Methods in Molecular Biology

Series Editor
John M. Walker
School of Life and Medical Sciences
University of Hertfordshire
Hatfield, Hertfordshire, UK

For further volumes:
http://www.springer.com/series/7651

For over 35 years, biological scientists have come to rely on the research protocols and methodologies in the critically acclaimed *Methods in Molecular Biology* series. The series was the first to introduce the step-by-step protocols approach that has become the standard in all biomedical protocol publishing. Each protocol is provided in readily-reproducible step-by-step fashion, opening with an introductory overview, a list of the materials and reagents needed to complete the experiment, and followed by a detailed procedure that is supported with a helpful notes section offering tips and tricks of the trade as well as troubleshooting advice. These hallmark features were introduced by series editor Dr. John Walker and constitute the key ingredient in each and every volume of the *Methods in Molecular Biology* series. Tested and trusted, comprehensive and reliable, all protocols from the series are indexed in PubMed.

MAIT Cells

Methods and Protocols

Edited by

Helen Kaipe

Department of Laboratory Medicine, Karolinska Institutet, Stockholm, Sweden

Clinical Immunology and Transfusion Medicine, Karolinska University Hospital, Stockholm, Sweden

Isabelle Magalhaes

Department of Oncology-Pathology, Karolinska Institutet, Stockholm, Sweden

Humana Press

Editors
Helen Kaipe
Department of Laboratory Medicine
Karolinska Institutet
Stockholm, Sweden

Clinical Immunology and Transfusion Medicine
Karolinska University Hospital
Stockholm, Sweden

Isabelle Magalhaes
Department of Oncology-Pathology
Karolinska Institutet
Stockholm, Sweden

ISSN 1064-3745 ISSN 1940-6029 (electronic)
Methods in Molecular Biology
ISBN 978-1-0716-0206-5 ISBN 978-1-0716-0207-2 (eBook)
https://doi.org/10.1007/978-1-0716-0207-2

This Humana imprint is published by the registered company Springer Science+Business Media, LLC, part of Springer Nature.
The registered company address is: 233 Spring Street, New York, NY 10013, U.S.A.

Preface

Mucosal-associated invariant T (MAIT) cells are an invariant type of T cells which recognizes bacterial derived riboflavin metabolites presented on the MHC-class I related (MR1) molecule. MAIT cells are characterized by a rapid innate-like response when activated, by mediating both cytotoxicity and production of inflammatory mediators. In this volume of *Methods in Molecular Biology*, we aim to describe methods for studying MAIT cells in several aspects.

The function and importance of MAIT cells in health and disease has only started to be unraveled, and the techniques described in this volume can be used to increase our knowledge in MAIT cell biology. The first part describes methods to isolate and characterize MAIT cells from human tissues, including liver, colon tumors, placenta and decidua, and endometrium. These chapters also contain descriptions of phenotypic and functional analysis by flow cytometry, including stochastic neighbor embedding (SNE) analysis for high-dimensional flow cytometry, as well as immunohistochemistry techniques to detect MAIT cells and MR1-expressing cells in tissues.

MAIT cells can be activated in an MR1-dependent manner by bacterial and fungal species with a functional riboflavin pathway, but they can also be partly activated by MR1-independent stimulation. The second part includes methods for studying activation of MAIT cells by different stimulatory agents. Protocols using anti-CD3 and -CD28 beads, Toll-like receptor agonists, bacterial *Escherichia coli* species, riboflavin intermediates, fungal *Aspergillus* and *Mucorales* species, influenza A virus, and inflammatory cytokines for stimulation are presented. Readouts for detection of responses include flow cytometry, ELISA, and cytotoxicity assays. An assay to study migration of MAIT cells in flow chambers is also presented.

MAIT cells are characterized by the expression of the invariant T cell receptor Vα7.2 and the C-type lectin CD161, but the use of MR1 tetramers has provided an even more reliable tool for the identification of MAIT cells. In the third part of this edition, the production of MR1 tetramers loaded with bacterial antigens that can be used for the detection of MAIT cells is described. A method for studying MAIT cells by quantitative proteomics is presented, which could be used for immune monitoring in different patient groups. This part also contains a protocol for the generation of MR1-restricted T cell clones that can be used for subsequent analysis and a description of reprogramming of MAIT cells to pluripotency and redifferentiation into MAIT cells, a tool that can be used for in vitro expansion of MAIT cells.

MAIT cells are highly conserved in mammalian species and are hence also present in mice. In the last part of this volume, the use of murine models for studying MAIT cells is presented. One chapter describes the use of a mouse model to study MAIT cells in type 1 diabetes, and one presents an in vivo model for studying antiviral responses mediated by MAIT cells. A description of the enrichment of murine MAIT cells for subsequent analysis is also presented.

Finally, we would like to thank all the authors of this book for their great contributions. We think that their efforts will provide great help for other researchers with an interest in MAIT cell biology. We would also like to thank the series editor, Professor John Walker, for giving us the opportunity to edit this volume and for helping us throughout the whole process of completing the book.

Stockholm, Sweden

Helen Kaipe
Isabelle Magalhaes

Contents

Contributors

ERIN J. ADAMS • *Department of Biochemistry and Molecular Biology, University of Chicago, Chicago, IL, USA*

HOSSAIN D. AKTHER • *Peter Medawar Building for Pathogen Research, Nuffield Department of Medicine, University of Oxford, Oxford, UK*

ALI AMINI • *Translational Gastroenterology Unit, Nuffield Department of Medicine, University of Oxford, Oxford, UK*

LUCIE BEAUDOIN • *INSERM U1016, Institut Cochin, Paris, France; CNRS, UMR8104, Paris, France; Université Paris Descartes, Laboratoire d'Excellence Inflamex, Sorbonne Paris Cité, Paris, France*

ISABEL BERNAL • *Cellular Proteomics, Helmholtz Centre for Infection Research, Braunschweig, Germany; Institute of Medical Microbiology and Hospital Hygiene, Infection Immunology, Otto-von-Guericke University Magdeburg, Magdeburg, Germany; ESF Graduate School ABINEP, Magdeburg, Germany*

MATTHEW BILTON • *Peter Medawar Building for Pathogen Research, Nuffield Department of Medicine, University of Oxford, Oxford, UK*

JONNA BISTER • *Center for Infectious Medicine, Department of Medicine Huddinge, Karolinska Institutet, Karolinska University Hospital Huddinge, Stockholm, Sweden*

NIKLAS K. BJÖRKSTRÖM • *Center for Infectious Medicine, Department of Medicine Huddinge, Karolinska Institutet, Karolinska University Hospital Huddinge, Stockholm, Sweden*

SARAH BOETTCHER • *Infections in Hematology and Oncology, Leibniz-Institute for Natural Product Research and Infection Biology, Jena, Germany*

CAROLINE BOULOUIS • *Center for Infectious Medicine, Department of Medicine, Karolinska Institutet, Stockholm, Sweden*

MEGHAN CANSLER • *Department of Pediatrics, Oregon Health & Science University, Portland, OR, USA*

VINCENZO CERUNDOLO • *MRC Human Immunology Unit, MRC Weatherall Institute of Molecular Medicine, University of Oxford, John Radcliffe Hospital, Oxford, UK*

ZHENJUN CHEN • *Department of Microbiology and Immunology, The University of Melbourne, The Peter Doherty Institute for Infection and Immunity, Melbourne, VIC, Australia*

ALEXANDRA J. CORBETT • *Department of Microbiology and Immunology, The University of Melbourne, The Peter Doherty Institute for Infection and Immunity, Melbourne, VIC, Australia*

GENNARO DE LIBERO • *Experimental Immunology, Department of Biomedicine, University Hospital and University of Basel, Basel, Switzerland*

DANIELA DI BLASI • *Experimental Immunology, Department of Biomedicine, University Hospital and University of Basel, Basel, Switzerland*

MATTHEW EDMANS • *Peter Medawar Building for Pathogen Research, Nuffield Department of Medicine, University of Oxford, Oxford, UK*

JOSHUA M. FARBER • *Inflammation Biology Section, Laboratory of Molecular Immunology, National Institute of Allergy and Infectious Diseases, National Institutes of Health, Bethesda, MD, USA*

CARA A. FROYD • *Department of Biochemistry and Molecular Biology, University of Chicago, Chicago, IL, USA*
LUCY C. GARNER • *Translational Gastroenterology Unit, Nuffield Department of Medicine, University of Oxford, Oxford, UK*
ANNA GIBBS • *Department of Medicine Solna, Division of Infectious Diseases, Center for Molecular Medicine, Karolinska Institutet, Karolinska University Hospital, Stockholm, Sweden*
LAIA GORCHS • *Department of Laboratory Medicine, Karolinska Institutet, Stockholm, Sweden*
MUHAMMAD YAASEEN GULAM • *Program in Emerging Infectious Diseases, Duke-National University of Singapore Medical School, Singapore, Singapore*
CARL-PHILIPP HACKSTEIN • *Peter Medawar Building for Pathogen Research, Nuffield Department of Medicine, University of Oxford, Oxford, UK; Translational Gastroenterology Unit, Nuffield Department of Medicine, University of Oxford, Oxford, UK*
JOACHIM P. HAGEL • *Peter Medawar Building for Pathogen Research, Nuffield Department of Medicine, University of Oxford, Oxford, UK*
MELANIE HARRIFF • *Department of Pulmonary and Critical Care Medicine, Oregon Health & Science University, Portland, OR, USA; Research and Development, VA Portland Health Care System, Portland, OR, USA*
TIMOTHY S. C. HINKS • *Respiratory Medicine Unit, Nuffield Department of Medicine Experimental Medicine, University of Oxford, Oxfordshire, UK*
MARTIN A. IVARSSON • *Center for Infectious Medicine, Department of Medicine Huddinge, Karolinska Institutet, Karolinska University Hospital Huddinge, Stockholm, Sweden*
LOTHAR JÄNSCH • *Cellular Proteomics, Helmholtz Centre for Infection Research, Braunschweig, Germany*
SUSANNE JAHREIS • *Klinik für Innere Medizin II, Hematology and Oncology, Jena University Hospital, Jena, Germany; Infections in Hematology and Oncology, Leibniz-Institute for Natural Product Research and Infection Biology, Jena, Germany*
HELEN KAIPE • *Department of Laboratory Medicine, Karolinska Institutet, Stockholm, Sweden; Clinical Immunology and Transfusion Medicine, Karolinska University Hospital, Stockholm, Sweden*
KATHERINE KEDZIERSKA • *Department of Microbiology and Immunology, The Peter Doherty Institute for Infection and Immunity, The University of Melbourne, Parkville, Australia*
PAUL KLENERMAN • *Peter Medawar Building for Pathogen Research, Nuffield Department of Medicine, University of Oxford, Oxford, UK; Translational Gastroenterology Unit, Nuffield Department of Medicine, University of Oxford, Oxford, UK; NIHR Biomedical Research Centre, University of Oxford, Oxford, UK*
MARIOS KOUTSAKOS • *Department of Microbiology and Immunology, The Peter Doherty Institute for Infection and Immunity, The University of Melbourne, Parkville, Australia*
ANDREAS KRUEGER • *Institute for Molecular Medicine, Goethe University Frankfurt, Frankfurt am Main, Germany*
ANDREA LAY HOON KWA • *Program in Emerging Infectious Diseases, Duke-National University of Singapore Medical School, Singapore, Singapore; Department of Pharmacy, Singapore General Hospital, Singapore, Singapore*
NICOLE A. LADD • *Department of Biochemistry and Molecular Biology, University of Chicago, Chicago, IL, USA*

EDWIN LEEANSYAH • *Program in Emerging Infectious Diseases, Duke-National University of Singapore Medical School, Singapore, Singapore; Center for Infectious Medicine, Department of Medicine, Karolinska Institutet, Stockholm, Sweden*

AGNÈS LEHUEN • *INSERM U1016, Institut Cochin, Paris, France; CNRS, UMR8104, Paris, France; Université Paris Descartes, Laboratoire d'Excellence Inflamex, Sorbonne Paris Cité, Paris, France*

TIANQI LENG • *Peter Medawar Building for Pathogen Research, Nuffield Department of Medicine, University of Oxford, Oxford, UK*

DAVID LEWINSOHN • *Department of Pediatrics, Oregon Health & Science University, Portland, OR, USA; Department of Pulmonary and Critical Care Medicine, Oregon Health & Science University, Portland, OR, USA; Research and Development, VA Portland Health Care System, Portland, OR, USA*

LIYEN LOH • *Department of Microbiology and Immunology, The Peter Doherty Institute for Infection and Immunity, The University of Melbourne, Parkville, Australia; Department of Immunology and Microbiology, University of Colorado Anschutz Medical Campus, Aurora, CO, USA*

ISABELLE MAGALHAES • *Department of Oncology-Pathology, Karolinska Institutet, Stockholm, Sweden*

ERIN MEERMEIER • *Department of Pulmonary and Critical Care Medicine, Oregon Health & Science University, Portland, OR, USA*

HEMA MEHTA • *Peter Medawar Building for Pathogen Research, Nuffield Department of Medicine, University of Oxford, Oxford, UK*

LUCIA MORI • *Experimental Immunology, Department of Biomedicine, University Hospital and University of Basel, Basel, Switzerland*

ISABELLE NEL • *INSERM U1016, Institut Cochin, Paris, France; CNRS, UMR8104, Paris, France; Université Paris Descartes, Laboratoire d'Excellence Inflamex, Sorbonne Paris Cité, Paris, France*

MEGAN NULL • *Department of Pediatrics, Oregon Health & Science University, Portland, OR, USA*

FARHAT PARWEEN • *Inflammation Biology Section, Laboratory of Molecular Immunology, National Institute of Allergy and Infectious Diseases, National Institutes of Health, Bethesda, MD, USA*

PRABHJEET PHALORA • *Peter Medawar Building for Pathogen Research, Nuffield Department of Medicine, University of Oxford, Oxford, UK*

NICHOLAS M. PROVINE • *Translational Gastroenterology Unit, Nuffield Department of Medicine, University of Oxford, Oxford, UK*

MARIANNE QUIDING JÄRBRINK • *Department of Microbiology and Immunology, Institute of Biomedicine, University of Gothenburg, Göteborg, Sweden*

WILLIAM RODIN • *Department of Microbiology and Immunology, Institute of Biomedicine, University of Gothenburg, Göteborg, Sweden*

MARIOLINA SALIO • *MRC Human Immunology Unit, MRC Weatherall Institute of Molecular Medicine, University of Oxford, John Radcliffe Hospital, Oxford, UK*

JOHAN K. SANDBERG • *Department of Medicine, Center for Infectious Medicine, Karolinska Institutet, Stockholm, Sweden*

WAN RONG SIA • *Program in Emerging Infectious Diseases, Duke-National University of Singapore Medical School, Singapore, Singapore*

MICHAL J. SOBKOWIAK • *Department of Medicine, Center for Infectious Medicine, Karolinska Institutet, Stockholm, Sweden*

MARTIN SOLDERS • *Department of Laboratory Medicine, Karolinska Institutet, Stockholm, Sweden; Clinical Immunology and Transfusion Medicine, Karolinska University Hospital, Stockholm, Sweden*
BENEDIKT STRUNZ • *Center for Infectious Medicine, Department of Medicine Huddinge, Karolinska Institutet, Karolinska University Hospital Huddinge, Stockholm, Sweden*
PATRIK SUNDSTRÖM • *Department of Microbiology and Immunology, Institute of Biomedicine, University of Gothenburg, Göteborg, Sweden*
GWENDOLYN SWARBRICK • *Department of Pediatrics, Oregon Health & Science University, Portland, OR, USA*
ANNELIE TJERNLUND • *Department of Medicine Solna, Division of Infectious Diseases, Center for Molecular Medicine, Karolinska Institutet, Karolinska University Hospital, Stockholm, Sweden*
ALESSANDRO VACCHINI • *Experimental Immunology, Department of Biomedicine, University Hospital and University of Basel, Basel, Switzerland*
MARCO VAN HAM • *Cellular Proteomics, Helmholtz Centre for Infection Research, Braunschweig, Germany*
BONNIE VAN WILGENBURG • *Peter Medawar Building for Pathogen Research and Translational Gastroenterology Unit, Nuffield Department of Clinical Medicine, University of Oxford, Oxford, UK*
MARIE VON LILIENFELD-TOAL • *Klinik für Innere Medizin II, Hematology and Oncology, Jena University Hospital, Jena, Germany; Infections in Hematology and Oncology, Leibniz-Institute for Natural Product Research and Infection Biology, Jena, Germany*
HIROSHI WAKAO • *Dokkyo Medical University, Tochigi, Japan*
HUIMENG WANG • *Department of Microbiology and Immunology, The University of Melbourne, The Peter Doherty Institute for Infection and Immunity, Melbourne, VIC, Australia*
CHRISTIAN B. WILLBERG • *Peter Medawar Building for Pathogen Research, Nuffield Department of Medicine, University of Oxford, Oxford, UK; NIHR Biomedical Research Centre, University of Oxford, Oxford, UK*
SAMANTHA J. WINTER • *Institute for Molecular Medicine, Goethe University Frankfurt, Frankfurt am Main, Germany*
HONGWEI H. ZHANG • *Inflammation Biology Section, Laboratory of Molecular Immunology, National Institute of Allergy and Infectious Diseases, National Institutes of Health, Bethesda, MD, USA*
CHRISTINE L. ZIMMER • *Center for Infectious Medicine, Department of Medicine Huddinge, Karolinska Institutet, Karolinska University Hospital Huddinge, Stockholm, Sweden*

Part I

MAIT Cells in Human Tissues and Blood

Chapter 1

MAIT Cells in Health and Disease

Isabelle Magalhaes, Martin Solders, and Helen Kaipe

Abstract

Mucosal-associated invariant T (MAIT) cells are a newly described subset of T cells that are found in the blood and are enriched in many tissues, particularly in the liver. MAIT cells express a semi-invariant T cell receptor restricted by the MHC class I-related (MR1) molecule. MAIT cells are activated in a MR1-dependent manner in response to microbial-derived riboflavin metabolites which leads to rapid effector functions, but they can also be activated in a MR1-independent manner by cytokines and viruses. The use of mice models and MR1 tetramers, among other recent methodological advances, have provided more insight into the development, mode of activation, characterization in different diseases and tissues of MAIT cells. In this chapter, we provide an overview of MAIT cells and yet remaining questions about their potential therapeutic role.

Key words MAIT cells, MR1, Bacteria, Virus, Autoimmune diseases, Cancer

1 Introduction

Mucosal-associated invariant T (MAIT) cells are a subset of unconventional innate-like T cells that respond to metabolites of riboflavin (vitamin B2) synthesis [1], which can be produced by a large number of microbes but not by humans. After presentation of riboflavin metabolites on the MHC class I-related (MR1) molecule to the semi-invariant T cell receptor (TCR) expressed by MAIT cells [2], they are triggered to promote a rapid immune response mediated by the release of cytotoxic molecules and inflammatory cytokines. MAIT cells have a propensity to localize in non-lymphoid tissues, such as the liver, gut, and lung, and are also present in peripheral blood [3, 4]. MAIT cells predominantly express the CD8 coreceptor, although some are negative for both CD8 and CD4, and a small portion expresses CD4 [5].

1.1 MAIT Cell TCR

Compared to conventional T cells, which can display an almost unlimited TCR repertoire, the specificities of MAIT cell TCRs are limited. The preferred TCR Vα7.2 chain is most often coupled with

Helen Kaipe and Isabelle Magalhaes (eds.), *MAIT Cells: Methods and Protocols*, Methods in Molecular Biology, vol. 2098,
https://doi.org/10.1007/978-1-0716-0207-2_1,

a TCR Jα33 chain, although the pairing with Jα12 or Jα20 can also be seen [6]. The Vβ chain diversity within the MAIT cell population is greater than that of the Vα chain, but the number of Vβ sequences found in human MAIT cells is still markedly fewer compared to conventional T cells [6]. The most commonly used Vβ chains are the Vβ13.2 and Vβ2 [6, 7], and MAIT cells utilizing these Vβ chains seem to promote a stronger response to *E. coli* stimulation compared to MAIT cells using rarer Vβ segments [7].

Due to their uniform expression of Vα7.2, MAIT cells can be identified by the expression of Vα7.2 together with high expression of the C-type lection CD161. These have been shown to provide reliable markers of MAIT cells in several settings [7, 8]. However, the expression of CD161 can be affected by environmental conditions [9, 10], and these markers also appear less reliable for identifying MAIT cells in newborns [11]. The development of MR1 tetramers loaded with 5-OP-RU, a derivate of the microbial riboflavin precursor 5-A-RU, has therefore provided a valuable tool for identifying MAIT cells more accurately based on their specificity [12, 13].

1.2 The MR1 Molecule and Its Function in MAIT Cell Activation

As described above, the defining property of MAIT cells is their ability to recognize metabolites of riboflavin synthesis [1] bound to the MR1 molecule [2]. The gene encoding for human MR1 is located on chromosome 1 (1q25.3) [14]. In contrast to the polymorphic HLA class I molecules A, B, and C and the oligomorphic HLA-E and G, the MR1 gene is monomorphic [14]. The MR1 gene is highly conserved between mammalian species, including non-human primates, mice, and pigs. Murine MR1 shares 80–90% of the MR1 sequence identity of the antigen-binding domain with human MR1 [14–16]. However, there are mammals lacking a functional MR1 gene, including the rabbit, pika, and armadillo [17]. MAIT cells are thus present in many mammals, such as humans and mice, but it should be noted that the number of MAIT cells in many laboratory mouse strains are very low. However, clean wild-derived inbred mice have a higher frequency of MAIT cells [18].

MR1 is widely expressed in different tissues and by both hematopoietic and non-hematopoietic cells, but at varying intensities depending on the location and the activation status of the cell [14, 15]. Extracellular expression of MR1 is rarely seen in resting cells [19–21]. Instead, MR1 molecules generally reside in the endoplasmic reticulum (ER) in an incompletely folded conformational state, dissociated from the β_2 microglobulin subunit [19]. After riboflavin metabolites produced by bacteria or fungi are taken up by the cell by yet unknown mechanisms, they are transported to the ER where the antigen is bound to the MR1 molecule. The antigen-loaded MR1 molecule associates with the β_2 microglobulin subunit, and the MR1–antigen complex is then folded into a functional shape. After passing through the ER and the Golgi apparatus, the complex is transported to the cell surface [19]. Most MR1 molecules are degraded upon reinternalization into the cells [19]. However, recycling of the MR1 molecule can occur since it has been shown that antigens can be loaded to the MR1 molecules in an ER-independent manner within endosomes in the context of *Mycobacterium tuberculosis* infection [22]. By blocking NF-κB signaling, the turnover of loaded extracellular MR1 molecules from the cell surface was greatly decreased, and NF-κB signaling has been shown to be required for MR1 signaling [20].

Several types of antigen presenting cells (APCs) can activate MAIT cells in an MR1 dependent manner, including dendritic cells [23, 24], monocytes [3, 24–26], macrophages [24, 27], and B cells from both blood [21] and liver [27]. Moreover, non-hematopoietic cells such as epithelial cells from lungs [23] and from bile ducts [27] have the ability to activate MAIT cells in vitro.

In humans and mice, MAIT cells have a broad anti-bacterial spectrum, including *Escherichia*, *Lactobacillus*, *Staphylococcus*, *Salmonella*, *Mycobacteria*, and *Clostridioides* species among others [1, 25, 28]. Bacterial species lacking the riboflavin synthesis pathway (e.g., *Streptococcus* group A, *Listeria monocytogenes*, and *Enterococcus faecalis*) consequentially lack the ability to activate MAIT cells through an MR1-dependant pathway [23, 25]. Fungal species, such as *Candida*, *Saccharomyces*, and *Aspergillus*, are also able to activate MAIT cells since they are equipped with riboflavin-producing capacities [25, 29, 30].

1.3 MR1-Independent Activation of MAIT Cells

In addition to MR1-dependent activation, MAIT cell can also be partially activated by cytokines, including IL-7, IL-12, IL-15, and IL-18 and type I interferons [24, 31–33]. Resting MAIT cells express the receptors for IL-18 and IL-12 [3, 25]. IL-15-induced production of IL-18 from monocytes leads to the expression of both IFN-γ and cytotoxic molecules in MAIT cells [32]. IL-12 together with IL-18, which both are produced from APCs in response to Toll-like receptor (TLR) ligands, are also potent stimulators of IFN-γ production from MAIT cells [31]. IL-7 can increase the cytotoxic effects of MAIT cells and potentiates the response to bacterial stimulation [33]. Similarly, IL-12 and IL-18 have been shown to be important for potentiating MR1-dependent bacterial MAIT cell activation [20, 32, 34].

The potential of inflammatory cytokines to activate MAIT cells broadens their ability to respond to other types of TLR-activating microbes beyond those with a functional riboflavin-producing capacity. Indeed, there is now emerging evidence that MAIT cells can be involved in immune responses to viruses [24, 35–37]. MAIT cells can be activated in vitro by influenza virus, dengue virus, and hepatitis C virus (HCV) in a cytokine-dependent manner, independent of TCR–MR1 interaction [24, 36]. IL-18 released by APCs after viral stimulation appears to be the most important cytokine in the viral-mediated activation of MAIT cells, but IL-12, IL-15, and type I interferons can further potentiate the response. Several studies have reported that the frequencies of peripheral and/or hepatic MAIT cells are reduced in patients with viral infections, including human immunodeficiency virus (HIV), influenza, hepatitis B (HBV), and hepatitis delta virus [24, 33, 36, 38–40]. The remaining MAIT cells displayed an activated phenotype, and the function of MAIT cells also appears to be affected by viral infections, as reflected by a decreased capacity to produce inflammatory cytokines and cytotoxic molecules in response to stimulation. Whether the MAIT cell-mediated inflammation and cytotoxicity are involved in eradicating the virus is not known, but MAIT cells have been shown to be important in clearing influenza A virus in a mouse model [35]. It is also possible that bystander activation by viral-induced cytokines may promote MAIT cells to react off target and hence affect healthy cells.

Superantigens derived from bacteria such as *Streptococcus pyogenes* and *Staphylococcus aureus* can activate MAIT cells in an MR1-independent manner by cross-linking the MAIT cell TCR with MHC class II on APCs [26]. Interestingly, the response of MAIT cells to superantigens is stronger than that of conventional T cells and other types of innate-like T cells. MAIT cells display an exhausted phenotype with high expression of LAG-3 after exposure to superantigens, which also diminish their capacity to mediate MR1-dependent responses to riboflavin-producing bacteria.

1.4 MAIT Cell Development and Changes During Life

The precise process for MAIT cell development is still not completely understood, but it requires a functional thymus. In mice, it has been shown that MR1-expressing $CD4^+CD8^+$ thymocytes select MAIT cells in the thymus [41]. Koay et al. proposed a MAIT cell development process in three stages by using MR1:5-OP-RU tetramers; $CD161^-CD27^-$ and $CD4^+$ or $CD4^+CD8^+$ (stage 1), $CD161^-CD27^+$ and $CD4^+$, $CD8^+$ or $CD4^+CD8^+$ (stage 2), and $CD161^+CD27^{+\text{ or low}}$ and $CD4^-CD8^-$ or $CD8^+$ (with a minor fraction of $CD4^+$) (stage 3) [42]. Immature stage 1 MAIT cells were restricted to the thymus, but stage 2 and 3 MAIT cells could also be found in the thymus. Stage 2 MAIT cells were present in cord blood and peripheral blood in children, but not in adults. About 80% and 90% of the MAIT cells in cord blood and peripheral blood in children, respectively, were in stage 3, whereas virtually all MAIT cells in adult peripheral blood were in stage 3. MAIT cells in human cord blood [34, 43] as well as thymic MAIT cells in mice display a naïve phenotype [44]. Leeansyah et al. made similar findings when investigating MAIT cells in human fetuses from second trimester using the Vα7.2 and CD161 expression [45]. MAIT cells isolated from primary and secondary lymphoid organs displayed a naïve phenotype and were poor producers of IFN-γ after bacterial stimulation. On the other hand, tissue-resident MAIT cells isolated from various fetal tissues expressed an effector memory phenotype and had a stronger IFN-γ response.

During thymic development, MAIT cells are programmed to express promyelocytic leukemia zinc finger (PLZF), a transcription factor which have been shown to be important for the differentiation into functional MAIT cells with a memory phenotype [11, 42]. MR1 ligands are likely also important for the development of MAIT cells, as germ-free mice lack functional MAIT cells [2]. Mature MAIT cells travel with the blood circulation, until a proportion of the MAIT cells eventually end up as tissue-resident cells in mucosal tissues, liver, and lungs. Tissue residency is associated with the upregulation of CD103 and CD69 and lack of expression of the lymph node homing markers CCR7 and CD62L. Interestingly, human MAIT cells have been shown to express low levels of CCR7 and CD62L already in the thymus, indicating that they are programmed for tissue residency and devoid of lymph node homing capacities already from the start [46].

MAIT cells in cord blood have a greater TCR Vβ chain diversity compared to MAIT cells from peripheral blood of adults, and, as mentioned above, display a naïve phenotype [11]. The IFN-γ response to MR1-dependent activation is low, whereas the proliferative effect upon phytohemagglutinin (PHA) stimulation is greater compared to MAIT cells from adults [11]. A substantial proportion of the $V\alpha7.2^{+}CD161^{high}$ T cells in cord blood do not bind to the MR1:5-OP-RU tetramers [42]. Thus, there is probably a gradual and MR1-dependent maturation of MAIT cells following birth, which will result in a relative increase in MAIT cells expressing Vβ chains with high affinity for MR1–ligand complexes.

In healthy adults, there is a great variance in the frequency of MAIT cells in blood, ranging between 0.2% and 20% with a median of 2% of all $CD3^{+}$ cells [3, 47, 48]. The MAIT cell frequency is low at birth, and an increase is seen over puberty and early adulthood [11]. A study comparing four monocygotic and 17 dizygotic twins shortly after birth showed no difference in MAIT cells frequency, indicating that environmental factors are more influential than genetics in this setting [11]. During their fertile age, women have higher MAIT cell frequencies compared to men, whereas no differences between sexes were seen after 50 years of age [47]. MAIT cells continue to increase from early adulthood, and the frequency reaches a peak when the individual is 30–40 years old. Thereafter, a gradual decrease in peripheral blood MAIT cells is seen until death [47–50]. Whether this is due to tissue compartmentalization or if it is due to depletion as a consequence of inflamm-ageing is not known.

The distribution of $CD4^+$, $CD8^+$, and $CD4^-CD8^-$ (DN) MAIT cells also changes over the lifespan of an individual. $CD8^+$ MAIT cells gradually decrease with age, whereas $CD4^+$ [48] or DN MAIT cells increase [47]. However, no differences were seen regarding the IFN-γ or IL-17 production of MAIT cells in response to phorbol myristate acetate (PMA)/ionomycin simulation when comparing younger and older individuals [48].

1.5 MAIT Cell Effector Functions

Resting MAIT cells express granulysin, granzyme A, and perforin. Following activation, there is an upregulation of the expression of the activation markers CD25 and CD69 and the degranulation marker CD107a [10, 25, 51]. There is also an increase in the expression of intracellular perforin and the MAIT cells start to express granzyme B [3, 33, 34, 52]. Perforin creates a channel in the target cell membrane, and MAIT cells then transfer granulysin and granzymes into the cytoplasm. Granulysin disrupts cellular membranes and aids in killing intracellular bacteria, whereas the granzymes induce apoptosis of the cell [53]. It is well established that infected cells are killed by MAIT cells in an MR1-dependent manner in co-cultures [33, 51, 52], and by using time-lapse microscopy it has also been shown that sorted MAIT cells induce apoptosis in infected target cells [51].

MAIT cells also respond with the production and secretion of cytokines upon stimulation. Peripheral MAIT cells are potent producers of IFN-γ and TNF-α, and liver-residing MAIT cells also have the capacity to produce IL-17, both after stimulation with PMA/Ionomycin and MR1-dependent activation [54]. When investigating MAIT cells from the female genital tract mucosa, Gibbs et al. found that the production of IFN-γ and TNF-α was lower compared to peripheral blood MAIT cells in response to *E. coli*, but that the production of both IL-17 and IL-22 was higher [55]. MAIT cells isolated from maternal blood in the intervillous space of the human placenta were also shown to express higher levels of IFN-γ, perforin, and granzyme B compared to peripheral blood in response to bacterial stimulation, whereas decidual MAIT cell responses were comparable to those of peripheral MAIT cells [34]. Together, this suggests that MAIT cells are influenced by their environment, at least with regard to their secretory functions. Any factor behind such a tissue-based specialization remains unknown, and it would be of great interest to determine these mechanisms.

1.6 MAIT Cells in Disease

1.6.1 MAIT Cells in Infectious Disease

The importance of MAIT cell in infections has been studied in vivo using murine models. MR1-knockout mice, which hence lack MAIT cells, that were injected with *E. coli* or *Mycobacterium abscessus* had higher splenic bacterial load compared to wild-type mice [25]. Such mice also had a higher mortality upon infection with *Klebsiella pneumonia* [56]. Furthermore, it has been shown that MAIT cells accumulate in the lungs after challenge with a live vaccine strain of *Francisella tularensis,* and that mice lacking MAIT cells displayed a delayed immune response with poor recruitment of conventional T cells [57]. Similarly, mice infected with *Salmonella typhimurium* presented with an enrichment of MAIT cells in the lungs shortly after infection [50]. MAIT cells have been shown to contribute to a protective effect against influenza in an MR1-independent manner [35].

There are also reports on how MAIT cells are affected by infections in humans. A prospective controlled human malaria infection study was performed by infecting healthy individuals with *Plasmodium falciparum* sporozoites [58]. They observed an initial drop in MAIT cell frequency in blood, which was followed by a subsequent long-lasting increase which exceeded base line proportions. A reduced frequency of MAIT cells in peripheral blood has been associated with a wide range of bacterial infections in humans, including *Shigella dysenteriae* [51], *Helicobacter pylori* [59], *Mycobacterium tuberculosis* [23, 25, 60], *Vibrio cholera* (in children but not adults) [61], and *Salmonella enterica* resulting in typhoid fever [62]. Patients with *Mycobacterium tuberculosis* infections have also been described to have an enrichment of MAIT cells in the lungs, suggesting that MAIT cells are recruited to the site of infection [23].

As mentioned above, HIV infection has also been correlated to both a decrease in MAIT cells and an impaired functionality of the remaining cells. In these studies, there were no measurable effects on MAIT cells after anti-retroviral therapy [39, 63]. A decrease in both frequency and functionality was also seen in patients coinfected by both HIV-1 and *Mycobacterium tuberculosis* [64]. In the case of both HCV and coinfection of HCV and HIV-1, patients had fewer MAIT cells overall, and their MAIT cells showed a more activated phenotype, both in peripheral blood and in the liver [24, 38, 65–67]. This differed from patients with HBV infection, where the MAIT cells did not markedly differ from healthy controls, except for a modest increase in CD38 expression [68]. Following infection with either Dengue fever or influenza, MAIT cell frequencies were also markedly lower compared to healthy controls [24, 36].

Thus, the indirect activation of MAIT cells during viral infections seems to be measurable in vivo, and the activation leads to a subsequent depletion of MAIT cells, which persists even after infection clearance. The mechanisms for this and the possible clinical consequences are yet unknown. Since there is a risk for secondary bacterial infections following influenza clearance, it can be speculated that the reduced MAIT cell numbers after viral infection could be one factor contributing to the morbidity and mortality of secondary bacterial and fungal infections seen post influenza [69].

1.6.2 Liver Diseases

MAIT cells are particularly enriched in the liver where they can represent up to 45% of liver lymphocytes [3], and also make up the majority of IL-17-producing cells [54]. MAIT cells reside mostly in peri-biliary regions of portal tracts, and can be activated, in an MR1-dependent fashion, by liver-infiltrating B cells and biliary epithelium cells exposed to *E. coli* [27]. In patients with primary sclerosis or primary cholangitis (characterized by chronic inflammation), MAIT cell frequency was lower in blood, but no changes were observed in the bile ducts [70]. Circulating MAIT cells in these patients showed an activated phenotype (e.g., CD69 expression) and impaired functions (e.g., decreased IFN-γ production). Interestingly, in patients with primary sclerosing cholangitis, MAIT cells were found to localize preferentially to fibrotic areas [71]. In patients with chronic hepatitis delta virus infection (a severe form of hepatitis, associated with high risk of *hepatocellular carcinoma*, HCC), peripheral blood MAIT cell frequency, and particularly the $CD8^+$ MAIT cell population, was decreased, and they exhibited an abnormal activated phenotype and functional impairment [40].

Lower circulating frequency of MAIT cells has been reported in patients with alcoholic or non-alcoholic fatty liver (NAFLD) disease-related cirrhosis, where MAIT cells were found to accumulate in liver fibrotic septa, and exhibited profibrogenic properties mediated by TNF production [72]. Another report in patients with NAFLD confirmed the decreased MAIT cell frequency, while showing increased MAIT cell numbers in the liver. In vitro-activated MAIT cells were shown to induce M2 macrophage differentiation, suggesting that MAIT cells may play a role in protecting against inflammation [73].

In patients with HCV infection, the frequency of circulating MAIT cells was decreased and not restored up to 18 months even after successful IFN-free treatment [74]. In addition, 9 months after treatment, MAIT cells displayed impaired effector functions. Longitudinal monitoring of patients coinfected with HIV and HCV after successful IFN-free treatment (6 months) and HCV clearance also reported circulating MAIT cell depletion [75]. Altogether, the studies of MAIT cells in patients with different kinds of liver diseases showed that impairment in both frequency and effector function of MAIT cell can be detected in circulating blood, and that MAIT cells in the liver can influence other neighboring cells by enhancing fibrogenesis and differentiation.

1.6.3 Solid Cancers

Very little is known regarding the role of MAIT cells in cancer [76]. As compared to healthy donors, there were no differences in circulating MAIT cells frequency in patients with colon adenocarcinoma [77]. On the other hand, decreased MAIT cell frequencies have been observed in blood in patients with cervical cancer [78], and in the blood and in tumor samples of patients with HCC [79]. In the HCC tumor, more MAIT cells expressed checkpoint inhibitors compared to peritumor or normal liver MAIT cells. Gene expression analysis showed downregulation of cytokine secreting and cytolysis functions, but upregulation of glucose and cholesterol metabolism, and a shift toward antitumor immunity. Lastly, high tumor infiltration of MAIT cells was correlated with worse outcome [79].

Other groups have reported that the frequency of circulating MAIT cells was decreased in patients with mucosal-associated cancers [80], or with colorectal cancer (CRC), and that MAIT cells accumulated in colorectal neoplasms [81], and in the colonic tumor tissue as compared to unaffected tissue [77]. In patients with CRC, circulating and tumor-infiltrating MAIT cells display a Th17-biased function. MAIT cells were shown to reduce the cell viability of the colon cancer cell line HCT116 in in vitro experiments in an MR1-dependent manner [81]. It is yet unknown whether tumor-infiltrating MAIT cells play a protective or deleterious role in CRC. It has recently been shown that a high frequency of tumor-infiltrating MAIT cells, as compared to unaffected tissue, is a marker of poor prognosis in patients with CRC [82]. In patients suffering from colorectal liver metastasis, tumor-infiltrating and tumor-margin MAIT cells failed and displayed a reduced capacity to produce IFN-γ, respectively, as compared to MAIT cells isolated from healthy tissue [83].

In patients with colon adenocarcinoma, tumor-infiltrating MAIT cells displayed cytotoxic capacity, suggesting that MAIT cells can potentially participate to antitumor responses [84]. Lastly, lower frequencies of MAIT cells in blood and lesions were reported in patients with Langerhans cell histiocytosis, while the proportion of MAIT cells expressing $CD4^{+}$ was increased in the blood of patients as compared to healthy controls [85].

1.6.4 Type 1 Diabetes and Other Autoimmune Disorders

A report that included a limited number of patients showed that MAIT cells could be detected in chronic pancreatitis but not in normal pancreatic tissue, and that in patients with chronic pancreatitis, MAIT cells accumulated in the pancreatic tissue as compared to blood [86]. In children at the onset of type 1 diabetes (T1D), but not in patients with established T1D, alterations of circulating MAIT cells were reported. The same authors, using a mouse model of T1D, showed that MAIT cells were present in the pancreatic islets of NOD mice and were more abundant than in control mice, and that their function was altered with diabetes development

[87]. In contrast, another report did not find differences in circulating MAIT cells frequency in a juvenile cohort of new onset T1D patients [88]. However, the time points differed between the two reports: within 10 days of the first insulin injection in the former (onset), and <12 months since diagnosis (new onset) in the latter, as well as the staining strategy to define MAIT cells (i.e., the anti-Vα7.2 antibody was not included in the latter). A decrease in MAIT cells frequency in the peripheral blood of subjects who had seroconverted to autoantibodies was reported [89], while in a smaller cohort subjects at risk of T1D showed MAIT cell frequencies comparable to controls [87]. No MAIT cells were detected in insulitic lesions in patients with recent onset of T1D, indicating that they may not play a role in β cell destruction in vivo [90], even if MAIT cells could kill human β cells in in vitro experiments [87].

An increased frequency of circulating MAIT cells was observed in pediatric-onset multiple sclerosis (MS) patients compared to healthy donors, and to children with non-MS monophasic inflammatory demyelinating disease [91]. The same study also showed that peripheral MAIT cells from children with MS produced higher levels of IFN-γ and IL-17 in response to PMA/ionomycin stimulation. In contrast, decreased MAIT cell frequencies have been seen in a wide range of other autoimmune disorders and other immune-mediated diseases, including systemic lupus erythematosus, rheumatoid arthritis, Crohn's disease, ulcerative colitis, coeliac disease, asthma, and chronic obstructive pulmonary disease and obesity [92].

1.6.5 Cardiometabolic Diseases

Decreased frequencies of MAIT cells in peripheral blood, and functional defects in MAIT cells in obese and/or with type 2 diabetic (T2D) patients have been reported [8, 93]. MAIT cell depletion was confirmed in an additional study, not only in patients with T2D and obesity but also in patients with metabolic syndrome or coronary heart disease (with or without congestive heart failure) [94]. Of note, circulating MAIT cell loss increased with the severity of cardiac dysfunction. Increased glucose concentration in vitro was shown to induce apoptosis in MAIT cells (but not conventional T cells), which could at least partially explain MAIT cell loss in patients with chronically high glucose levels. Interestingly, a recent report showed that in obese patients, activated MAIT cells displayed decreased amino acid uptake and mTORC1 activity, and subsequent defective glycolytic metabolism, which in turn impacted on IFN-γ production [95].

1.6.6 Hematopoietic Stem Cell Transplantation

In patients with MS who received autologous stem cell transplantation (SCT), a reduced frequency of circulating $CD161^{high}CD8^{+}$ T cells (presumably mainly MAIT cells) after non-myeloablative conditioning [96] and $CCR6^{+}$ MAIT cells after myeloablative chemotherapy [97] has been reported up to 2 years and 1 year, respectively. In patients with hematological malignancies, after myeloablative chemotherapy and autologous stem cell transplant, circulating MAIT cells recovered to pretreatment levels, and early recovery was associated with younger age [98], suggesting that thymic input contributes to MAIT cells reconstitution.

In patients (mainly with malignant diseases) who received allogeneic hematopoietic SCT and myeloablative or non-myeloablative conditioning, no normalization of MAIT cells frequency was seen 3 months after transplantation. Low numbers of MAIT cells at day 60 post transplant was associated with grade I–II and grade II–IV graft versus host disease (GVHD) [99]. One report showed that MAIT cell numbers were not normalized 1 year following transplantation [100]. Cord blood transplant and post-HCT cyclophosphamide treatment for GVHD had profound impairment of MAIT cell reconstitution.

Another report showed that peripheral MAIT cell frequency was decreased up to 2 years after transplantation, and although an impairment in functionality were detected 2–6 months post transplantation, MAIT cell responses to bacterial stimulation were restored 2 years later [101]. No correlation was found between MAIT cell frequency and acute GVHD. In a mouse model, recipient MAIT cells have been shown to promote gastrointestinal tract integrity and protect against acute GVHD in the colon [102]. High expression by MAIT cells of the ATP-binding cassette-multi-drug efflux protein 1 (MDR1) [3, 103] allows rapid efflux of xenobiotics and increased resistance to immunosuppressive MDR1 substrates [104]. Cyclosporine A inhibits MDR1, and in the presence of cyclosporine A, efflux of rhodamine 123 by MAIT cells was inhibited [3]. MAIT cells have been shown to be more sensitive than conventional T cells to cyclosporine A and sirolimus in vitro [101]. Lastly, the relationship between MAIT cells and gastrointestinal tract microbiota in the setting of SCT indicated that peripheral MAIT cell reconstitution correlated with higher abundance of certain gut microbial species (*Bifidobacterium longum* and *Blautia* spp.) [100], and that IL-17 production by MAIT cells regulated microbiome diversity [102].

1.7 Conclusion and Future Directions

The function of MAIT cells in human health and disease is only starting to be unraveled, but it is evident that MAIT cells are affected by various immune-mediated disorders and play a role in infectious diseases. MAIT cells are reduced in the circulation in several types of infectious diseases as well as in other inflammatory disorders and malignancies, which could suggest that they localize to inflamed tissue. However, this does not always seem to be the case since MAIT cell frequencies in the liver of hepatitis patients also have been shown to be reduced. MAIT cells display a dysregulated function in several types of diseases, suggesting that inflammatory signals can promote MAIT cell exhaustion and depletion in certain settings. It also appears that the loss of circulating MAIT cells is long lasting, since in the different settings reported, circulating MAIT cell frequency was not restored to normal levels months or even years after treatment. This could suggest that de novo MAIT cell development in adults is impaired. Whether this is due to diminished thymic function in adults and/or microbiota composition remains to be clarified. How MAIT cells discriminate between pathogenic and commensal microbes is also largely unexplored. The studies of MAIT cells have demonstrated that they can, upon direct TCR-mediated activation or indirect activation, via cytokine and cytotoxic granules production, participate in inflammation, have regulatory properties, and exert cytotoxic functions. The exact role of MAIT cells as protective, regulatory, or pathogenic in tumors and other non-infectious diseases still remains to be elucidated. More studies are therefore needed to decipher whether MAIT cells may be desirable targets for therapeutic manipulation and/or relevant clinical biomarkers.

Acknowledgments

HK was supported by the Swedish Research Council, the Swedish Childhood Cancer Foundation, the Cancer Society in Stockholm, the Swedish Cancer Foundation, and Karolinska Institutet. MS was supported by Karolinska Institutet and Stockholm County Council. IM was supported by Karolinska Institutet, *Clas Groschinskys Minnesfond*, and Dr. *Åke Olsson Foundation.*

References

1. Kjer-Nielsen L, Patel O, Corbett AJ, Le Nours J, Meehan B, Liu L, Bhati M, Chen Z, Kostenko L, Reantragoon R, Williamson NA, Purcell AW, Dudek NL, McConville MJ, O'Hair RA, Khairallah GN, Godfrey DI, Fairlie DP, Rossjohn J, McCluskey J (2012) MR1 presents microbial vitamin B metabolites to MAIT cells. Nature 491 (7426):717–723. https://doi.org/10.1038/nature11605
2. Treiner E, Duban L, Bahram S, Radosavljevic M, Wanner V, Tilloy F, Affaticati P, Gilfillan S, Lantz O (2003) Selection of evolutionarily conserved mucosal-associated invariant T cells by MR1. Nature 422(6928):164–169. https://doi.org/10.1038/nature01433
3. Dusseaux M, Martin E, Serriari N, Peguillet I, Premel V, Louis D, Milder M, Le Bourhis L, Soudais C, Treiner E, Lantz O (2011) Human MAIT cells are xenobiotic-resistant, tissue-targeted, CD161hi IL-17-secreting T cells. Blood 117(4):1250–1259. https://doi.org/10.1182/blood-2010-08-303339
4. Kurioka A, Walker LJ, Klenerman P, Willberg CB (2016) MAIT cells: new guardians of the liver. Clin Transl Immunol 5(8):e98. https://doi.org/10.1038/cti.2016.51
5. Dias J, Boulouis C, Gorin JB, van den Biggelaar R, Lal KG, Gibbs A, Loh L, Gulam MY, Sia WR, Bari S, Hwang WYK, Nixon DF, Nguyen S, Betts MR, Buggert M, Eller MA, Broliden K, Tjernlund A, Sandberg JK, Leeansyah E (2018) The CD4(−)CD8(−) MAIT cell subpopulation is a functionally distinct subset developmentally related to the main CD8(+) MAIT cell pool. Proc Natl Acad Sci U S A 115 (49):E11513–E11522. https://doi.org/10.1073/pnas.1812273115
6. Lepore M, Kalinichenko A, Colone A, Paleja B, Singhal A, Tschumi A, Lee B,

Poidinger M, Zolezzi F, Quagliata L, Sander P, Newell E, Bertoletti A, Terracciano L, De Libero G, Mori L (2014) Parallel T-cell cloning and deep sequencing of human MAIT cells reveal stable oligoclonal TCRbeta repertoire. Nat Commun 5:3866. https://doi.org/10.1038/ncomms4866

7. Dias J, Leeansyah E, Sandberg JK (2017) Multiple layers of heterogeneity and subset diversity in human MAIT cell responses to distinct microorganisms and to innate cytokines. Proc Natl Acad Sci U S A 114: E5434–E5443. https://doi.org/10.1073/pnas.1705759114
8. Magalhaes I, Pingris K, Poitou C, Bessoles S, Venteclef N, Kiaf B, Beaudoin L, Da Silva J, Allatif O, Rossjohn J, Kjer-Nielsen L, McCluskey J, Ledoux S, Genser L, Torcivia A, Soudais C, Lantz O, Boitard C, Aron-Wisnewsky J, Larger E, Clement K, Lehuen A (2015) Mucosal-associated invariant T cell alterations in obese and type 2 diabetic patients. J Clin Invest 125 (4):1752–1762. https://doi.org/10.1172/JCI78941
9. Sharma PK, Wong EB, Napier RJ, Bishai WR, Ndung'u T, Kasprowicz VO, Lewinsohn DA, Lewinsohn DM, Gold MC (2015) High expression of CD26 accurately identifies human bacteria-reactive MR1-restricted MAIT cells. Immunology 145(3):443–453. https://doi.org/10.1111/imm.12461
10. Dias J, Sobkowiak MJ, Sandberg JK, Leeansyah E (2016) Human MAIT-cell responses to Escherichia coli: activation, cytokine production, proliferation, and cytotoxicity. J Leukoc Biol 100(1):233–240. https://doi.org/10.1189/jlb.4TA0815-391RR
11. Ben Youssef G, Tourret M, Salou M, Ghazarian L, Houdouin V, Mondot S, Mburu Y, Lambert M, Azarnoush S, Diana JS, Virlouvet AL, Peuchmaur M, Schmitz T, Dalle JH, Lantz O, Biran V, Caillat-Zucman S (2018) Ontogeny of human mucosal-associated invariant T cells and related T cell subsets. J Exp Med 215:459–479. https://doi.org/10.1084/jem.20171739
12. Rahimpour A, Koay HF, Enders A, Clanchy R, Eckle SB, Meehan B, Chen Z, Whittle B, Liu L, Fairlie DP, Goodnow CC, McCluskey J, Rossjohn J, Uldrich AP, Pellicci DG, Godfrey DI (2015) Identification of phenotypically and functionally heterogeneous mouse mucosal-associated invariant T cells using MR1 tetramers. J Exp Med 212 (7):1095–1108. https://doi.org/10.1084/jem.20142110
13. Corbett AJ, Eckle SB, Birkinshaw RW, Liu L, Patel O, Mahony J, Chen Z, Reantragoon R, Meehan B, Cao H, Williamson NA, Strugnell RA, Van Sinderen D, Mak JY, Fairlie DP, Kjer-Nielsen L, Rossjohn J, McCluskey J (2014) T-cell activation by transitory neo-antigens derived from distinct microbial pathways. Nature 509(7500):361–365. https://doi.org/10.1038/nature13160
14. Hashimoto K, Hirai M, Kurosawa Y (1995) A gene outside the human MHC related to classical HLA class I genes. Science 269 (5224):693–695
15. Riegert P, Wanner V, Bahram S (1998) Genomics, isoforms, expression, and phylogeny of the MHC class I-related MR1 gene. J Immunol 161(8):4066–4077
16. Xiao X, Liu B, Ma X, Yang S, Cai J (2019) Molecular cloning and characterization of the pig MHC class-related MR1 gene. Dev Comp Immunol 96:58–67. https://doi.org/10.1016/j.dci.2019.02.020
17. Boudinot P, Mondot S, Jouneau L, Teyton L, Lefranc MP, Lantz O (2016) Restricting nonclassical MHC genes coevolve with TRAV genes used by innate-like T cells in mammals. Proc Natl Acad Sci U S A 113(21): E2983–E2992. https://doi.org/10.1073/pnas.1600674113
18. Cui Y, Franciszkiewicz K, Mburu YK, Mondot S, Le Bourhis L, Premel V, Martin E, Kachaner A, Duban L, Ingersoll MA, Rabot S, Jaubert J, De Villartay JP, Soudais C, Lantz O (2015) Mucosal-associated invariant T cell-rich congenic mouse strain allows functional evaluation. J Clin Invest 125(11):4171–4185. https://doi.org/10.1172/JCI82424
19. McWilliam HE, Eckle SB, Theodossis A, Liu L, Chen Z, Wubben JM, Fairlie DP, Strugnell RA, Mintern JD, McCluskey J, Rossjohn J, Villadangos JA (2016) The intracellular pathway for the presentation of vitamin B-related antigens by the antigen-presenting molecule MR1. Nat Immunol 17 (5):531–537. https://doi.org/10.1038/ni.3416
20. Ussher JE, van Wilgenburg B, Hannaway RF, Ruustal K, Phalora P, Kurioka A, Hansen TH, Willberg CB, Phillips RE, Klenerman P (2016) TLR signaling in human antigen-presenting cells regulates MR1-dependent activation of MAIT cells. Eur J Immunol 46 (7):1600–1614. https://doi.org/10.1002/eji.201545969
21. Salerno-Goncalves R, Rezwan T, Sztein MB (2014) B cells modulate mucosal associated invariant T cell immune responses. Front

Immunol 4:511. https://doi.org/10.3389/fimmu.2013.00511

22. Harriff MJ, Karamooz E, Burr A, Grant WF, Canfield ET, Sorensen ML, Moita LF, Lewinsohn DM (2016) Endosomal MR1 trafficking plays a key role in presentation of Mycobacterium tuberculosis ligands to MAIT cells. PLoS Pathog 12(3):e1005524. https://doi.org/10.1371/journal.ppat.1005524
23. Gold MC, Cerri S, Smyk-Pearson S, Cansler ME, Vogt TM, Delepine J, Winata E, Swarbrick GM, Chua WJ, Yu YY, Lantz O, Cook MS, Null MD, Jacoby DB, Harriff MJ, Lewinsohn DA, Hansen TH, Lewinsohn DM (2010) Human mucosal associated invariant T cells detect bacterially infected cells. PLoS Biol 8(6):e1000407. https://doi.org/10.1371/journal.pbio.1000407
24. van Wilgenburg B, Scherwitzl I, Hutchinson EC, Leng T, Kurioka A, Kulicke C, de Lara C, Cole S, Vasanawathana S, Limpitikul W, Malasit P, Young D, Denney L, Consortium S-H, Moore MD, Fabris P, Giordani MT, Oo YH, Laidlaw SM, Dustin LB, Ho LP, Thompson FM, Ramamurthy N, Mongkolsapaya J, Willberg CB, Screaton GR, Klenerman P (2016) MAIT cells are activated during human viral infections. Nat Commun 7:11653. https://doi.org/10.1038/ncomms11653
25. Le Bourhis L, Martin E, Peguillet I, Guihot A, Froux N, Core M, Levy E, Dusseaux M, Meyssonnier V, Premel V, Ngo C, Riteau B, Duban L, Robert D, Huang S, Rottman M, Soudais C, Lantz O (2010) Antimicrobial activity of mucosal-associated invariant T cells. Nat Immunol 11(8):701–708. https://doi.org/10.1038/ni.1890
26. Shaler CR, Choi J, Rudak PT, Memarnejadian A, Szabo PA, Tun-Abraham ME, Rossjohn J, Corbett AJ, McCluskey J, McCormick JK, Lantz O, Hernandez-Alejandro R, Haeryfar SMM (2017) MAIT cells launch a rapid, robust and distinct hyperinflammatory response to bacterial superantigens and quickly acquire an anergic phenotype that impedes their cognate antimicrobial function: defining a novel mechanism of superantigen-induced immunopathology and immunosuppression. PLoS Biol 15(6): e2001930. https://doi.org/10.1371/journal.pbio.2001930
27. Jeffery HC, van Wilgenburg B, Kurioka A, Parekh K, Stirling K, Roberts S, Dutton EE, Hunter S, Geh D, Braitch MK, Rajanayagam J, Iqbal T, Pinkney T, Brown R, Withers DR, Adams DH, Klenerman P, Oo YH (2016) Biliary epithelium and liver B cells exposed to bacteria activate intrahepatic MAIT cells through MR1. J Hepatol 64(5):1118–1127. https://doi.org/10.1016/j.jhep.2015.12.017
28. Bernal I, Hofmann JD, Bulitta B, Klawonn F, Michel AM, Jahn D, Neumann-Schaal M, Bruder D, Jansch L (2018) Clostridioides difficile activates human mucosal-associated invariant T cells. Front Microbiol 9:2532. https://doi.org/10.3389/fmicb.2018.02532
29. Jahreis S, Bottcher S, Hartung S, Rachow T, Rummler S, Dietl AM, Haas H, Walther G, Hochhaus A, von Lilienfeld-Toal M (2018) Human MAIT cells are rapidly activated by Aspergillus spp. in an APC-dependent manner. Eur J Immunol 48(10):1698–1706. https://doi.org/10.1002/eji.201747312
30. Meermeier EW, Harriff MJ, Karamooz E, Lewinsohn DM (2018) MAIT cells and microbial immunity. Immunol Cell Biol 96 (6):607–617. https://doi.org/10.1111/imcb.12022
31. Ussher JE, Bilton M, Attwod E, Shadwell J, Richardson R, de Lara C, Mettke E, Kurioka A, Hansen TH, Klenerman P, Willberg CB (2014) CD161++ CD8+ T cells, including the MAIT cell subset, are specifically activated by IL-12+IL-18 in a TCR-independent manner. Eur J Immunol 44(1):195–203. https://doi.org/10.1002/eji.201343509
32. Sattler A, Dang-Heine C, Reinke P, Babel N (2015) IL-15 dependent induction of IL-18 secretion as a feedback mechanism controlling human MAIT-cell effector functions. Eur J Immunol 45(8):2286–2298. https://doi.org/10.1002/eji.201445313
33. Leeansyah E, Svard J, Dias J, Buggert M, Nystrom J, Quigley MF, Moll M, Sonnerborg A, Nowak P, Sandberg JK (2015) Arming of MAIT cell Cytolytic antimicrobial activity is induced by IL-7 and defective in HIV-1 infection. PLoS Pathog 11(8):e1005072. https://doi.org/10.1371/journal.ppat.1005072
34. Solders M, Gorchs L, Erkers T, Lundell AC, Nava S, Gidlof S, Tiblad E, Magalhaes I, Kaipe H (2017) MAIT cells accumulate in placental intervillous space and display a highly cytotoxic phenotype upon bacterial stimulation. Sci Rep 7(1):6123. https://doi.org/10.1038/s41598-017-06430-6
35. Wilgenburg BV, Loh L, Chen Z, Pediongco TJ, Wang H, Shi M, Zhao Z, Koutsakos M, Nussing S, Sant S, Wang Z, D'Souza C, Jia X, Almeida CF, Kostenko L, Eckle SBG, Meehan BS, Kallies A, Godfrey DI, Reading PC,

Corbett AJ, McCluskey J, Klenerman P, Kedzierska K, Hinks TSC (2018) MAIT cells contribute to protection against lethal influenza infection in vivo. Nat Commun 9 (1):4706. https://doi.org/10.1038/s41467-018-07207-9

36. Loh L, Wang Z, Sant S, Koutsakos M, Jegaskanda S, Corbett AJ, Liu L, Fairlie DP, Crowe J, Rossjohn J, Xu J, Doherty PC, McCluskey J, Kedzierska K (2016) Human mucosal-associated invariant T cells contribute to antiviral influenza immunity via IL-18-dependent activation. Proc Natl Acad Sci U S A 113(36):10133–10138. https://doi.org/10.1073/pnas.1610750113

37. Ussher JE, Willberg CB, Klenerman P (2018) MAIT cells and viruses. Immunol Cell Biol 96 (6):630–641. https://doi.org/10.1111/imcb.12008

38. Eberhard JM, Kummer S, Hartjen P, Hufner A, Diedrich T, Degen O, Lohse AW, van Lunzen J, Schulze Zur Wiesch J (2016) Reduced CD161(+) MAIT cell frequencies in HCV and HIV/HCV co-infection: Is the liver the heart of the matter? J Hepatol 65 (6):1261–1263. https://doi.org/10.1016/j.jhep.2016.07.031

39. Cosgrove C, Ussher JE, Rauch A, Gartner K, Kurioka A, Huhn MH, Adelmann K, Kang YH, Fergusson JR, Simmonds P, Goulder P, Hansen TH, Fox J, Gunthard HF, Khanna N, Powrie F, Steel A, Gazzard B, Phillips RE, Frater J, Uhlig H, Klenerman P (2013) Early and nonreversible decrease of CD161++ / MAIT cells in HIV infection. Blood 121 (6):951–961. https://doi.org/10.1182/blood-2012-06-436436

40. Dias J, Hengst J, Parrot T, Leeansyah E, Lunemann S, Malone DFG, Hardtke S, Strauss O, Zimmer CL, Berglin L, Schirdewahn T, Ciesek S, Marquardt N, von Hahn T, Manns MP, Cornberg M, Ljunggren HG, Wedemeyer H, Sandberg JK, Bjorkstrom NK (2019) Chronic hepatitis delta virus infection leads to functional impairment and severe loss of MAIT cells. J Hepatol 71:301–312. https://doi.org/10.1016/j.jhep.2019.04.009

41. Seach N, Guerri L, Le Bourhis L, Mburu Y, Cui Y, Bessoles S, Soudais C, Lantz O (2013) Double-positive thymocytes select mucosal-associated invariant T cells. J Immunol 191 (12):6002–6009. https://doi.org/10.4049/jimmunol.1301212

42. Koay HF, Gherardin NA, Enders A, Loh L, Mackay LK, Almeida CF, Russ BE, Nold-Petry CA, Nold MF, Bedoui S, Chen Z, Corbett AJ, Eckle SB, Meehan B, d'Udekem Y, Konstantinov IE, Lappas M, Liu L, Goodnow CC, Fairlie DP, Rossjohn J, Chong MM, Kedzierska K, Berzins SP, Belz GT, McCluskey J, Uldrich AP, Godfrey DI, Pellicci DG (2016) A three-stage intrathymic development pathway for the mucosal-associated invariant T cell lineage. Nat Immunol 17(11):1300–1311. https://doi.org/10.1038/ni.3565

43. Gold MC, Eid T, Smyk-Pearson S, Eberling Y, Swarbrick GM, Langley SM, Streeter PR, Lewinsohn DA, Lewinsohn DM (2013) Human thymic MR1-restricted MAIT cells are innate pathogen-reactive effectors that adapt following thymic egress. Mucosal Immunol 6(1):35–44. https://doi.org/10.1038/mi.2012.45

44. Martin E, Treiner E, Duban L, Guerri L, Laude H, Toly C, Premel V, Devys A, Moura IC, Tilloy F, Cherif S, Vera G, Latour S, Soudais C, Lantz O (2009) Stepwise development of MAIT cells in mouse and human. PLoS Biol 7(3):e54. https://doi.org/10.1371/journal.pbio.1000054

45. Leeansyah E, Loh L, Nixon DF, Sandberg JK (2014) Acquisition of innate-like microbial reactivity in mucosal tissues during human fetal MAIT-cell development. Nat Commun 5:3143. https://doi.org/10.1038/ncomms4143

46. Salou M, Legoux F, Gilet J, Darbois A, du Halgouet A, Alonso R, Richer W, Goubet AG, Daviaud C, Menger L, Procopio E, Premel V, Lantz O (2019) A common transcriptomic program acquired in the thymus defines tissue residency of MAIT and NKT subsets. J Exp Med 216(1):133–151. https://doi.org/10.1084/jem.20181483

47. Novak J, Dobrovolny J, Novakova L, Kozak T (2014) The decrease in number and change in phenotype of mucosal-associated invariant T cells in the elderly and differences in men and women of reproductive age. Scand J Immunol 80(4):271–275. https://doi.org/10.1111/sji.12193

48. Lee OJ, Cho YN, Kee SJ, Kim MJ, Jin HM, Lee SJ, Park KJ, Kim TJ, Lee SS, Kwon YS, Kim N, Shin MG, Shin JH, Suh SP, Ryang DW, Park YW (2014) Circulating mucosal-associated invariant T cell levels and their cytokine levels in healthy adults. Exp Gerontol 49:47–54. https://doi.org/10.1016/j.exger.2013.11.003

49. Walker LJ, Tharmalingam H, Klenerman P (2014) The rise and fall of MAIT cells with age. Scand J Immunol 80(6):462–463. https://doi.org/10.1111/sji.12237

50. Chen Z, Wang H, D'Souza C, Sun S, Kostenko L, Eckle SB, Meehan BS, Jackson DC, Strugnell RA, Cao H, Wang N, Fairlie DP, Liu L, Godfrey DI, Rossjohn J, McCluskey J, Corbett AJ (2017) Mucosal-associated invariant T-cell activation and accumulation after in vivo infection depends on microbial riboflavin synthesis and co-stimulatory signals. Mucosal Immunol 10 (1):58–68. https://doi.org/10.1038/mi.2016.39
51. Le Bourhis L, Dusseaux M, Bohineust A, Bessoles S, Martin E, Premel V, Core M, Sleurs D, Serriari NE, Treiner E, Hivroz C, Sansonetti P, Gougeon ML, Soudais C, Lantz O (2013) MAIT cells detect and efficiently lyse bacterially-infected epithelial cells. PLoS Pathog 9(10):e1003681. https://doi.org/10.1371/journal.ppat.1003681
52. Kurioka A, Ussher JE, Cosgrove C, Clough C, Fergusson JR, Smith K, Kang YH, Walker LJ, Hansen TH, Willberg CB, Klenerman P (2015) MAIT cells are licensed through granzyme exchange to kill bacterially sensitized targets. Mucosal Immunol 8 (2):429–440. https://doi.org/10.1038/mi.2014.81
53. Voskoboinik I, Whisstock JC, Trapani JA (2015) Perforin and granzymes: function, dysfunction and human pathology. Nat Rev Immunol 15(6):388–400. https://doi.org/10.1038/nri3839
54. Tang XZ, Jo J, Tan AT, Sandalova E, Chia A, Tan KC, Lee KH, Gehring AJ, De Libero G, Bertoletti A (2013) IL-7 licenses activation of human liver intrasinusoidal mucosal-associated invariant T cells. J Immunol 190 (7):3142–3152. https://doi.org/10.4049/jimmunol.1203218
55. Gibbs A, Leeansyah E, Introini A, Paquin-Proulx D, Hasselrot K, Andersson E, Broliden K, Sandberg JK, Tjernlund A (2016) MAIT cells reside in the female genital mucosa and are biased towards IL-17 and IL-22 production in response to bacterial stimulation. Mucosal Immunol 10:35–45. https://doi.org/10.1038/mi.2016.30
56. Georgel P, Radosavljevic M, Macquin C, Bahram S (2011) The non-conventional MHC class I MR1 molecule controls infection by Klebsiella pneumoniae in mice. Mol Immunol 48(5):769–775. https://doi.org/10.1016/j.molimm.2010.12.002
57. Meierovics A, Yankelevich WJ, Cowley SC (2013) MAIT cells are critical for optimal mucosal immune responses during in vivo pulmonary bacterial infection. Proc Natl Acad Sci U S A 110(33):E3119–E3128. https://doi.org/10.1073/pnas.1302799110
58. Mpina M, Maurice NJ, Yajima M, Slichter CK, Miller HW, Dutta M, McElrath MJ, Stuart KD, De Rosa SC, McNevin JP, Linsley PS, Abdulla S, Tanner M, Hoffman SL, Gottardo R, Daubenberger CA, Prlic M (2017) Controlled human malaria infection leads to long-lasting changes in innate and innate-like lymphocyte populations. J Immunol 199(1):107–118. https://doi.org/10.4049/jimmunol.1601989
59. Booth JS, Salerno-Goncalves R, Blanchard TG, Patil SA, Kader HA, Safta AM, Morningstar LM, Czinn SJ, Greenwald BD, Sztein MB (2015) Mucosal-associated invariant T cells in the human gastric mucosa and blood: role in helicobacter pylori infection. Front Immunol 6:466. https://doi.org/10.3389/fimmu.2015.00466
60. Jiang J, Wang X, An H, Yang B, Cao Z, Liu Y, Su J, Zhai F, Wang R, Zhang G, Cheng X (2014) Mucosal-associated invariant T-cell function is modulated by programmed death-1 signaling in patients with active tuberculosis. Am J Respir Crit Care Med 190(3):329–339. https://doi.org/10.1164/rccm.201401-0106OC
61. Leung DT, Bhuiyan TR, Nishat NS, Hoq MR, Aktar A, Rahman MA, Uddin T, Khan AI, Chowdhury F, Charles RC, Harris JB, Calderwood SB, Qadri F, Ryan ET (2014) Circulating mucosal associated invariant T cells are activated in Vibrio cholerae O1 infection and associated with lipopolysaccharide antibody responses. PLoS Negl Trop Dis 8 (8):e3076. https://doi.org/10.1371/journal.pntd.0003076
62. Salerno-Goncalves R, Luo D, Fresnay S, Magder L, Darton TC, Jones C, Waddington CS, Blohmke CJ, Angus B, Levine MM, Pollard AJ, Sztein MB (2017) Challenge of humans with wild-type Salmonella enterica Serovar Typhi elicits changes in the activation and homing characteristics of mucosal-associated invariant T cells. Front Immunol 8:398. https://doi.org/10.3389/fimmu.2017.00398
63. Leeansyah E, Ganesh A, Quigley MF, Sonnerborg A, Andersson J, Hunt PW, Somsouk M, Deeks SG, Martin JN, Moll M, Shacklett BL, Sandberg JK (2013) Activation, exhaustion, and persistent decline of the antimicrobial MR1-restricted MAIT-cell population in chronic HIV-1 infection. Blood 121 (7):1124–1135. https://doi.org/10.1182/blood-2012-07-445429

64. Wong EB, Akilimali NA, Govender P, Sullivan ZA, Cosgrove C, Pillay M, Lewinsohn DM, Bishai WR, Walker BD, Ndung'u T, Klenerman P, Kasprowicz VO (2013) Low levels of peripheral CD161++CD8+ mucosal associated invariant T (MAIT) cells are found in HIV and HIV/TB co-infection. PLoS One 8(12):e83474. https://doi.org/10.1371/journal.pone.0083474
65. Bolte FJ, O'Keefe AC, Webb LM, Serti E, Rivera E, Liang TJ, Ghany M, Rehermann B (2017) Intra-hepatic depletion of mucosal-associated invariant T cells in hepatitis C virus-induced liver inflammation. Gastroenterology 153(5):1392–1403. e1392. https://doi.org/10.1053/j.gastro.2017.07.043
66. Beudeker BJB, van Oord GW, Arends JE, Schulze Zur Wiesch J, van der Heide MS, de Knegt RJ, Verbon A, Boonstra A, Claassen MAA (2017) Mucosal-associated invariant T-cell frequency and function in blood and liver of HCV mono- and HCV/HIV co-infected patients with advanced fibrosis. Liver Int 38:458–468. https://doi.org/10.1111/liv.13544
67. Barathan M, Mohamed R, Vadivelu J, Chang LY, Saeidi A, Yong YK, Ravishankar Ram M, Gopal K, Velu V, Larsson M, Shankar EM (2016) Peripheral loss of CD8(+) CD161(++) TCRValpha7.2(+) mucosal-associated invariant T cells in chronic hepatitis C virus-infected patients. Eur J Clin Invest 46 (2):170–180. https://doi.org/10.1111/eci.12581
68. Boeijen LL, Montanari NR, de Groen RA, van Oord GW, van der Heide-Mulder M, de Knegt RJ, Boonstra A (2017) Mucosal-associated invariant T cells are more activated in chronic hepatitis B, but not depleted in blood: reversal by antiviral therapy. J Infect Dis 216(8):969–976. https://doi.org/10.1093/infdis/jix425
69. Paget C, Trottein F (2019) Mechanisms of bacterial Superinfection post-influenza: a role for unconventional T cells. Front Immunol 10:336. https://doi.org/10.3389/fimmu.2019.00336
70. von Seth E, Zimmer CL, Reuterwall-Hansson M, Barakat A, Arnelo U, Bergquist A, Ivarsson MA, Bjorkstrom NK (2018) Primary sclerosing cholangitis leads to dysfunction and loss of MAIT cells. Eur J Immunol 48(12):1997–2004. https://doi.org/10.1002/eji.201847608
71. Berglin L, Bergquist A, Johansson H, Glaumann H, Jorns C, Lunemann S, Wedemeyer H, Ellis EC, Bjorkstrom NK (2014) In situ characterization of intrahepatic non-parenchymal cells in PSC reveals phenotypic patterns associated with disease severity. PLoS One 9(8):e105375. https://doi.org/10.1371/journal.pone.0105375
72. Hegde P, Weiss E, Paradis V, Wan J, Mabire M, Sukriti S, Rautou PE, Albuquerque M, Picq O, Gupta AC, Ferrere G, Gilgenkrantz H, Kiaf B, Toubal A, Beaudoin L, Letteron P, Moreau R, Lehuen A, Lotersztajn S (2018) Mucosal-associated invariant T cells are a profibrogenic immune cell population in the liver. Nat Commun 9(1):2146. https://doi.org/10.1038/s41467-018-04450-y
73. Li Y, Huang B, Jiang X, Chen W, Zhang J, Wei Y, Chen Y, Lian M, Bian Z, Miao Q, Peng Y, Fang J, Wang Q, Tang R, Gershwin ME, Ma X (2018) Mucosal-associated invariant T cells improve nonalcoholic fatty liver disease through regulating macrophage polarization. Front Immunol 9:1994. https://doi.org/10.3389/fimmu.2018.01994
74. Hengst J, Strunz B, Deterding K, Ljunggren HG, Leeansyah E, Manns MP, Cornberg M, Sandberg JK, Wedemeyer H, Bjorkstrom NK (2016) Nonreversible MAIT cell-dysfunction in chronic hepatitis C virus infection despite successful interferon-free therapy. Eur J Immunol 46(9):2204–2210. https://doi.org/10.1002/eji.201646447
75. Cannizzo ES, Cerrone M, Merlini E, van Wilgenburg B, Swadling L, Ancona G, De Bona A, d'Arminio Monforte A, Klenerman P, Marchetti G (2019) Successful direct-acting antiviral therapy in HIV/HCV co-infected patients fails to restore circulating mucosal-associated invariant T cells. Eur J Immunol 49:1127–1129. https://doi.org/10.1002/eji.201948152
76. Howson LJ, Salio M, Cerundolo V (2015) MR1-restricted mucosal-associated invariant T cells and their activation during infectious diseases. Front Immunol 6:303. https://doi.org/10.3389/fimmu.2015.00303
77. Sundstrom P, Ahlmanner F, Akeus P, Sundquist M, Alsen S, Yrlid U, Borjesson L, Sjoling A, Gustavsson B, Wong SB, Quiding-Jarbrink M (2015) Human mucosa-associated invariant T cells accumulate in colon adenocarcinomas but produce reduced amounts of IFN-gamma. J Immunol 195(7):3472–3481. https://doi.org/10.4049/jimmunol.1500258
78. Huang WC, Hsiao YC, Wu CC, Hsu YT, Chang CL (2019) Less circulating mucosal-associated invariant T cells in patients with cervical cancer. Taiwan J Obstet Gynecol 58

(1):117–121. https://doi.org/10.1016/j.tjog.2018.11.022

79. Duan M, Goswami S, Shi JY, Wu LJ, Wang XY, Ma JQ, Zhang Z, Shi Y, Ma LJ, Zhang S, Xi RB, Cao Y, Zhou J, Fan J, Zhang XM, Gao Q (2019) Activated and exhausted MAIT cells foster disease progression and indicate poor outcome in hepatocellular carcinoma. Clin Cancer Res 25(11):3304–3316. https://doi.org/10.1158/1078-0432.CCR-18-3040
80. Park Y-W, Kee S-J (2015) Mucosal-associated invariant T cells: a new player in innate immunity. J Rheum Dis 22(6):337–345
81. Ling L, Lin Y, Zheng W, Hong S, Tang X, Zhao P, Li M, Ni J, Li C, Wang L, Jiang Y (2016) Circulating and tumor-infiltrating mucosal associated invariant T (MAIT) cells in colorectal cancer patients. Sci Rep 6:20,358. https://doi.org/10.1038/srep20358
82. Zabijak L, Attencourt C, Guignant C, Chatelain D, Marcelo P, Marolleau JP, Treiner E (2015) Increased tumor infiltration by mucosal-associated invariant T cells correlates with poor survival in colorectal cancer patients. Cancer Immunol Immunother 64(12):1601–1608. https://doi.org/10.1007/s00262-015-1764-7
83. Shaler CR, Tun-Abraham ME, Skaro AI, Khazaie K, Corbett AJ, Mele T, Hernandez-Alejandro R, Haeryfar SMM (2017) Mucosa-associated invariant T cells infiltrate hepatic metastases in patients with colorectal carcinoma but are rendered dysfunctional within and adjacent to tumor microenvironment. Cancer Immunol Immunother 66:1563–1575. https://doi.org/10.1007/s00262-017-2050-7
84. Sundstrom P, Szeponik L, Ahlmanner F, Sundquist M, Wong JSB, Lindskog EB, Gustafsson B, Quiding-Jarbrink M (2019) Tumor-infiltrating mucosal-associated invariant T (MAIT) cells retain expression of cytotoxic effector molecules. Oncotarget 10(29):2810–2823. https://doi.org/10.18632/oncotarget.26866
85. Mitchell J, Kvedaraite E, von Bahr Greenwood T, Henter JI, Pellicci DG, Berzins SP, Kannourakis G (2018) Altered populations of unconventional T cell lineages in patients with Langerhans cell Histiocytosis. Sci Rep 8(1):16506. https://doi.org/10.1038/s41598-018-34873-y
86. Jupp J, Mansour S, Johnson CD, Sanderson J, Fine D, Gadola S (2015) T-cell populations in chronic pancreatitis. Pancreatology 15(4):311–312. https://doi.org/10.1016/j.pan.2015.04.009
87. Rouxel O, Da Silva J, Beaudoin L, Nel I, Tard C, Cagninacci L, Kiaf B, Oshima M, Diedisheim M, Salou M, Corbett A, Rossjohn J, McCluskey J, Scharfmann R, Battaglia M, Polak M, Lantz O, Beltrand J, Lehuen A (2017) Cytotoxic and regulatory roles of mucosal-associated invariant T cells in type 1 diabetes. Nat Immunol 18(12):1321–1331. https://doi.org/10.1038/ni.3854
88. Harms RZ, Lorenzo KM, Corley KP, Cabrera MS, Sarvetnick NE (2015) Altered CD161 bright CD8+ mucosal associated invariant T (MAIT)-like cell dynamics and increased differentiation states among juvenile type 1 diabetics. PLoS One 10(1):e0117335. https://doi.org/10.1371/journal.pone.0117335
89. Harms RZ, Lorenzo-Arteaga KM, Ostlund KR, Smith VB, Smith LM, Gottlieb P, Sarvetnick N (2018) Abnormal T cell frequencies, including cytomegalovirus-associated expansions, distinguish seroconverted subjects at risk for type 1 diabetes. Front Immunol 9:2332. https://doi.org/10.3389/fimmu.2018.02332
90. Kuric E, Krogvold L, Hanssen KF, Dahl-Jorgensen K, Skog O, Korsgren O (2018) No evidence for presence of mucosal-associated invariant T cells in the insulitic lesions in patients recently diagnosed with type 1 diabetes. Am J Pathol 188(8):1744–1748. https://doi.org/10.1016/j.ajpath.2018.04.009
91. Mexhitaj I, Nyirenda MH, Li R, O'Mahony J, Rezk A, Rozenberg A, Moore CS, Johnson T, Sadovnick D, Collins DL, Arnold DL, Gran B, Yeh EA, Marrie RA, Banwell B, Bar-Or A (2019) Abnormal effector and regulatory T cell subsets in paediatric-onset multiple sclerosis. Brain 142(3):617–632. https://doi.org/10.1093/brain/awz017
92. Hinks TS (2016) Mucosal-associated invariant T cells in autoimmunity, immune-mediated diseases and airways disease. Immunology 148(1):1–12. https://doi.org/10.1111/imm.12582
93. Carolan E, Tobin LM, Mangan BA, Corrigan M, Gaoatswe G, Byrne G, Geoghegan J, Cody D, O'Connell J, Winter DC, Doherty DG, Lynch L, O'Shea D, Hogan AE (2015) Altered distribution and increased IL-17 production by mucosal-associated invariant T cells in adult and childhood obesity. J Immunol 194(12):5775–5780. https://doi.org/10.4049/jimmunol.1402945
94. Touch S, Assmann KE, Aron-Wisnewsky J, Marquet F, Rouault C, Fradet M,

Mosbah H, Consortium M, Isnard R, Helft G, Lehuen A, Poitou C, Clement K, Andre S, MetaCardis Consortium (2018) Mucosal-associated invariant T (MAIT) cells are depleted and prone to apoptosis in cardiometabolic disorders. FASEB J: fj201800052RR. https://doi.org/10.1096/fj.201800052RR

95. O'Brien A, Loftus RM, Pisarska MM, Tobin LM, Bergin R, Wood NAW, Foley C, Mat A, Tinley FC, Bannan C, Sommerville G, Veerapen N, Besra GS, Sinclair LV, Moynagh PN, Lynch L, Finlay DK, O'Shea D, Hogan AE (2019) Obesity reduces mTORC1 activity in mucosal-associated invariant T cells, driving defective metabolic and functional responses. J Immunol 202(12):3404–3411. https://doi.org/10.4049/jimmunol.1801600

96. Abrahamsson SV, Angelini DF, Dubinsky AN, Morel E, Oh U, Jones JL, Carassiti D, Reynolds R, Salvetti M, Calabresi PA, Coles AJ, Battistini L, Martin R, Burt RK, Muraro PA (2013) Non-myeloablative autologous haematopoietic stem cell transplantation expands regulatory cells and depletes IL-17 producing mucosal-associated invariant T cells in multiple sclerosis. Brain 136 (Pt 9):2888–2903. https://doi.org/10.1093/brain/awt182

97. Moore JJ, Massey JC, Ford CD, Khoo ML, Zaunders JJ, Hendrawan K, Barnett Y, Barnett MH, Kyle KA, Zivadinov R, Ma KC, Milliken ST, Sutton IJ, Ma DDF (2019) Prospective phase II clinical trial of autologous haematopoietic stem cell transplant for treatment refractory multiple sclerosis. J Neurol Neurosurg Psychiatry 90(5):514–521. https://doi.org/10.1136/jnnp-2018-319446

98. Novak J, Dobrovolny J, Brozova J, Novakova L, Kozak T (2016) Recovery of mucosal-associated invariant T cells after myeloablative chemotherapy and autologous peripheral blood stem cell transplantation. Clin Exp Med 16(4):529–537. https://doi.org/10.1007/s10238-015-0384-z

99. Kawaguchi K, Umeda K, Hiejima E, Iwai A, Mikami M, Nodomi S, Saida S, Kato I, Hiramatsu H, Yasumi T, Nishikomori R, Kondo T, Takaori-Kondo A, Heike T, Adachi S (2018) Influence of post-transplant mucosal-associated invariant T cell recovery on the development of acute graft-versus-host disease in allogeneic bone marrow transplantation. Int J Hematol 108(1):66–75. https://doi.org/10.1007/s12185-018-2442-2

100. Bhattacharyya A, Hanafi LA, Sheih A, Golob JL, Srinivasan S, Boeckh MJ, Pergam SA, Mahmood S, Baker KK, Gooley TA, Milano F, Fredricks DN, Riddell SR, Turtle CJ (2018) Graft-derived reconstitution of mucosal-associated invariant T cells after allogeneic hematopoietic cell transplantation. Biol Blood Marrow Transplant 24 (2):242–251. https://doi.org/10.1016/j.bbmt.2017.10.003

101. Solders M, Erkers T, Gorchs L, Poiret T, Remberger M, Magalhaes I, Kaipe H (2017) Mucosal-associated invariant T cells display a poor reconstitution and altered phenotype after allogeneic hematopoietic stem cell transplantation. Front Immunol 8:1861. https://doi.org/10.3389/fimmu.2017.01861

102. Varelias A, Bunting MD, Ormerod KL, Koyama M, Olver SD, Straube J, Kuns RD, Robb RJ, Henden AS, Cooper L, Lachner N, Gartlan KH, Lantz O, Kjer-Nielsen L, Mak JY, Fairlie DP, Clouston AD, McCluskey J, Rossjohn J, Lane SW, Hugenholtz P, Hill GR (2018) Recipient mucosal-associated invariant T cells control GVHD within the colon. J Clin Invest 128(5):1919–1936. https://doi.org/10.1172/JCI91646

103. Turtle CJ, Swanson HM, Fujii N, Estey EH, Riddell SR (2009) A distinct subset of self-renewing human memory CD8+ T cells survives cytotoxic chemotherapy. Immunity 31 (5):834–844. https://doi.org/10.1016/j.immuni.2009.09.015

104. Fergusson JR, Ussher JE, Kurioka A, Klenerman P, Walker LJ (2018) High MDR-1 expression by MAIT cells confers resistance to cytotoxic but not immunosuppressive MDR-1 substrates. Clin Exp Immunol 194(2):180–191. https://doi.org/10.1111/cei.13165

Chapter 2

Isolation and Characterization of MAIT Cells from Human Tissue Biopsies

Daniela Di Blasi, Alessandro Vacchini, Gennaro De Libero, and Lucia Mori

Abstract

Human mucosal-associated invariant T (MAIT) cells are unconventional T cells highly enriched in tissues exposed to microbial antigens including the oral, gastrointestinal and genital mucosae, liver, and lung. Here we describe a protocol for isolation and characterization of peripheral blood and tissue-infiltrating MAIT cells by using multicolor flow cytometry. This technology allows the analysis of multiple markers in a single sample at a single-cell level. Study of human samples requires particular care since the sample amount is often limited. We present a protocol optimized for the isolation and characterization of human MAIT cells and the identification of MAIT cell populations detected by simultaneous expression of multiple activation markers and inhibitory receptors.

Key words Human MR1 tetramer, Immune phenotype, MAIT cells, Multicolor flow cytometry, Tissue-resident cells

1 Introduction

Flow cytometry is a powerful methodology used in clinical practice and in research to perform complex cellular analyses quickly and efficiently. Thanks to its versatility by analyzing several parameters simultaneously, it has made a tremendous impact in immunology. The level of information obtained from a single sample can be greatly expanded by using multiple fluorescent antibodies directed against different targets. A second important advantage of this technology is the possibility to sort unique populations of cells that can be used in functional studies.

This chapter intends to introduce the basic theory, material and suggestions for appropriate protocols needed to plan and run a multidimensional analysis of MAIT cells.

MAIT cells are unconventional T cells with both innate and adaptive immune properties [1]. They represent up to 10% of circulating T cells in humans and are more abundant in tissues

Helen Kaipe and Isabelle Magalhaes (eds.), *MAIT Cells: Methods and Protocols*, Methods in Molecular Biology, vol. 2098,
https://doi.org/10.1007/978-1-0716-0207-2_2,

constantly exposed to microbial antigens including the gastrointestinal and genital mucosae, liver, and lung [2–5].

MAIT cells recognize as antigens small molecular-weight molecules generated in the riboflavin (vitamin B2) metabolic pathway, which are presented by the major histocompatibility complex (MHC)-related protein 1 (MR1) [6].

Like MHC class I molecules, MR1 is ubiquitously expressed, although MR1 protein levels are almost undetectable on the cell surface under physiological conditions [7]. Surface expression increases in cells infected by microbes as a result of MR1 protein stabilization by the specific ligands [8]. The MR1–antigen complexes establish cognate interactions with the TCR of MAIT cells without the requirements for co-stimulatory molecules. Antigen recognition by MAIT cells is mediated by semi-invariant αβ T-cell receptors (TCRs) with the TCRα chains made of the Vα7.2 (TRAV1-2) region, which is evolutionarily conserved in mammals [9], joined to Jα33 or Jα12 or Jα20 gene segments [5, 10, 11] and paired with a limited number of TCRVβ chains [1]. MAIT cells are typically $CD8^+$ or $CD4^-$ $CD8^-$ double negative, although some $CD4^+$ MAIT cells have been described [5], and they express high levels of the NK cell-related marker CD161 [12].

MAIT cells have typically been identified by using combined anti-TCR Vα7.2 and anti-CD161 mAbs. Recently, antigen-loaded MR1 tetramers have been developed for the unequivocal identification of human MAIT cells [13, 14].

We have previously studied MAIT cells in the gastrointestinal lamina propria and in the liver, thus we propose the following protocol for the isolation of MAIT cells from both gut and liver human biopsies.

In optimizing this protocol, we compared the efficiency of two different techniques for isolating MAIT cells from the two types of tissue biopsies: (1) mechanical dissociation of the tissue above using a cell strainer, and (2) enzymatic digestion using a mixture of collagenase and DNase enzymes. While the enzymatic digestion is necessary for MAIT cell isolation from gut biopsies, due to the localization of the T cells into the lamina propria, mechanical dissociation is sufficient for MAIT cell isolation from liver biopsies and allows the recovery of a large number of viable cells. The resulting lymphocyte suspensions contain viable lymphocytes (including MAIT cells), epithelial cells and debris. In order to improve the purity of MAIT cell isolation, after tissue disruption, lymphocytes are separated from epithelial cells and debris by using density gradient centrifugation.

We also describe a high dimensional data analysis using multicolor flow cytometry followed by dimensional reduction by t-SNE analysis. This type of analysis allows precise characterization and comparison of MAIT cells isolated from human tissues and peripheral blood.

2 Materials

All work is conducted with sterile material and under sterile conditions in a biosafety cabinet. Sterile conditions have to be maintained during the preparation of reagents and during cell culture assays.

2.1 Isolation of Immune Cells from Peripheral Blood and Human Biopsies

1. Ficoll Density Gradient Media (Density 1.077 ± 0.001 g/mL, osmolarity 290 ± 15 mOsm).
2. Dulbecco's PBS without Ca^{2+} and Mg^{2+} (DPBS).
3. Polypropylene conical tubes, 15 and 50 mL.
4. Tissue culture flasks, 175 cm^2.
5. Surgical forceps.
6. Thermostatic water bath.
7. Cell strainers, 40 μm nylon.
8. Petri dishes.
9. Pasteur pipettes, 230 mm, glass, pre-plugged.
10. Carbon steel scalpels.
11. Disposable 2 mL syringes.
12. Hemocytometer Neubauer chamber.
13. Cryopreservation medium: 90% fetal calf serum (FCS)/10% dimethyl sulfoxide (DMSO).
14. Digestion medium: RPMI-1640 medium containing 25 mM Hepes, 100 U/mL DNAse I, 200 U/mL Collagenase IV, 2.5 μg/mL Amphotericin B (Fungizone), 5 μg/mL Vancomycin, 26.6/3.3 μg/mL Piperacillin/Tazobactam, 10 μg/mL Ciprofloxacin (Ciproxin infusion 0.2 g/100 mL solution).
15. Cryovials, 1.8 mL.

2.2 *In Vitro* Stimulation

1. Antigen presenting cells (APCs): Human A375 melanoma cells β_2 microglobulin (β_2m)-deficient and overexpressing human MR1 (A375-MR1, *see* **Note 1**) [5].
2. Bacterial metabolite antigens or bacterial lysates. *E. coli* DH5α lysates corresponding to 10^7 CFU/mL can be used as well as 5-OP-RU at 50 nM [15].
3. Tissue culture, 24-well plates.
4. Cell culture medium: RPMI 1640 medium supplemented with 10% heat-inactivated FCS, 100 U/mL Kanamycin, 2 mM stable glutamine, 1 mM sodium pyruvate, and 1% MEM non-essential amino acids (RPMI 10% FCS).
5. Phorbol 12-myristate 13-acetate (PMA), stock solution 1 mg/mL in DMSO. Store at −70 °C.

6. Calcium-ionophore A23187, stock solution 2 mg/mL in DMSO. Store at −70 °C.
7. Brefeldin A (BFA), stock solution 5 mg/mL in ethanol. Store at −70 °C.

2.3 Multicolor Flow Cytometry

1. FACS buffer used for staining: DPBS containing 0.5% bovine serum albumin (BSA Fraction V, pH 7.0), 0.02% NaN_3. Store at 4 °C.
2. FcR-blocking buffer: DPBS containing 50% human serum.
3. Saponin-washing buffer: DPBS containing 0.1% saponin.
4. Fixation buffer: DPBS containing 4% paraformaldehyde, pH 7.4. Prepare fresh each time.
5. LIVE/DEAD™ Fixable Dead Cell Stain Sampler Kit.
6. 4′, 6-diamidino-2-phenylindole (DAPI).
7. Antibodies of interest labeled with appropriately selected fluorochromes (Tables 1, 2, and 3).
8. 5-(2-oxopropylideneamino)-6-D-ribitylaminouracil (5-OP-RU)-loaded human MR1 tetramer labeled with phycoerythrin (PE).
9. VersaComp Antibody Capture Bead Kit.
10. Flow cytometry calibration particles: Rainbow Calibration Particles, 8 peaks.
11. Tissue culture, round-bottomed 96-well plates.
12. Polypropylene round-bottom tubes, 5 mL.
13. Cell strainer (35 μm nylon mesh) cupped polystyrene round-bottom tubes, 5 mL.
14. Flow Cytometer equipped with 355 nm (UV), 405 nm (Violet), 488 nm (Blue), 561 (Yellow), and 640 (Red) lasers.
15. Flow cytometry data analysis software (i.e., FlowJo, LLC/BD).

3 Methods

3.1 Isolation of Immune Cells from Peripheral Blood or Buffy Coat

1. Dilute fresh anticoagulant-supplemented peripheral blood with an equal amount of DPBS (1:1) in a 50 mL sterile tube. If a buffy coat is used, dilute it by adding two times the volume of room temperature DPBS (1:3) in 175 cm^2 tissue culture flasks (*see* **Note 2**).
2. For each gradient use a sterile tube with an appropriate volume of room temperature Ficoll such that the ratio of DPBS:blood: Ficoll is 1:1:1 (or DPBS:buffy coat:Ficoll 2:1:1).

Table 1
Antibody panel used for cell surface staining of MAIT cells from gut biopsies

Marker	Clone	Dye
CD3	OKT3	Brilliant Violet 711™
TCR Vα7.2	3C10	PE
CD161	HP-3G10	Brilliant Violet 605™
CD25	BC96	Brilliant Violet 650™
CD69	FN50	Brilliant Violet 421™
CD137	4B4-1	PE/Dazzle™ 594
CD38	HB-7	APC/Cy7
HLA-DR	L243	Alexa Fluor® 700
CD224 (2B4)	C1.7	FITC
CD223 (LAG-3)	3DS223H	APC/Cy7™
CD279 (PD-1)	EH12.2H7	Brilliant Violet 785™
CD152 (CTLA-4)	14D3	PE/Cy7™
TIGIT	MBSA43	PerCP-eFluor™ 710
Live/dead		DAPI[a]

[a]Excitation using UV-355, Filter 450/50

Table 2
Antibody panel used for cell surface staining followed by intracellular staining of MAIT cells from gut biopsies

Marker	Clone	Dye
CD3	OKT3	Brilliant Violet 711™
MR1-5-OP-RU-tetramer		PE
CD161	HP-3G10	Brilliant Violet 605™
CD25	BC96	Brilliant Violet 650™
IFN-γ	4S.B3	Brilliant Violet 421™
TNF-α	MAb11	PE/Cy7™
IL-17A	BL168	PE/Dazzle™ 594
Granzyme B	GB11	FITC
RORγt	AFKJS-9	APC
T-bet	4B10	Brilliant Violet 785™
IL-22	2G12A41	PerCP/Cy5.5™
Live/dead		UV[a]

[a]LIVE/DEAD™ Fixable Dead Cell Stain Sampler Kit, Excitation using UV-355, Filter 450/50

Table 3
Antibody panel used for cell surface staining of MAIT cells from liver biopsies

Marker	Clone	Dye
CD3	UCHT1	PE/Cy7™
MR1-5-OP-RU-tetramer		PE
CD8	RPA-T8	BUV496™
CD137	4B4-1	PE/Dazzle™ 594
CD150 (SLAM)	A12 (7D4)	FITC
CD38	HB-7	Brilliant Violet 650™
ICOS	C398.4A	Brilliant Violet 510™
HLA-DR	L243	Alexa Fluor® 700
CD69	FN50	APC/Cy7™
CD103	Ber-ACT8	APC
TIGIT	MBSA43	PerCP-eFluor™ 710
CD279 (PD-1)	EH12.2H7	Brilliant Violet 785™
KLRG1	2F1/KLRG1	Brilliant Violet 605™
Live/dead		DAPI[a]

[a]Excitation using UV-355, Filter 450/50

3. For a standard separation, carefully layer 30 mL of diluted blood over 15 mL of Ficoll in a 50 mL conical tube. Hold the tube already containing Ficoll almost horizontal, add the diluted blood slowly, letting it flow down the side of the tube and distribute on the surface of the Ficoll, then carefully return the tube to vertical position as more blood is layered.
4. Centrifuge at room temperature at 800 × *g* for 20 min with minimal acceleration and the brake off since the deceleration disrupts the density gradient.
5. Remove upper plasma layer, take the white ring containing PBMCs carefully and transfer no more than 10 mL to a fresh tube.
6. Wash once with a large excess of room temperature DPBS to remove any residues of Ficoll or plasma.
7. Count the cells using a hemocytometer.
8. Pellet the cells for 10 min at 400 × *g*.
9. Freeze the cells in 1 mL of cryopreservation medium for further use. Freeze between 10^6 and 10^7 cells/tube. If necessary, make several aliquots.

3.2 Isolation of Immune Cells from Human Tissue Biopsies

1. Biopsies are collected in a tube containing sterile DPBS.
2. Place the tube with the biopsy on ice. Some cells can be released into the DPBS.
3. To process liver biopsies, small cylinders of liver tissue can easily be mechanically dissociated by using the piston of a 2 mL syringe above a 40 μm cell strainer without previous dissociation and enzymatic digestion. Following this, proceed directly to **step 11**.
4. To dissociate the gut tissue, transfer the biopsy into a Petri dish using sterile forceps.
5. Add 1 mL DPBS and dissociate the tissue with a sterile scalpel in small pieces (2–3 mm) taking care not to spread the individual pieces.
6. Rinse the Petri dish with 2 mL of DPBS using a sterile glass Pasteur pipette and gently pipette up and down several times to further dissociate the tissue debris.
7. Collect both debris and DPBS in a conical 15 mL tube.
8. Centrifuge 5 min at 400 × *g*.
9. Re-suspend the pellet in digestion medium pre-warmed at 37 °C, by pipetting up and down multiple times using a sterile glass Pasteur pipette.
10. Place the tube into the water bath at 37 °C for 3–4 h and gently mix a few times during the incubation in order to regularly re-suspend precipitating material.
11. With a sterile glass Pasteur pipette, flush the cell suspension (or cylinders of liver biopsies from **step 3**) through a 40 μM cell strainer placed on top of a 50 mL tube, further mechanically dissociate the tissue on the cell strainer using a piston of a 2 mL syringe and filter tissue debris.
12. Rinse the cell strainer once with 1 mL cold DPBS and then place the tube on ice.
13. Centrifuge for 5 min at 400 × *g* the tube with the cells from dissociated tissue as well as the tube in which the biopsy was originally collected as it may contain cells released from the biopsy.
14. We suggest that the cells isolated from the two tubes are kept separately.
15. Count the cells, pellet them for 10 min at 400 × *g*, and freeze in 1 mL of cryopreservation medium. Freeze 10^6–10^7 cells/tube. If necessary, make several aliquots.

3.3 In Vitro Stimulation of MAIT Cells

MAIT cells present in PBMCs or in the cell suspensions obtained from tissue biopsies can be stimulated in vitro in an antigen-specific manner using MR1-expressing APCs and bacterial extracts or synthetic antigens. A375-MR1 cells were preferentially used as APCs because they lack β_2m-associated antigen-presenting molecules and, following single-chain transfection of human MR1 linked to β_2m, they overexpress MR1 (*see* **Note 1**). Thus, they allow minimal background in a variety of T-cell activation assays and optimal MR1 loading and presentation. We have previously shown that various metabolite antigens of bacterial origin stimulate MAIT cells differently [15]. Alternatively, to study the effector function profile of MAIT cells, a combination of PMA and Ca-ionophore A23187 can be used. PMA/Ca-ionophore stimulation allows the activation of MAIT cells in a TCR-independent manner by increasing the level of intracellular Ca^{2+} and activating PKC.

3.3.1 Antigen-Specific Stimulation

1. Incubate freshly isolated PBMCs or tissue-infiltrating T cells (3×10^6/mL) with A375-MR1 cells (1×10^6/mL) and antigen (for example, 5-OP-RU, 50 nM) for 24 h at 37 °C in 24-well cell culture plates.
2. Add BFA (10 μg/mL) during the last 4 h of stimulation to accumulate cytokine inside cells.

3.3.2 Antigen-Independent Stimulation

1. Incubate freshly isolated PBMCs or tissue-infiltrating T cells with PMA (25 ng/mL) and Calcium-ionophore (1 μg/mL) in the presence of BFA (10 μg/mL) for 4 h at 37 °C in 24-well cell culture plates.

Following these alternative stimulation methods, proceed with the staining to evaluate MAIT cell activation by flow cytometry.

3.4 Antibody Panel Selection

1. Determine the antibody panel for the detection of cell surface and intracellular markers of interest. We propose the following criteria, which consider antigen density (the number of molecules expressed on cells) and if antigen expression levels are constant or variable. It is convenient to use bright fluorochromes (e.g., PE or APC) to stain dimly expressed antigens, and a dimmer fluorochrome (e.g., FITC or Alexa Fluor™ 700) for highly expressed antigens.
2. In Tables 1, 2, and 3, we highlight the panels we used to characterize MAIT cells from gut [15] and liver biopsies [16] and compare them to circulating MAIT cells.

3.5 Optimization of Antibody Concentrations for Staining

1. Although all antibody manufacturers suggest a working dilution of antibody, we recommend performing a titration experiment to optimize antibody concentrations for their application. Several dilutions (e.g., 1:10, 1:50, 1:100, and 1:200) of each antibody and tetramer can be tested on the cells of interest (e.g., PBMCs from healthy donors). For optimal detection of intracellular targets (e.g., cytokines and transcription factors), we stimulated PBMCs with PMA/Calcium-

ionophore in the presence of BFA for 4 h to allow intracellular accumulation.

2. Once optimal antibody concentrations are found, perform a trial experiment using the full antibody panel to stain PBMCs from healthy donors and evaluate panel performance.

3.6 Cell Surface Staining

All steps must be done on ice or at 4 °C, except when using MR1 tetramers, which are incubated at room temperature.

1. Wash cells twice with DPBS at 400 × *g* for 10 min at room temperature (*see* **Note 3**), discard the supernatant, and transfer cells to a 96-well U-bottom plate (max cell number 10^6/well). Leave empty wells all around the filled ones, to avoid contamination during the procedure.
2. Prepare proper control for the cell surface staining (*see* **Note 4**).
3. Re-suspend cell pellets in FcR-blocking buffer for 30 min at room temperature.
4. As first staining reagent use fluorochrome-labeled MR1 5-OP-RU tetramers (4 μg/mL) in FACS buffer for 40 min at room temperature in the dark (*see* **Note 5**).
5. Wash the cells twice at 600 × *g* for 5 min with cold FACS buffer.
6. Proceed with the staining by using a cocktail of mAbs for surface antigens for 20 min at 4 °C in the dark.
7. Wash the cells twice at 600 × *g* for 5 min with cold FACS buffer.
8. Dilute the DAPI to ~3 μM in FACS buffer and re-suspend the cells in FACS buffer plus DAPI just before data acquisition.

3.7 Intracellular Staining

1. After in vitro stimulation, wash the cells twice at 600 × *g* for 5 min with room temperature DPBS.
2. Prepare LIVE/DEAD™ Fixable Dead Cell dye by dissolving the dry fluorescent reactive dye in 50 μL of DMSO (*see* **Note 6**).
3. Count the cells and adjust the cell density with room temperature DPBS to 10^6 cells in 1 mL.
4. Stain cells in a 5 mL round-bottom tube with the reconstituted fluorescent reactive dye diluted 1:1000 in DPBS and incubate for 30 min at room temperature in the dark.
5. Add 1 mL of RPMI 10% FCS to block the staining reaction and centrifuge at 600 × *g* for 5 min.
6. Wash the cells at 600 × *g* for 5 min with cold DPBS.
7. Re-suspend the cells in 100 μL of FcR-blocking buffer and transfer the cells to U-bottom 96-well plates and incubate for 30 min at room temperature.

8. Incubate with mAb cocktail for extracellular staining (final volume is 50 μL), 20 min at 4 °C in the dark.
9. Fix cells with freshly prepared paraformaldehyde 2% in DPBS, for 20 min at room temperature in the dark (*see* **Note 7**).
10. Wash the cells three times with saponin-washing buffer at 600 × *g* for 5 min (*see* **Note 8**).
11. Incubate the cells with saponin-washing buffer together with mAbs for 25 min at room temperature (*see* **Note 9**).
12. Wash three times with saponin-washing buffer.
13. Re-suspend cells in cold FACS buffer and filter sample through the cell strainer cup of a 5 mL tube before proceeding to flow cytometry (*see* **Note 10**).

3.8 Compensation Controls

1. Add 5 μL of negative beads and 5 μL of positive beads to each single stained compensation control tube (*see* **Note 11**).
2. Stain the beads by adding fluorescent antibodies into each control tube (use the same antibody concentration used for staining your sample) and mix well.
3. Incubate the antibody capture beads for 20 min at 4 °C in the dark.
4. Wash the beads with 1 mL of cold FACS buffer at 300 × *g* for 5 min.

Re-suspend the beads in 500 μL of FACS buffer and use as flow cytometry compensation controls.

3.9 Cytometer Standardization

Cytometer standardization allows comparison of data obtained at different time points or between different cytometers.

Prepare the beads:

1. Vortex the beads and add one drop of particles to 1 mL of FACS buffer.
2. Start to acquire the unstained sample and the compensation control using appropriate voltage.
3. Acquire calibration particles.
4. Proceed with sample acquisition.
5. Use the Mean Fluorescence Intensity (MFI) obtained from the acquisition of the beads during the first experiment as absolute reference to normalize data acquired at later time points.
6. Calculate the normalization ratio (β):

$$\beta = \text{MFI (day 2)}/\text{MFI (day 1)}.$$

7. The relative normalization of all the parameters of interest can be performed using a flow cytometry analysis software, by

calculating the "derive parameter" for each parameter (e.g., FITC):

$$\text{MFI (normalized)} = \text{MFI (day 2)} \times \beta.$$

3.10 Data Analysis for Identification and Characterization of Unique MAIT Cell Subsets

1. Analyze flow cytometry data by using an appropriate software analysis program.
2. Gate MAIT cells according to the expression of cell surface markers. MAIT cells can be identified as $CD3^+CD161^{bright}$MR1-5-OP-RU tetramer$^+$ or $CD3^+CD161^{bright}V\alpha7.2^+$. Figure 1 shows one example of gating strategy based on the expression of CD3 and CD161, and staining with MR1-5-OP-RU tetramers.
3. Note scale values discriminating positive and negative populations for each marker in the panel. These "cutoffs" will be used to center the data prior to further analysis.
4. Export data from the selected population as a new CSV file.
5. Select the files in the workspace you would like to export.

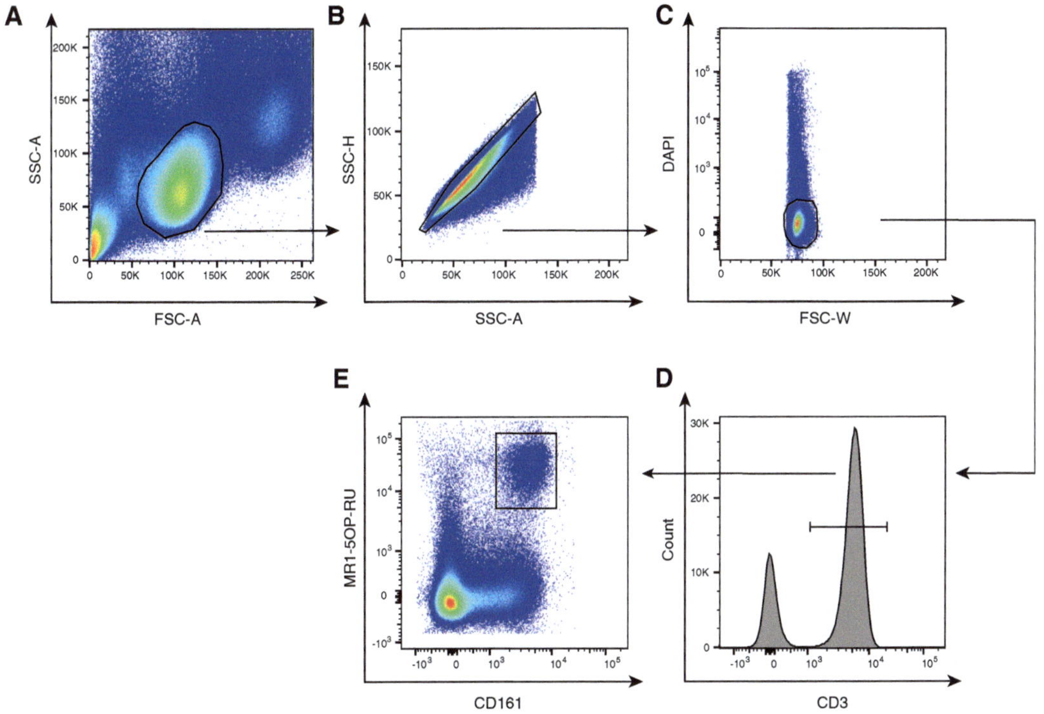

Fig. 1 Gating strategy to identify MAIT cells in PBMCs using 5-OP-RU-loaded MR1 tetramers. Shown is a representative MAIT gating strategy, whereby, within the lymphocyte population, gated populations (**a**–**e** from left to right) are indicated defining viable (DAPI negative) $CD3^+$ T cells. MAIT cells were discriminated on the basis of staining with MR1-5-OP-RU tetramer and expression of CD161

6. Select the parameter and the number of events per sample (i.e., 5000 events) to export for further analyses (*see* **Note 12**).
7. Transform the data using a centered inverse hyperbolic sine function (Eq. 1) and the cutoffs defined in **step 3**.

$$y_{ij} = \sinh^{-1}\left(\frac{x_{ij} - c_j}{b}\right) \quad (1)$$

Transformed value y for event i and marker j is the inverse hyperbolic sine product of the raw event x_{ij} centered by the cutoff value c_j, proportional to the cofactor b

8. Apply dimensionality reduction of transformed data with the Barnes-Hutt implementation of t-SNE (Fig. 2) [17].

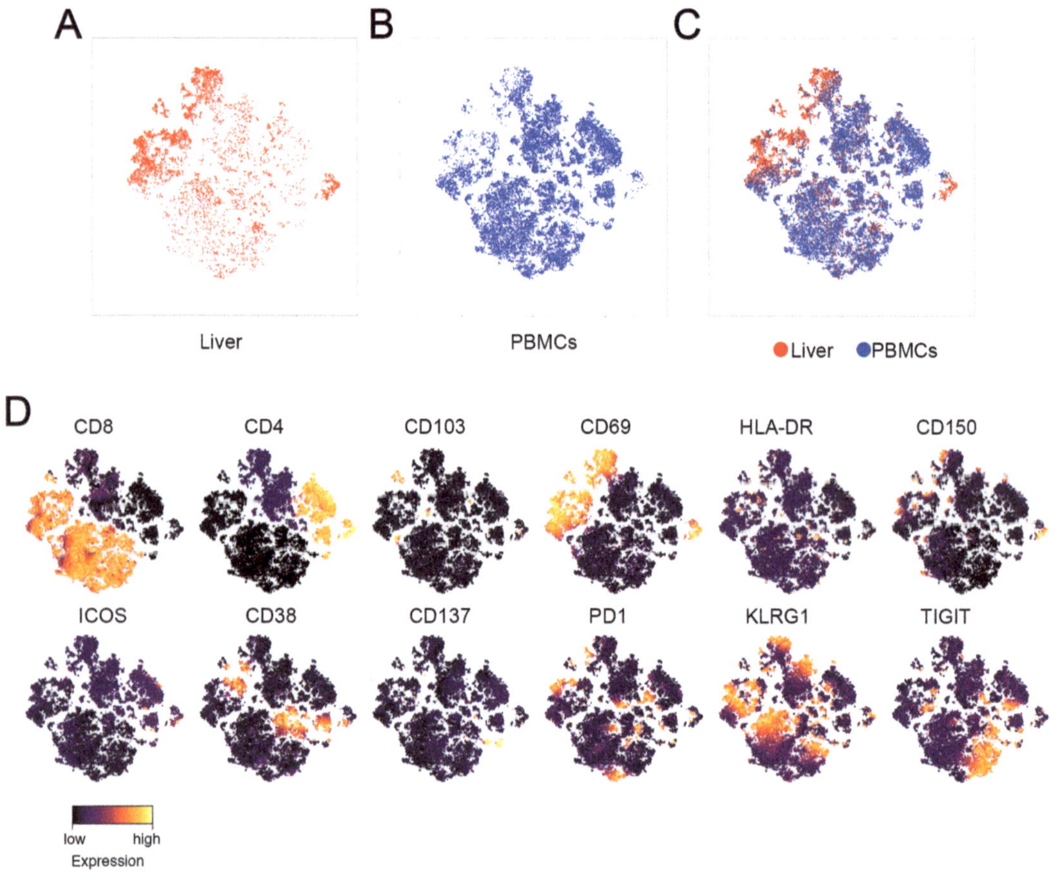

Fig. 2 MAIT cells in the liver are phenotypically different from MAIT cells in the blood. t-SNE analysis performed using dimensionality reduction of transformed data with the Barnes-Hutt implementation of t-SNE on MAIT cells gated from samples stained with antibodies listed in Table 3. (**a**) t-SNE map showing the distribution of MAIT cell subpopulations from liver-infiltrating lymphocytes. (**b**) t-SNE map of MAIT cells showing the distribution of MAIT cell subpopulations in PBMCs. (**c**) Overlapped t-SNE maps of MAIT cells from liver (red) and PBMCs (blue). (**d**) t-SNE maps of MAIT cells from liver and PBMCs show the expression level and the distribution of the indicated markers for activation molecules and inhibitory receptors

Multidimensional analysis of all markers using the t-SNE algorithm allows all of the data to be reduced to two dimensions. The cells are distributed in two dimensions according to the simultaneous expression of all markers at the single-cell level. This analysis identifies unique profiles of MAIT cells by combining the expression of multiple markers. Cells with a similar profile are grouped together and separated from other cell clusters.

4 Notes

1. To generate optimal APCs for antigen presentation to MAIT cells, the human melanoma cell line A375 was genetically engineered to become deficient for MHC class I, CD1, and all class I-like molecules, and overexpress MR1. Briefly, we knocked out β2m gene by transfecting A375 cells with a pool of five plasmids containing Cas9 and five independent chimeric single-guide RNA targeting the β2m gene (Horizon). Four days after transfection, cells were stained with the BBM.1 mAbs against human β2m (ATCC HB-28), and negative cells were enriched by FACS sorting before being cloned by limiting dilution. The absence of β2m expression was confirmed on selected clones by western blot and absence of MHC class I cell surface expression by flow cytometry using anti-human HLA-A,B,C mAbs (clone W6/32). A375-β2m-deficient cells were then transduced with a lentiviral vector containing a human MR1A cDNA construct linked to β2m via a flexible Gly-Ser linker as previously described [5, 18]. Surface expression of MR1 was assessed by flow cytometry using the MR1-specific mAb 26.5, and positive cells were FACS sorted.
2. The peripheral blood or buffy coat samples should not be older than 10 h, and they should be supplemented with anticoagulants (e.g., heparin or EDTA). If samples cannot be immediately processed, store them at room temperature prior to the isolation of PBMCs.
3. Previous studies have suggested addition of DNase I to the medium used to wash freshly thawed PBMCs. This step prevents cell-to-cell adhesion and clumping by digesting the DNA that is released by dead and dying cells, thus avoiding risks of obstruction of the flow cytometer during data acquisition and increasing the number of single cells available for the staining and analysis.
4. Control wells should always include (1) unstained cells, (2) single color controls for setting and compensation, and (3) cells stained with isotype-matched irrelevant mAbs.
5. It is recommended to use tetrameric reagents in the first staining steps because of possible steric hindrance issues when using antibodies specific for the TCR/CD3 complex.

6. Reconstituted fixable viability dye can be stored at −20 °C. We recommend preparation of 2 μL aliquots to avoid repeated freeze–thaw cycles. Aliquots should be used within 1 month of preparation.
7. Solutions containing paraformaldehyde in DPBS can be titrated for optimal performance depending on the cell type. Optimal cell fixation of human PBMCs is obtained by using freshly prepared 2% paraformaldehyde solution. The solution can be stored for a few days at 4 °C in the dark. Paraformaldehyde is toxic and suspected carcinogen.
8. After fixation, cells reduce their adherence to the plastic and can be readily lost during washing. Therefore, for samples containing small numbers of cells (i.e., MAIT cells isolated from human biopsies), irrelevant tumor cells can be added before the first washing step to avoid sample loss. To facilitate gating out of these irrelevant tumor cells during data analysis, they can be fixed and stained with live/dead fixable dye. In this manner, they are gated out as dead cells.
9. If unlabeled or biotinylated antibodies are used, after the first intracellular staining cells are washed three times with saponin-washing buffer and then are incubated with secondary Abs or fluorochrome-labeled streptavidin diluted in saponin 0.1% (total cocktail volume is 50 μL) for further 25 min at 4 °C in the dark.
10. A procedure recommended for all cell samples before acquisition by flow cytometer is to dilute stained cells in FACS buffer and filter them through a cell strainer-cupped tube. This procedure limits blockage of the instrument and consequential sample loss.
11. The best compensation controls are obtained using antibody capture beads. For each fluorochrome, the same labeled mAbs present in the panel is used. If some mAbs in the panel are IgM and thus cannot be captured by beads, an IgG antibody labeled with the same fluorochrome of the IgM mAb is used.
12. Before performing the dimensional reduction, you can randomly down sample the data to an equal number of events (i.e., 5000 events) after data transformation.

Acknowledgments

We thank Paula Cullen and Julian Spagnuolo for critical reading of the manuscript, and the DBM Flow Facility staff. The MR1 tetramer technology was developed jointly by Dr. James McCluskey, Dr. Jamie Rossjohn, and Dr. David Fairlie. The MR1 tetramer was produced by the NIH Tetramer Core Facility as permitted to be distributed by the University of Melbourne. The author's studies are supported by SNF 310030-173240, KFS-3730-08-2015, and PMB-02-17 Grants.

References

1. Tilloy F, Treiner E, Park SH, Garcia C, Lemonnier F, de la Salle H, Bendelac A, Bonneville M, Lantz O (1999) An invariant T cell receptor alpha chain defines a novel TAP-independent major histocompatibility complex class Ib-restricted alpha/beta T cell subpopulation in mammals. J Exp Med 189:1907–1921
2. Treiner E, Duban L, Bahram S, Radosavljevic M, Wanner V, Tilloy F, Affaticati P, Gilfillan S, Lantz O (2003) Selection of evolutionarily conserved mucosal-associated invariant T cells by MR1. Nature 422:164–169
3. Gold MC, Cerri S, Smyk-Pearson S, Cansler ME, Vogt TM, Delepine J, Winata E, Swarbrick GM, Chua WJ, Yu YY, Lantz O, Cook MS, Null MD, Jacoby DB, Harriff MJ, Lewinsohn DA, Hansen TH, Lewinsohn DM (2010) Human mucosal associated invariant T cells detect bacterially infected cells. PLoS Biol 8: e1000407
4. Tang XZ, Jo J, Tan AT, Sandalova E, Chia A, Tan KC, Lee KH, Gehring AJ, De Libero G, Bertoletti A (2013) IL-7 licenses activation of human liver intrasinusoidal mucosal-associated invariant T cells. J Immunol 190:3142–3152
5. Lepore M, Kalinichenko A, Colone A, Paleja B, Singhal A, Tschumi A, Lee B, Poidinger M, Zolezzi F, Quagliata L, Sander P, Newell E, Bertoletti A, Terracciano L, De Libero G, Mori L (2014) Parallel T-cell cloning and deep sequencing of human MAIT cells reveal stable oligoclonal TCRbeta repertoire. Nat Commun 5:3866
6. Kjer-Nielsen L, Patel O, Corbett AJ, Le Nours J, Meehan B, Liu L, Bhati M, Chen Z, Kostenko L, Reantragoon R, Williamson NA, Purcell AW, Dudek NL, McConville MJ, O'Hair RA, Khairallah GN, Godfrey DI, Fairlie DP, Rossjohn J, McCluskey J (2012) MR1 presents microbial vitamin B metabolites to MAIT cells. Nature 491:717–723
7. Miley MJ, Truscott SM, Yu YY, Gilfillan S, Fremont DH, Hansen TH, Lybarger L (2003) Biochemical features of the MHC-related protein 1 consistent with an immunological function. J Immunol 170:6090–6098
8. Huang S, Gilfillan S, Kim S, Thompson B, Wang X, Sant AJ, Fremont DH, Lantz O, Hansen TH (2008) MR1 uses an endocytic pathway to activate mucosal-associated invariant T cells. J Exp Med 205:1201–1211
9. Treiner E, Duban L, Moura IC, Hansen T, Gilfillan S, Lantz O (2005) Mucosal-associated invariant T (MAIT) cells: an evolutionarily conserved T cell subset. Microbes Infect 7:552–559
10. Gold MC, McLaren JE, Reistetter JA, Smyk-Pearson S, Ladell K, Swarbrick GM, Yu YY, Hansen TH, Lund O, Nielsen M, Gerritsen B, Kesmir C, Miles JJ, Lewinsohn DA, Price DA, Lewinsohn DM (2014) MR1-restricted MAIT cells display ligand discrimination and pathogen selectivity through distinct T cell receptor usage. J Exp Med 211:1601–1610
11. Reantragoon R, Kjer-Nielsen L, Patel O, Chen Z, Illing PT, Bhati M, Kostenko L, Bharadwaj M, Meehan B, Hansen TH, Godfrey DI, Rossjohn J, McCluskey J (2012) Structural insight into MR1-mediated recognition of the mucosal associated invariant T cell receptor. J Exp Med 209:761–774
12. Dusseaux M, Martin E, Serriari N, Peguillet I, Premel V, Louis D, Milder M, Le Bourhis L, Soudais C, Treiner E, Lantz O (2011) Human MAIT cells are xenobiotic-resistant, tissue-targeted, CD161hi IL-17-secreting T cells. Blood 117:1250–1259
13. Reantragoon R, Corbett AJ, Sakala IG, Gherardin NA, Furness JB, Chen Z, Eckle SB, Uldrich AP, Birkinshaw RW, Patel O, Kostenko L, Meehan B, Kedzierska K, Liu L, Fairlie DP, Hansen TH, Godfrey DI, Rossjohn J, McCluskey J, Kjer-Nielsen L (2013) Antigen-loaded MR1 tetramers define T cell receptor heterogeneity in mucosal-associated invariant T cells. J Exp Med 210:2305–2320
14. Corbett AJ, Eckle SB, Birkinshaw RW, Liu L, Patel O, Mahony J, Chen Z, Reantragoon R, Meehan B, Cao H, Williamson NA, Strugnell RA, Van Sinderen D, Mak JY, Fairlie DP, Kjer-Nielsen L, Rossjohn J, McCluskey J (2014) T-cell activation by transitory neo-antigens derived from distinct microbial pathways. Nature 509:361–365

15. Schmaler M, Colone A, Spagnuolo J, Zimmermann M, Lepore M, Kalinichenko A, Bhatia S, Cottier F, Rutishauser T, Pavelka N, Egli A, Azzali E, Pieroni M, Costantino G, Hruz P, Sauer U, Mori L, De Libero G (2018) Modulation of bacterial metabolism by the microenvironment controls MAIT cell stimulation. Mucosal Immunol 11:1060–1070
16. Di Blasi D, Boldanova T, Mori L, Terracciano L, Heim MH, De Libero G (2019) Cell Mol Gastroenterol Hepatol, https://doi.org/10.1016/j.jcmgh.2019.08.004
17. van der Maaten L (2014) Accelerating t-SNE using tree-based algorithms. J Mach Learn Res 15:3221–3245
18. Lepore M, Kalinichenko A, Calogero S, Kumar P, Paleja B, Schmaler M, Narang V, Zolezzi F, Poidinger M, Mori L, De Libero G (2017) Functionally diverse human T cells recognize non-microbial antigens presented by MR1. Elife 6:e24476

Chapter 3

Isolation and Characterization of MAIT Cells from Tumor Tissues

William Rodin, Patrik Sundström, and Marianne Quiding Järbrink

Abstract

Mucosal-associated invariant T (MAIT) cell infiltration has been demonstrated in colorectal and hepatocellular carcinoma, and their ability to produce Th1- and Th17-associated cytokines, as well as their cytotoxic function, suggests that MAIT cells may have important functions in both reducing and promoting protective tumor immunity. Here, we describe enzymatic methods to isolate intraepithelial and lamina propria lymphocyte single cell suspensions from colon tissue and tumors containing viable MAIT cells, which can be used for further purification, flow cytometry analysis, or culture.

Key words MAIT cell, Colon cancer, Intraepithelial lymphocytes, Lamina propria lymphocytes, CD161, Cytokines, Cytotoxicity

1 Introduction

Mucosal-associated invariant T (MAIT) cells have several features suggesting that they may be potent effector cells in anti-tumor immunity, such as their rapid production of the Th1-type cytokines IFN-γ and TNF and their cytotoxic potential. However, MAIT cell cytokine profiles are different in different organs, and Th17-type cytokines with tumor-promoting effects are preferentially produced by MAIT cells in the lungs, liver, and female genital tract, while gastrointestinal and circulating MAIT cells secrete Th1-type cytokines [1]. The difference in MAIT cell function between different anatomical locations makes it important to study MAIT cells isolated from the organ of interest and not just circulating cells, even if they derive from patients with relevant symptoms or disease. In cancer disease, it is also preferable to compare the functions of tumor-infiltrating MAIT cells with those isolated from the corresponding unaffected tissue.

The large majority of MAIT cells express CD8, but there are also smaller populations of $CD4^+$ and $CD4^-CD8^-$ double-negative MAIT cells. Most $CD8^+$ and double-negative MAIT cells

Helen Kaipe and Isabelle Magalhaes (eds.), *MAIT Cells: Methods and Protocols*, Methods in Molecular Biology, vol. 2098, https://doi.org/10.1007/978-1-0716-0207-2_3,

contain the canonical Vα7.2 T cell receptor alpha chain, and bind to MR1 tetramers presenting the MAIT cell antigen 5-(2-oxopropylideneamino)-6-D-ribitylaminouracil (5-OP-RU). The $CD4^+$ MAIT cells, on the other hand, have a more variable T cell receptor, and only about one-third of them react with 5-OP-RU presenting tetramers [2]. It is now becoming increasingly evident that these three MAIT cell populations have partly different functional properties, and that they may represent distinct differentiation stages [2–4]. However, their precise functions in the tumor microenvironment have not yet been established.

So far, MAIT cell infiltration has been assessed in colorectal cancer (CRC) [5–8], lymphoma [9], hepatocellular carcinoma [10], and CRC metastases in the liver [11]. In addition, an early study showed the presence of transcripts for the canonical MAIT cell T cell receptor Vα7.2 joined with Jα33 in brain and kidney tumors [12]. MAIT cells are clearly enriched in CRC, as assessed by both flow cytometry and immunofluorescence [5–8], while there is no accumulation of MAIT cells in the bone marrow of patients with multiple myeloma [9]. In contrast, hepatocellular carcinoma and CRC metastases harbor fewer MAIT cells than the surrounding unaffected liver tissue [10, 11]. MAIT cells in colon cancer preferentially produce Th1-associated cytokines, IFN-γ and TNF. However, frequencies of IFN-γ-producing MAIT cells are reduced in tumor tissue compared to macroscopically unaffected tissues [5], while cytotoxic effector functions are maintained in tumor-infiltrating MAIT cells [13]. In liver metastases of CRC, there is a similar but even more pronounced reduction of IFN-γ production by MAIT cells, but no concomitant increase in IL-17 production [11]. Also in hepatocellular carcinomas, IFN-γ-production is reduced in the infiltrating MAIT cells, and interestingly also granzyme B and Perforin, in comparison to MAIT cells from the healthy liver tissue [10]. Furthermore, IL-8 secretion was significantly increased in tumor-infiltrating MAIT cells, and a high IL-8 signature in the tumor actually correlated to reduced patient survival [10]. Two studies have assessed the impact of total MAIT cell infiltration on patient outcome, and both find that a high MAIT cell infiltration into tumor tissue is correlated with a lower relapse free survival [6, 10]. Taken together, the current literature suggests that MAIT cells loose important effector molecules implicated in anti-tumor immunity when located in the tumor microenvironment, and they may thus contribute to an unfavorable clinical outcome.

In this chapter, we describe enzymatic methods to isolate intraepithelial and lamina propria lymphocyte single cell suspensions from unaffected colon tissue and tumors containing viable MAIT cells, which can be used for flow cytometry analysis or stimulation. Single cell suspensions can also be further processed to allow MAIT cell isolation using magnetic beads or flow cytometric cell sorting, to allow functional studies of tumor-infiltrating MAIT cells.

2 Materials

Prepare and store all solutions at room temperature (unless otherwise indicated). Keep equipment, reagents, and solutions sterile. It is strongly recommended to perform all work in a laminar flow cabinet/biosafety cabinet.

2.1 Intraepithelial Lymphocyte Isolation

1. EDTA medium (EDTA-M): gH_20, 10% Hank's Balanced Salt Solution $(HBSS)^{-mg-ca}$, 2% fetal calf serum (FCS), 1 mM EDTA, 20 mM HEPES (*see* **Note 1**).
2. Wash medium (WM): RPMI containing 10% FCS, 50 μM β-mercaptoethanol, 1 mM HEPES, 1 mM sodium pyruvate, 100 U/mL penicillin, 100 ng/mL streptomycin, 2 mM L-glutamine, 10 ng/mL gentamicin (*see* **Note 2**).
3. Digestion medium (DM): WM containing 180 μg/mL DNase I, 71.4 μg/mL Liberase TM, 5 mM $CaCl_2$ (*see* **Note 3**).
4. Phosphate-buffered saline (PBS).
5. 70% Percoll solution: PBS containing 70% Percoll stock solution (*see* **Note 4**).
6. 40% Percoll solution: PBS containing 40% Percoll stock solution, 5 mM EDTA (*see* **Note 5**).
7. 50 mL polypropylene centrifuge tubes.
8. 30 mL flat-bottomed polypropylene tubes.
9. Temperature-regulated centrifuge.
10. Fine tweezers and scissors.
11. Cell medium (CM): RPMI containing 10% FCS, 50 ng/mL gentamicin, 25 mM HEPES, 100 U/mL penicillin, 100 ng/mL streptomycin, 2 mM L-glutamine (*see* **Note 6**).
12. Balance ranging from 0.01 to 10 g.
13. Magnetic stirrer and magnetic stir bars.
14. Heating cabinet set at 37 °C.
15. Ice.

2.2 Lamina Propria Lymphocyte Isolation

1. Wash medium (WM): Same as **item 2** in Subheading 2.1.
2. Digestion medium (DM): Same as **item 3** in Subheading 2.1.
3. 50 mL polypropylene centrifuge tubes.
4. 30 mL flat-bottomed polypropylene tubes.
5. Temperature-regulated centrifuge.
6. 50 μm nylon filters.
7. Cell medium (CM): Same as **item 11** in Subheading 2.1.

Table 1
Suggestion for monoclonal antibodies to identify MAIT cells

Marker	Fluorochrome	Clone
CD45	AF700	H130
CD45	FITC	H130
CD3	APC/H7	SK7
CD4	BV650	RPA-T4
CD8	BUV395	RPA-T8
Vα7.2	APC	3C10
CD161	eF450	HP-3G10

2.3 MAIT Cell Staining for Flow Cytometry Analyses

1. 96-well plates with conical bottom.
2. PBS containing 2% FCS.
3. RPMI medium containing 1% FCS.
4. eBioscience™ Foxp3/Transcription Factor Staining Buffer Set.
5. LIVE/DEAD™ Fixable cell stain kit or BD Horizon™ Fixable viability stain.
6. Fluorochrome conjugated antibodies (*see* Table 1 for suggestion of antibodies to use).
7. MR1 tetramers presenting 5-OP-RU (*see* **Note 7**).
8. Intracellular flow cytometry staining kit containing fixation and permeabilization buffers.

2.4 Magnetic Bead Isolation of Vα7.2+ Cells

1. PBS/EDTA buffer: PBS containing 0.1% FCS, 2 mM EDTA.
2. Cell medium (CM): Same as **item 11** in Subheading 2.1.
3. Dynabeads™ CD8 Positive Isolation Kit, containing beads coated with CD8 and DETACHaBEAD® solution.
4. EasySep™ Human APC Positive Selection Kit II (*see* **Note 8**).
5. Anti-Vα7.2+—APC antibody.
6. 50 mL polypropylene centrifuge tubes.
7. Round-bottomed 5 mL polystyrene tubes.
8. Tube-holding magnets (*see* **Note 9**).

2.5 Cell Sorting for Isolation of MAIT Cells

1. Cell medium (CM): Same as **item 11** in Subheading 2.1.
2. PBS.
3. LIVE/DEAD™ Fixable cell stain kit or BD Horizon™ Fixable viability stain.

4. Fluorochrome conjugated antibodies (*see* Table 1 for suggestion of antibodies to use).
5. 5 mL polystyrene round-bottomed tubes.
6. Temperature-regulated centrifuge.
7. Ice.
8. FACS instrument.

3 Methods

When working with human material, be aware that the sample could be infectious, and all waste should be disposed according to human/contaminated hazardous waste protocols. We recommend to carry out all work in a laminar flow cabinet/biosafety cabinet.

3.1 Intraepithelial Lymphocyte Isolation

1. Carefully cut away extensive tissues, such as fat, connective tissue, and blood vessels from the colon mucosa sample, using tweezers and scissors (*see* **Notes 10** and **11**).
2. Weigh the tissue samples. If necessary, adjust the protocol according to the sample weight (*see* **Notes 12** and **13**).
3. Cut the tissue into 5–10 mm pieces and transfer to a 30 mL flat-bottomed plastic tube containing a magnetic stir bar.
4. Add 10 mL of preheated (37 °C) EDTA-M to the tissue and stir for 15 min at 37 °C in a heating cabinet (*see* **Note 14**).
5. Aspirate all the buffer using a sterile transfer pipette, leaving the tissue pieces in the original tube, and transfer the buffer to a sterile 50 mL tube. Keep the buffer containing the epithelial fraction on ice.
6. Repeat **steps 4** and **5** for a total of four treatments with EDTA-M (*see* **Notes 15** and **16**).

For isolation of intraepithelial lymphocytes (iEL) proceed with **step** 7 below, and for isolation of only lamina propria lymphocytes (LPL) proceed with **step 1** under Subheading 3.2.

7. Centrifuge the epithelial fraction at 500 × *g* for 6 min at 4 °C and discard the supernatant (*see* **Note 17**).
8. Re-suspend the pellet in 7 mL of preheated (37 °C) DM and gently stir for 30 min at 37 °C in a 30 mL flat-bottomed tube in a heating cabinet (*see* **Note 18**).
9. Add 7 mL of preheated (37 °C) EDTA-M to stop the enzymatic digestion and transfer the cells to a 50 mL tube.
10. Centrifuge at 500 × *g* for 6 min in a cold centrifuge and discard the supernatant.
11. Re-suspend the pellet in 9 mL of 40% Percoll solution.

12. Add 15 mL of 70% Percoll solution to a new 50 mL tube.
13. Carefully overlay the sample in 40% Percoll solution, on top of the 70% Percoll solution using a sterile transfer pipette (*see* **Note 19**).
14. Centrifuge at 600 × *g* for 20 min at room temperature (20–25 °C) (*see* **Note 20**).
15. Carefully transfer the iEL at the interface between the two Percoll solutions, using a sterile transfer pipette, to a new 50 mL tube (*see* **Note 21**).
16. Add cold PBS up to a total volume of 50 mL and spin at 600 × *g* for 6 min in a cold centrifuge and discard the supernatant.
17. Re-suspend the cells in the desired volume of CM, either preheated (37 °C) or cold (4 °C), depending on downstream application (*see* **Note 22**).

3.2 Lamina Propria Lymphocyte Isolation

(Continued from Subheading 3.1, **step 6**.)

1. Add 10 mL of preheated (37 °C) WM to the tissue pieces and stir for 15 min at 37 °C in a heating cabinet.
2. Aspirate the buffer, using a sterile transfer pipette, and discard it.
3. Repeat **steps 1** and **2**.
4. Add 7 mL of preheated (37 °C) DM to the tissue and gently stir for 2 h at 37 °C in a heating cabinet (*see* **Note 23**).
5. Add 7 mL of EDTA-M to the tissue to stop the enzymatic digestion.
6. Filter the sample through a 50 μm filter. Wash the tube and rinse through the filter with 10 mL cold PBS, repeat twice.
7. Centrifuge the sample at 500 × *g* for 6 min in a cold centrifuge and discard the supernatant.
8. Re-suspend the cells in the desired volume of CM, either preheated (37 °C) or cold (4 °C), depending on downstream application (*see* **Notes 24** and **25**).

3.3 MAIT Cell Staining for Flow Cytometry Analyses

1. Plate 1 × 10^6 cells per well in a 96-well plate with conical bottom (*see* **Note 26**)
2. Prepare a 1:1000 dilution of a cell viability dye in PBS (e.g., LIVE/DEAD™ Fixable cell stain kit or BD Horizon™ FVS) and add 50 μL of cell viability dye to each well.
3. Incubate the sample at room temperature in the dark for 25 min.

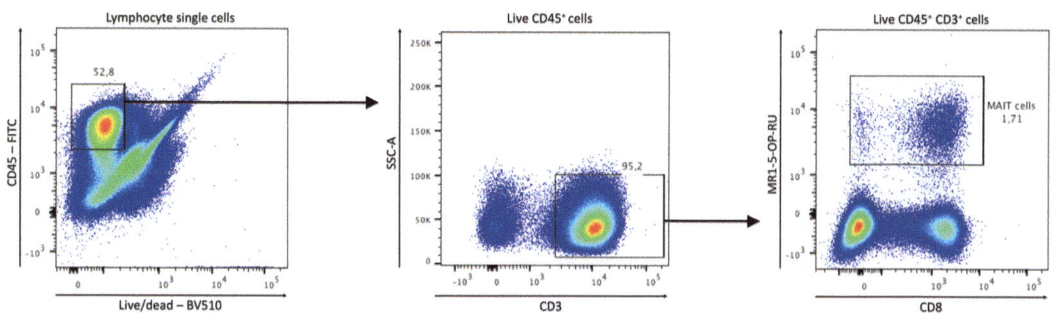

Fig. 1 Gating strategy to identify MAIT cells using MR1 tetramers. The figure shows FACS plots of a tumor lamina propria single-cell solution prepared as described in Subheadings 3.1 and 3.2, and the gating strategy used to identify MAIT cells among CD3$^+$ T cells. MAIT cells are identified as the CD45$^+$CD3$^+$MR1$^+$ subset

4. Wash off any excess cell viability dye by adding 100–200 μL of PBS and centrifuge the cells at 380 × *g* for 3 min in a cold centrifuge and discard the supernatant (*see* **Note 27**).
5. Prepare the MR1 tetramer at an appropriate dilution for your staining conditions (generally between 1:00 and 1:1000 in PBS containing 2% FCS) (*see* **Note 28**). See Fig. 1 for a representative staining of tumor-derived cells using MR1 tetramers to identify MAIT cells. Intraepithelial CD3$^+$ T cell suspensions from the unaffected mucosa usually contain 0.1–1% of MAIT cells and lamina propria 0.2–2%. MAIT cell accumulate in colon tumors, and cell suspensions from the tumor usually contain 0.3–3% of MAIT cells among CD3$^+$ T cells in the epithelial fraction and 0.4–3% in the lamina propria.
6. Add 40 μL of tetramer to each well, mix, and incubate the sample for 40 min at room temperature in the dark.
7. Wash off any excess tetramer as in **step 4**. After centrifugation, take away supernatant.
8. Prepare a mix of antibodies specific to relevant extracellular markers for your MAIT cell population, in PBS containing 2% FCS (*see* **Note 29**). Add the antibody mix (max volume 100 μL) to the cells and incubate 20 min at 4° C in the dark. Wash off any excess antibody as in **step 4**. After centrifugation, take away supernatant.
9. Add 200 μL of fixation buffer to each well and incubate the sample for 1 h in the dark at room temperature (*see* **Note 30**).
10. Wash off any excess fixation buffer as in **step 4**, but use permeabilization buffer instead of PBS. After centrifugation, take away supernatant.
11. Prepare a mix of antibodies specific to relevant intracellular markers for your MAIT cell population, in permeabilization buffer.

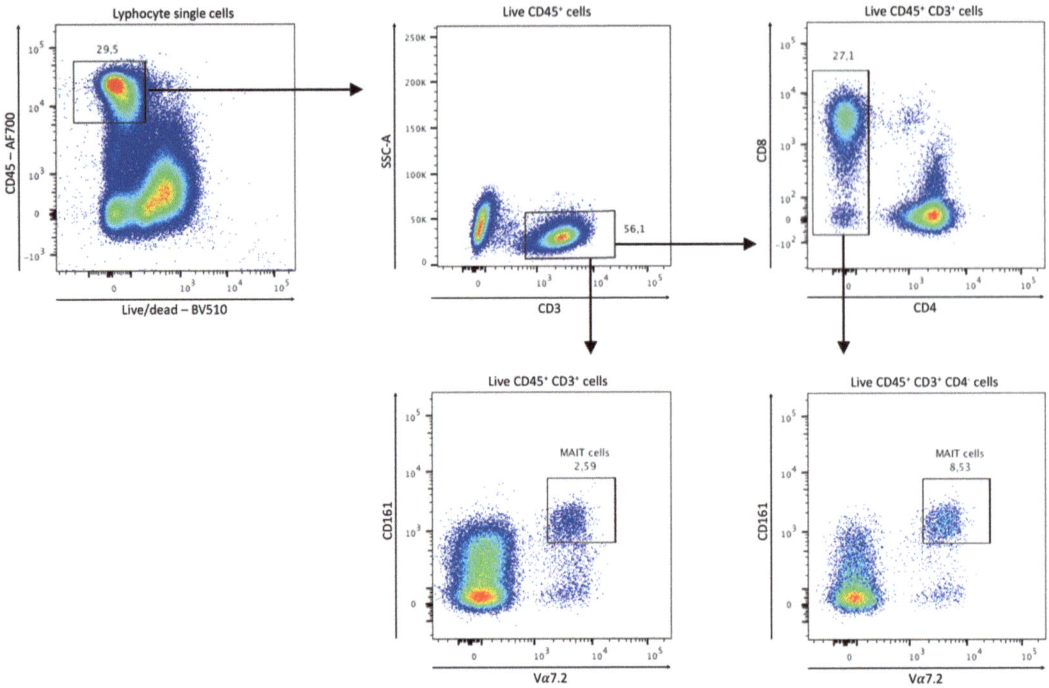

Fig. 2 Gating strategy to identify MAIT cells using CD161 and Vα7.2 T cell receptor. The figure shows FACS plots of a tumor lamina propria single-cell solution prepared as described in Subheadings 3.1 and 3.2, and the gating strategy used to identify MAIT cells among $CD8^+$ and double-negative cells, or among all $CD3^+$ T cells. MAIT cells are identified as the $CD45^+CD3^+V\alpha7.2^+CD161^{high}$ subset

12. Add 50 μL of the antibody mixture to each well and incubate for 30–60 min at room temperature in the dark (*see* **Note 30**).
13. Wash off any excess antibodies as in **step 10**. After centrifugation, take away supernatant.
14. Re-suspend the cells in 100–300 μL of PBS–2% FCS and keep them on ice until analysis (*see* **Note 31**). See Fig. 2 for a representative staining of tumor-derived cells using Vα7.2 and CD161 to identify MAIT cells.

3.4 Magnetic Bead Isolation of Vα7.2⁺ Cells

1. Count and dilute cells to 1×10^7 cells/mL in PBS/EDTA buffer in round-bottomed 5 mL tubes.
2. Wash the required amount of CD8 Dynabeads™ (1:1 bead to cell ratio) in, at least, 1 mL PBS/EDTA buffer by adding the buffer and letting the tube sit in a magnet for 1 min and then gently, without disturbing the pellet, aspirate and discard the buffer. Re-suspend the beads in PBS/EDTA to the original volume.
3. Add the CD8 Dynabeads™ to the cells and incubate for 20 min, while slowly tilting/rotating the tube, at 4 °C.

4. Place the tube in a magnet for 5 min and carefully remove the supernatant (*see* **Note 32**).
5. Remove the tube from the magnet and re-suspend the pellet containing $CD8^+$ cells in 100 μL of RPMI–1% FCS/10^7 cells. Cell numbers refer to the number of cells in the original suspension in **step 1**.
6. Add 10 μL of DETACHaBEAD® solution/100 μL cell suspension and incubate 45 min at room temperature with gentle mixing using a shaking table.
7. Add 3 mL RPMI–1% FCS to the tube and gently re-suspend the beads and cells. Place the tube in a magnet for 1 min and transfer the supernatant, containing the $CD8^+$ cells, to a 50 mL tube.
8. Repeat **step 7** twice. Centrifuge collected cells for 6 min at $400 \times g$, aspirate, and discard the supernatant to remove any residual DETACHaBEADs®.
9. Re-suspend the cells in PBS/EDTA buffer and count the cells.
10. Dilute the cells to a concentration of 2×10^8 cells/mL in PBS/EDTA buffer (*see* **Note 33**).
11. Add the species-specific Fc-receptor blocking antibody supplied in the EasySep® kit at 100 μL/mL of cells.
12. Add primary APC-conjugated antibody at a concentration of 0.5 μg/mL, mix well, and incubate for 15 min at room temperature and in the dark (*see* **Notes 34** and **35**).
13. Prepare a uniform suspension of the EasySep® magnetic nanoparticles by pipetting up and down at least five times. Add the nanoparticles at 50 μL/mL of cells. Mix well and incubate for 10 min at room temperature in the dark.
14. Add CM to a total volume of 2.5 mL and mix the cells by gently pipetting up and down three times, place the tube without cap in the EasySep® magnet for 5 min.
15. Remove the supernatant by inverting the magnet with the tube inside in one continuous motion. Keep the magnet and tube inverted for 3 s and then return it to its upright position (*see* **Note 36**).
16. Remove the tube from the magnet and repeat **steps 14** and **15** twice for a total of three washes in the magnet.
17. Re-suspend the $V\alpha7.2^+$ cells remaining in the tube in 100 μL of CM, count the cells, and dilute to the desired concentration depending on downstream application.

3.5 Cell Sorting for Isolation of MAIT Cells

If you are working with samples where the MAIT cells make up a small fraction, it may be beneficial to isolate the $CD8^+$ T cell fraction before proceeding with FACS sorting of MAIT cells. In that case perform **steps 1–9** in Subheading 3.4 before starting with this protocol.

1. Transfer cells to a 5 mL polystyrene tube and centrifuge at 380 × *g* for 6 min in a cold centrifuge and discard the supernatant.
2. Prepare a 1:1000 dilution of a cell viability dye in PBS (e.g., LIVE/DEAD™ Fixable cell stain kit or BD Horizon™ FVS) and add 100 μL/10 × 10^6 cells. Incubate the sample at room temperature in the dark for 25 min.
3. Wash off any excess cell viability dye by adding the same volume of CM and centrifuge the cells at 380 × *g* for 6 min in a cold centrifuge and discard the supernatant.
4. Prepare a mix of antibodies containing relevant markers for your MAIT cell population, in CM (*see* **Note 37**).
5. Add the antibody mix to your cells and incubate 20 min at 4° C in the dark.
6. Centrifuge the cells at 380 × *g* for 6 min in a cold centrifuge and discard the supernatant.
7. Wash off any excess antibodies as in **step 3**.
8. Re-suspend the cells at 5 × 10^6 cells/mL in CM and keep the cells on ice until you are ready to sort.
9. Acquire the sample on a FACS instrument (*see* **Note 38**). See Fig. 3 for a suggestion for gating strategy and a representative staining and purity of MAIT cells and Vα7.2$^+$CD161$^-$ conventional T cells.

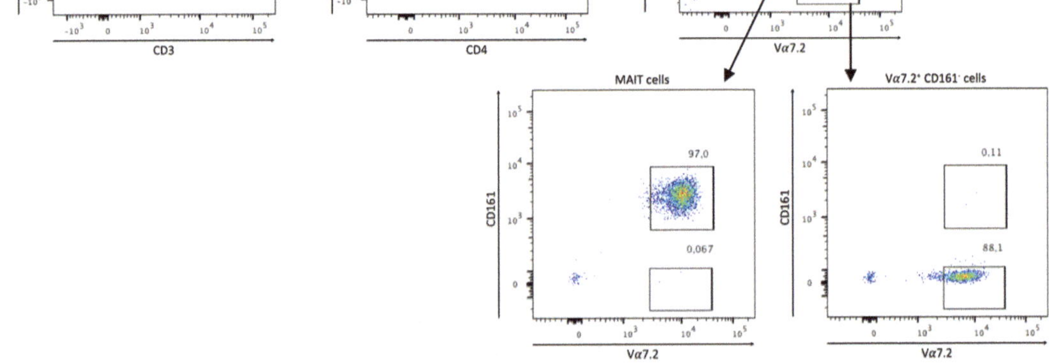

Fig. 3 Gating strategy for MAIT cells sorting. The figure shows FACS plots of a single cell solution from blood, and the gating strategy used to identify MAIT cells among CD8$^+$ cells isolated as in **steps 1–9** in Subheading 3.4. MAIT cells are identified as the CD45$^+$CD3$^+$Vα7.2$^+$CD161^{high} subset and if needed, CD45$^+$CD3$^+$Vα7.2$^+$CD161$^{-/low}$ conventional T cells can also be sorted

4 Notes

1. Store the EDTA-M at 4 °C if it is not going to be used within 4 h, otherwise store the EDTA-M at 37 °C. 110 mL of EDTA-M is required for two tissue samples.
2. Store the WM at 4 °C if it is not going to be used within 4 h, otherwise store the WM at 37 °C. 70 mL of WM is required for two tissue samples.

   β-Mercaptoethanol reduces the activity of the enzymes in the digestion buffer and can be excluded. However, including β-Mercaptoethanol is recommended if you plan to culture lymphocytes, as it prevents formation of reactive oxygen species.
3. Prepare DM by adding DNase, Liberase TM, and $CaCl_2$ to WM just before use. This is to avoid loss of enzyme activity. 28 mL of DM is required for four tissue samples (used for both LPL and iEL isolation). Addition of $CaCl_2$ is optional. RPMI does not naturally contain a calcium concentration high enough for optimal activity of the liberase TM. However, increased enzymatic activity can cause a loss of viability and/or unspecific cleavage of epitopes.
4. Percoll stock solution is prepared by mixing one part 1.5 mM NaCl and nine parts Percoll. Percoll stock solution should be stored at 4 °C.
5. Adding EDTA to the 40% Percoll is optional but improves the dissociation of lymphocytes from the epithelial cells
6. Store the CM at 4 °C.
7. MAIT-cell-specific MR1 tetramers presenting 5-OP-RU, and negative controls are available from the NIH tetramer facility (http://tetramer.yerkes.emory.edu/).
8. It is possible to use a kit with a different fluorochrome specificity and a different antibody if necessary.
9. Magnets used for the magnetic bead isolation of $CD8^+$ and $V\alpha7.2^+$ cells need to be suitable for the vials and beads used. See recommendations by bead manufacturer.
10. Work with the sample on a sterile paper cloth and prevent the sample from drying out by regularly washing it in cold PBS. Avoid perforating the tissue by cutting to deep.

    If it is difficult to determine when all the blood vessels, connective tissue, and fat are removed, place the sample in a sterile Petri dish with cold PBS; it will remain flat and not "curl" into a tube, and the connective tissue will appear slightly more whitish than the mucosa.

11. **Step 1** is not applicable for tumor tissue. When isolating cells from a disorganized tissue, such as a colon tumor, it is hard to sort out blood vessels, connective tissue, etc. that are present. For tumor tissues; wash the tissue in PBS, dry off any excess liquid and mucus on a sterile cloth and proceed to **step 2**.
12. Pre-weigh an empty sterile Petri dish or tube. Dry of any excess PBS on the tissue on a sterile paper cloth and weigh the sample in the Petri dish (or tube).
13. This protocol is recommended for tissue samples weighing up to 1 g. For larger samples, either split it into two or more tubes, isolate the cells in parallel and pool the cells when finished, or adjust buffer volumes and incubation times throughout the protocol.
14. The speed of the magnetic stirrer should be 300–400 rpm.
15. All four epithelial fractions can be collected in the same 50 mL tube as these will be pooled before proceeding with isolation of the iELs
16. Prior to aspirating the last round of EDTA-M, it is suggested to vortex the tube for a few seconds to further dissociate the epithelial layer from the underlying lamina propria.
17. It may be necessary to repeat this step if the cells do not form a distinct pellet. If so, spin the tube again as before.
18. The speed of the magnetic stirrer should be a maximum of 200 rpm. Increased mechanical force at higher speed will reduce the cell yield.

    This step will increase iEL recovery.
19. The gradient can also be created by carefully placing the 70% Percoll solution underneath the 40% Percoll solution using a sterile transfer pipette.
20. The Percoll gradient centrifuging must be performed at room temperature. To make sure that all reagents are at room temperature, it is recommended to prepare the 40% and 70% Percoll solutions and store in sterile tubes (at room temperature) when preparing the other buffers.
21. After gradient centrifugation, the iEL can be found as a white "band" at the interface between the Percoll solutions. Epithelial cells float on top of the 40% Percoll solution. Percoll separation will result in some loss of iEL, but at the same time remove the bulk of epithelial cells and thereby improve downstream analyses and/or culture. If the layer on top of the 40% Percoll solution is very thick and sticky, it may help to remove it, using a sterile transfer pipette, before recovering the iEL.

22. This procedure usually results in 1 to 8 million mononuclear cells in the epithelial fraction per gram of healthy tissue, and 2 to 9 million per gram tumor tissue.
23. The speed of the magnetic stirrer should be maximum 200 rpm. A too large mechanical force will reduce the cell yield.
24. The enzymatic digestion will result in loss of some surface molecules. To examine if a particular cell surface marker of interest is affected, one may treat for example mononuclear cells from blood as indicated in the protocol and assess potential loss of surface markers. With regard to MAIT cell function, CD127 (IL-7 receptor alpha) expression is lost during the isolation procedure. Culture overnight in cell culture medium in a humidified CO_2 incubator will restore CD127 expression.
25. This procedure usually results in 25–60 million mononuclear cells from the lamina propria per gram of healthy tissue. In the tumors, lymphocyte yields vary more but usually range from 20 to 130 million per gram tissue.
26. It is possible to stain in a U-bottomed plate, as well as U- and V-bottomed tubes, but it may be necessary to adjust centrifuge force and is not recommended for samples containing few cells.
27. Many cell viability dyes contain Sodium azide (highly toxic) which must be discarded according to hazardous chemical waste protocols.
28. It may be necessary to include negative control with 6-FP-loaded MR1 tetramer, Fluorescence Minus One (FMO) or isotype controls as part of your experiment to facilitate downstream analysis.
29. If you are using several antibodies based on the Brilliant Violet fluorochrome, it is possible to replace the PBS containing 2% FCS with Brilliant stain buffer (BD Biosciences) in order to minimize the unspecific fluorescence from combining several Brilliant violet-based dyes.
30. Adjusting the fixation and/or incubation time may improve staining result.
31. The cells can be kept up to 1 week at 4 °C following fixation, depending on the antibody stability (double-conjugated antibodies may degrade faster than non-conjugated antibodies); however, it is recommended to test this for your specific antibodies if you do not intend to analyze the samples immediately.
32. It is important to remove the supernatant while the tube is in the magnet (either using transfer pipette or by inverting the magnet and tube). Do not repeat this step as excessive handling could cause the beads to detach from the cells and decrease the yield.

If required, save the supernatant containing CD8-cells for purity assessment.

33. For samples containing fewer than 20×10^6 cells, re-suspend the cells in 100 μL of PBS/EDTA buffer and proceed with **step 11**.

    Make sure to use a polystyrene tube that fits in the Easy-Sep® magnet.

34. This concentration is suitable for APC-labeled anti-Vα7.2 (clone 3C10). If using another antibody preparation, titrate the antibody for optimal purity and recovery, increasing the antibody concentration generally increases cell yield but may reduce purity.

35. This procedure can also be carried out using APC-labeled MR1 tetramers presenting 5-OP-RU. This will yield pure MAIT cells without the conventional T cells using Vα7.2, but it will also activate the MAIT cells considerably and lead to maximal expression of granzyme B and degranulation.

36. The magnetically labeled cells will remain inside the tube during this process; however, it is important not to shake or blot off any drops hanging from the tube as these might contain magnetically labeled cells.

    If needed, collect the supernatant to check the isolation procedure, the supernatant would contain all Vα7.2⁻ cells.

37. When preparing the antibody mixture, it is important that you take into account the numbers of cells you intend to sort and adjust the volume of antibody accordingly. Titration of the antibodies may be necessary in order to optimize cell yield.

38. In order to optimize the cell yield and viability, keep the sample at 4 °C during the sort and only use a gentle setting for sample agitation (100–300 rpm).

    Make sure to match the event rate (events/second) with the droplet rate for the nozzle you are using so that one cell is present in every third drop. This will optimize the yield. It is recommended to sort using a 70 or 85 μm nozzle.

Acknowledgment

This work was supported by the Swedish Research Council and the Swedish Cancer Foundation.

References

1. Dias J, Boulouis C, Sobkowiak MJ, Lal KG, Emgard J, Buggert M, Parrot T, Gorin JB, Leeansyah E, Sandberg JK (2018) Factors influencing functional heterogeneity in human mucosa-associated invariant T cells. Front Immunol 9:1602. https://doi.org/10.3389/fimmu.2018.01602
2. Gherardin NA, Souter MN, Koay HF, Mangas KM, Seemann T, Stinear TP, Eckle SB, Berzins SP, d'Udekem Y, Konstantinov IE, Fairlie DP, Ritchie DS, Neeson PJ, Pellicci DG, Uldrich AP, McCluskey J, Godfrey DI (2018) Human blood MAIT cell subsets defined using MR1 tetramers. Immunol Cell Biol 96(5):507–525. https://doi.org/10.1111/imcb.12021
3. Gherardin NA, Keller AN, Woolley RE, Le Nours J, Ritchie DS, Neeson PJ, Birkinshaw RW, Eckle SBG, Waddington JN, Liu L, Fairlie DP, Uldrich AP, Pellicci DG, McCluskey J, Godfrey DI, Rossjohn J (2016) Diversity of T cells restricted by the MHC class I-related molecule MR1 facilitates differential antigen recognition. Immunity 44(1):32–45
4. Dias J, Boulouis C, Gorin JB, van den Biggelaar R, Lal KG, Gibbs A, Loh L, Gulam MY, Sia WR, Bari S, Hwang WYK, Nixon DF, Nguyen S, Betts MR, Buggert M, Eller MA, Broliden K, Tjernlund A, Sandberg JK, Leeansyah E (2018) The CD4(−)CD8(−) MAIT cell subpopulation is a functionally distinct subset developmentally related to the main CD8(+) MAIT cell pool. Proc Natl Acad Sci U S A 115(49):E11513–E11522. https://doi.org/10.1073/pnas.1812273115
5. Sundstrom P, Ahlmanner F, Akeus P, Sundquist M, Alsen S, Yrlid U, Borjesson L, Sjoling A, Gustavsson B, Wong SBJ, Quiding-Jarbrink M (2015) Human mucosa-associated invariant T cells accumulate in Colon adenocarcinomas but produce reduced amounts of IFN-gamma. J Immunol 195(7):3472–3481. https://doi.org/10.4049/jimmunol.1500258
6. Zabijak L, Attencourt C, Guignant C, Chatelain D, Marcelo P, Marolleau JP, Treiner E (2015) Increased tumor infiltration by mucosal-associated invariant T cells correlates with poor survival in colorectal cancer patients. Cancer Immunol Immun 64(12):1601–1608. https://doi.org/10.1007/s00262-015-1764-7
7. Won EJ, Ju JK, Cho YN, Jin HM, Park KJ, Kim TJ, Kwon YS, Kee HJ, Kim JC, Kee SJ, Park YW (2016) Clinical relevance of circulating mucosal-associated invariant T cell levels and their anti-cancer activity in patients with mucosal-associated cancer. Oncotarget 7 (46):76274–76290. https://doi.org/10.18632/oncotarget.11187
8. Ling LM, Lin YY, Zheng WW, Hong S, Tang XQ, Zhao PW, Li M, Ni JS, Li CG, Wang L, Jiang YF (2016) Circulating and tumor-infiltrating mucosal associated invariant T (MAIT) cells in colorectal cancer patients. Sci Rep-UK 6:ARTN 20358, https://doi.org/10.1038/srep20358
9. Gherardin NA, Loh L, Admojo L, Davenport AJ, Richardson K, Rogers A, Darcy PK, Jenkins MR, Prince HM, Harrison SJ, Quach H, Fairlie DP, Kedzierska K, McCluskey J, Uldrich AP, Neeson PJ, Ritchie DS, Godfrey DI (2018) Enumeration, functional responses and cytotoxic capacity of MAIT cells in newly diagnosed and relapsed multiple myeloma. Sci Rep 8 (1):4159. https://doi.org/10.1038/s41598-018-22130-1
10. Duan M, Goswami S, Shi JY, Wu LJ, Wang XY, Ma JQ, Zhang Z, Shi Y, Ma LJ, Zhang S, Xi R, Cao Y, Zhou J, Fan J, Zhang XM, Gao Q (2019) Activated and exhausted MAIT cells foster disease progression and indicate poor outcome in hepatocellular carcinoma. Clin Cancer Res. https://doi.org/10.1158/1078-0432.CCR-18-3040
11. Shaler CR, Tun-Abraham ME, Skaro AI, Khazaie K, Corbett AJ, Mele T, Hernandez-Alejandro R, Haeryfar SMM (2017) Mucosa-associated invariant T cells infiltrate hepatic metastases in patients with colorectal carcinoma but are rendered dysfunctional within and adjacent to tumor microenvironment. Cancer Immunol Immunother. https://doi.org/10.1007/s00262-017-2050-7
12. Peterfalvi A, Gomori E, Magyarlaki T, Pal J, Banati M, Javorhazy A, Szekeres-Bartho J, Szereday L, Illes Z (2008) Invariant Valpha7.2-Jalpha33 TCR is expressed in human kidney and brain tumors indicating infiltration by mucosal-associated invariant T (MAIT) cells. Int Immunol 20 (12):1517–1525. https://doi.org/10.1093/intimm/dxn111
13. Sundström P et al (2019) Tumor-infiltrating mucosal-associated invariant T (MAIT) cells retain expression of cytotoxic effector molecules. Oncotarget 10(29):2810–2823. https://doi.org/10.18632/oncotarget.26866

Chapter 4

Isolation of Immune Cells from Placental Tissues and Phenotypic and Functional Analysis of MAIT Cells

Martin Solders, Laia Gorchs, and Helen Kaipe

Abstract

The placenta is an immunological paradox since maternal immune cells infiltrating placental tissues need to be tolerant toward the fetus but still retain immunity against potential infections. This makes the placenta an interesting tissue for studying immunological processes. Mucosal-associated invariant T (MAIT) cells are a subset of T cells that respond to bacterially derived metabolites of riboflavin synthesis. Upon activation, MAIT cells respond by secretion of inflammatory cytokines and by directed killing of infected cells by the use of granzymes and perforin. In this protocol, we describe methods for the isolation of immune cells from the placental intervillous space and adjacent tissues such as the umbilical cord, decidua parietalis, and decidua basalis. We further describe how to stimulate MAIT cells in mixed cell suspensions of mononuclear cells with bacteria, and how to analyze the phenotypic and functional responses with flow cytometry.

Key words MAIT cells, Placenta, Intervillous blood, Decidua

1 Introduction

The placenta is an essential and temporary organ of human reproduction. It has a joint development with the fetus and is composed of both fetal and maternal cells. The fetus and amniotic fluid are enclosed by three membranes, the fetal membranes amnion and chorion, and the maternal decidua. The fetal membranes run on the fetal side of the placenta, whereas the decidua covers both the fetal membranes and the placenta. The part of the decidua covering the placenta is called decidua basalis, whereas the part joined with the chorion and amnion is termed decidua parietalis [1]. A schematic representation of a term placenta can be seen in Fig. 1.

The fetus relies on a continuous supply of oxygen and nutrients from the mother, and from the second trimester until birth, the fetus is supplied by maternal arterial blood. Spiral arteries penetrate the decidua basalis, and drains into the intervillous space of the placenta. From the fetal side, fetal villi form tree branch-like structures which protrude into the maternal blood. The fetal blood

Helen Kaipe and Isabelle Magalhaes (eds.), *MAIT Cells: Methods and Protocols*, Methods in Molecular Biology, vol. 2098, https://doi.org/10.1007/978-1-0716-0207-2_4,

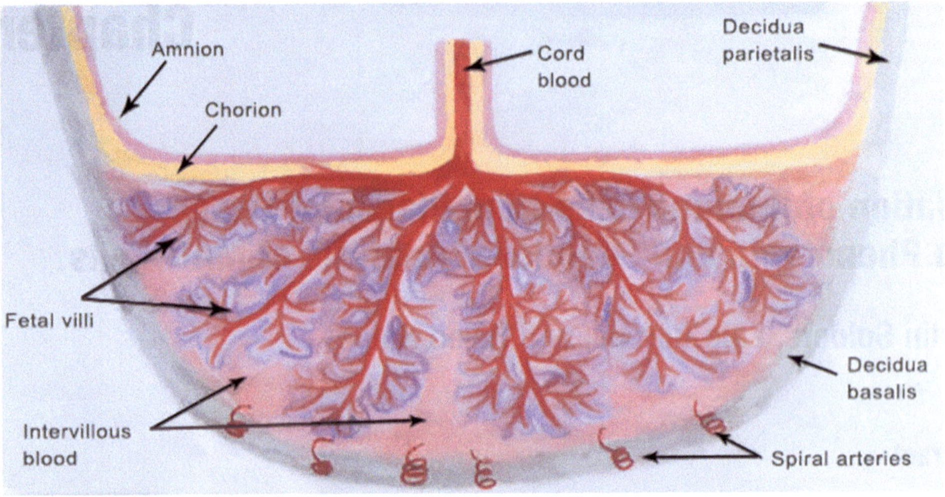

Fig. 1 A schematic representation of a term placenta. The placenta is covered by the maternal decidua. The part covering the placenta is called decidua basalis, and the part covering the fetus is termed decidua parietalis. The decidua parietalis encloses the fetal membranes chorion and amnion, which surround the fetus. Spiral arteries penetrate the decidua basalis into the placenta, and deliver blood to the intervillous space, which is then called intervillous blood. Fetal villi containing umbilical cord vessels then protrude down into the intervillous blood, and an exchange of gas and nutrients takes place. The smaller umbilical cord vessels then join together and form the umbilical cord vein and artery, supplying the fetus with gas and nutrients from the second trimester

vessels run inside the villi, and gas and nutrients are exchanged over a thin membrane of syncytiotrophoblasts and cytotrophoblasts. These blood vessels connect and end up forming the umbilical vein in the umbilical cord [1].

The sites where maternal immune cells come into direct contact with fetal cells are referred to as the fetal–maternal interface. Maternal immune cells infiltrate both the decidua parietalis and basalis [2, 3], and the maternal blood entering the intervillous space contains maternal immune cells that are in immediate contact with syncytiotrophoblasts [4]. In vitro, maternal immune cells detect and reject fetal cells in accordance with the mechanisms known from transplantation immunology [5, 6], but in the case of successful pregnancies, this reaction is quelled at the fetal–maternal interface. The fetal syncytiotrophoblast cells lining the villi do not express HLA, whereas the extravillous trophoblasts invading the decidua express HLA-C, but not HLA-A or -B. However, they express the non-classical, non-polymorphic HLA-E, -G, and -F molecules, which inhibit both NK cells and T cells [7, 8].

Mucosal-associated invariant T (MAIT) cells are a subset of invariant T cells that respond to metabolites from the synthesis of riboflavin [9]. Humans cannot synthesize riboflavin, and they depend on dietary sources for this essential vitamin. On the other hand, many species of bacteria have the ability to synthesize riboflavin, and the resulting metabolites make them a MAIT cell target. The riboflavin metabolites are presented by the non-polymorphic MHC class I-related molecule (MR1) [9], and MAIT cells bind the MR1 with its T cell receptor made up of a Vα7.2-segment coupled with either Jα33, 12, or 20, paired with a limited set of β chains [10–12].

The importance of MAIT cells in pregnancy is not known. However, we have recently shown that placental intervillous blood is enriched in MAIT cells when compared to matched peripheral blood at healthy term pregnancy [4]. Placenta-derived chemokines are involved in the recruitment or retention of MAIT cells in the intervillous space [13]. MAIT cells in intervillous blood also had a stronger reaction to bacterial stimulation than those in peripheral blood, even though no increase in activation markers was seen prior to the stimulation [4]. MAIT cells were present in the decidua, and the proportion of MAIT cells was higher in decidua basalis compared to decidua parietalis [3]. MAIT cells are few in cord blood compared to adult intervillous blood, and they also display a naïve phenotype [4, 14]. In line with lack of expression of HLA molecules on syncytiotrophoblasts, they are also negative for MR1, whereas fetal macrophages in the villi express MR1 [4].

The following protocols provide methods for the isolation of immune cells from the fetal–maternal interface, as well as phenotypic and functional studies of MAIT cells activated in cell cultures of mixed mononuclear cells.

2 Materials

2.1 Tools

1. Metal container, at least 20 × 30 × 5 cm.
2. Forceps.
3. Scissors.
4. Hemostatic clamps.
5. Scalpels.
6. Petri dishes, various sizes.
7. Metal nets, 100 μm pore size.

2.2 Plastic

1. 15 mL tubes.
2. 50 mL tubes.
3. Cell strainers, 40 μm pore size.
4. Cell strainers, 70 μm pore size.
5. Cell scrapers.
6. Heparin tubes.
7. Pipettes, 10 mL and 25 mL.

8. Aspirator pipettes, 2 mL.
9. GentleMacs dissociator tubes.
10. Flat-bottomed 96-well plate.
11. V-bottomed 96-well plate.

2.3 Reagents

1. Phosphate buffered saline (PBS).
2. Complete medium: RPMI containing 10% heat-inactivated fetal calf serum, 100 U/mL penicillin, and 100 μg/mL streptomycin.
3. RPMI containing 0.1 mM ethylenediamine tetraacetic acid (EDTA).
4. Phorbol 12-myristate 13-acetate (PMA), 25 ng/mL.
5. Ionomycin, 1 μg/mL.
6. Heparin, 100 U/mL.
7. FACS-buffer: PBS containing 0.1% bovine serum albumin (BSA) and 1 mM EDTA.
8. Freezing medium: complete RPMI containing 20% DMSO. Store at 4 °C.
9. BD Cytofix/Cytoperm™ kit or equivalent.
10. Brefeldin-A (BFA), 10 μg/mL.
11. Ficoll.
12. *Escherichia coli* (*E. coli*) or other riboflavin-producing bacteria.
13. 1% paraformaldehyde (PFA).

2.4 Antibodies

1. Anti-CD28mAb (clone CD28.2).
2. Flow cytometry antibodies (*see* Table 1).

2.5 Equipment

1. Laminar flow hood.
2. Automatic cell counter (Sysmex or equivalent) or hemocytometer.
3. Centrifuge.
4. Multicolor flow cytometer.
5. GentleMACS Dissociator.
6. Vacuum aspirator.
7. Freezing container.
8. Software for flow cytometry analysis (FlowJo or equivalent).

Table 1
Flow cytometry antibodies

Surface markers	Fluorochrome	Clone	Source
Base staining			
CD3	V450, FITC, PE-Cy7	UCHT1	BD
CD4	V500	RPA-T4	BD
CD8	A700, APC	RPA-T8	BD
CD161	PE	HP-3G10	BioLegend
TCR Vα7.2	APC-Cy7	3C10	BioLegend
—	7AAD	—	BD
Activation markers			
CD25	BV421	M-A251	BD
CD69	FITC, APC, PE-Cy7	L78	BD
HLA-DR	FITC	G46-6	BD
Co-inhibitory markers			
CD279 (PD-1)	BV421	EH12.1	BD
CD366 (TIM-3)	APC	F38-2E2	Miltenyi
Intracellular markers			
Granzyme B	FITC	GB11	BD
IFN-γ	PE-Cy7	4S.B3	BD
IL-17	A488	N49-653	BD
IL-22	APC	REA466	Miltenyi
Perforin	A647	δG9	BD

3 Methods

3.1 Collection of a Human Placenta and Peripheral Blood

1. Placentas can be obtained either from caesarian sections or from vaginal deliveries after informed consent.
2. Peripheral blood from the same donor should be collected in heparin tubes shortly before delivery, as the systemic factors involved in labor is likely to affect the immune cells both in the placenta and in the systemic circulation. If labor is induced by misoprostol, peripheral blood samples should be taken 30–60 min after the administration of the drug (*see* **Note 1**).
3. The gynecologist or midwife should try to keep the placenta as sterile as possible, and directly place it in a sterile container.
4. Transport the container directly to the lab.

3.2 Isolation of Placental Tissues and Cells

Most of this work can be performed alone, but some steps require aid from a colleague (**steps 4**, **5**, **10**, and **11** in Subheading 3.2.1).

3.2.1 Collection of Intervillous Blood and Umbilical Cord Blood

1. Work in a flow hood and use sterile gloves.
2. Place the placenta with the umbilical cord facing up. Inspect the placental surface and the umbilical cord for cuts, rifts, or other damages. If the umbilical cord is intact, use a hemostatic clamp to clamp the distal end of the umbilical cord. If the cord is damaged, clamp it proximally of the damage.
3. Turn the placenta so that the umbilical cord faces down. Using scissors and forceps, cut the joined "fetal membranes" (amnion, chorion, and decidua parietalis) $\geq$1 cm from the edge of the placenta. Place the tissue on a large Petri dish, cover with PBS, and put the Petri dish aside in the flow hood. Using PBS and a vacuum aspirator, wash the surface of the placenta (*see* **Note 2**). Inspect the placental surface for cuts, rifts, or other damages. If the placenta is damaged on either side, discard it as biological waste (*see* **Note 3**).
4. Using your hands, lift the placenta with the umbilical cord facing up. Intervillous blood will then seep, drip, or flow from the maternal side.
5. Get help from a colleague to collect the blood directly into heparin tubes and quickly turn them to mix the blood with heparin. Alternatively, collect the blood on a Petri dish and immediately transfer it to heparin tubes with a pipette. Work as quickly as possible as the intervillous blood quickly coagulate outside of the placenta (*see* **Note 4**).
6. A term placenta generally yields between 10 and 100 mL of intervillous blood without mechanical manipulation.
7. Place the placenta down with the umbilical cord facing up.
8. Change gloves. Wash the umbilical cord and the fetal side of the placenta with PBS and use the aspirator to remove the waste.
9. Remove the hemostatic clamp, and use scissors to cut off the damaged distal portion of the umbilical cord to create a fresh cut.
10. Using your hands, lift the placenta with the umbilical cord facing down.
11. With the help of a colleague collect the blood seeping or dripping from the umbilical cord into heparin tubes. If needed, press the umbilical cord vessels visible on the fetal side of the placenta to assist the transfer of blood into the cord, and then into the heparin tubes.

3.2.2 Isolation of Mononuclear Cells from Peripheral, Intervillous, and Umbilical Cord Blood

1. Dilute the blood 1:2 with PBS. Add 15 mL Ficoll to a 50 mL tube and carefully add 30 mL of diluted blood on top using a pipette. Alternatively, adapt according to the acquired blood volume so that the final ratio between Ficoll and diluted blood is 1:2. Centrifuge the tubes at room temperature, 400 × *g* for 30 min, 1 acceleration, 0 brake.
2. Transfer the white cell layer to new tubes, and fill them up with PBS. Wash the cells by three centrifugations at 4 °C, 600 × *g* for 8 min. After the third wash step, re-suspend the cells in complete medium.
3. Count the cells by using a hemocytometer or by using an automatic cell counter.

3.2.3 Collection of Decidua Basalis Tissue

1. Place the placenta with the umbilical cord facing down.
2. Wash the maternal side of the placenta with PBS using an aspirator.
3. The top of the placenta is covered by a thin gray membrane, the decidua basalis. Place a scalpel at an angle of 45°, and carefully scrape the thin membrane from the underlying tissue.
4. A term placenta generally yields between 5 and 20 mL of decidua basalis tissue.
5. Place the tissue in a 50 mL tube, no more than 10 mL of tissue per tube and fill the tubes with PBS. Continue to Subheading 3.2.5.

3.2.4 Collection of Decidua Parietalis Tissue

1. Take the Petri dish with the fetal membranes from **step 3** in Subheading 3.2.1. Using an aspirator pipette, remove the PBS covering the fetal membranes.
2. Using forceps, turn the membranes so that the decidua parietalis faces up. This is the side that has a rougher appearance, whereas the fetal side covered by the amnion is a sheer membrane.
3. Remove large blood clots mechanically with the forceps, and wash the decidua with PBS and the aspirator. Continue until the PBS is no longer colored by remaining blood. By using the aspirator, remove the PBS.
4. By using forceps and a cell scraper, carefully scrape the decidua parietalis from the chorion and amnion. Use the forceps to fixate the tissue in one direction, place the cell scraper at an angle of 45°, and scrape in the opposite direction. The decidua parietalis sometimes comes off as a largely intact membrane, and sometimes in small pieces. After the whole decidua parietalis is removed, discard of the chorion and amnion.
5. Using forceps together with a scalpel, mechanically dissect the decidua into as small pieces as possible.

6. A term placenta generally yields between 10 and 30 mL of decidua parietalis tissue.
7. Place the decidua parietalis tissue in 50 mL tubes, no more than 10 mL of tissue per tube and fill the tubes with PBS.

3.2.5 Isolation of Mononuclear Cells from Decidual Tissues

1. Take the tubes with decidua parietalis and basalis, and mix the tissue with the PBS.
2. Centrifuge the tubes at 4 °C, 600 × *g* for 30 s.
3. Remove the PBS using the aspirator and fill the tubes with new PBS.
4. Repeat **steps 1–3** until the PBS is no longer colored by contaminating blood, and you are left with decidua parietalis and basalis tissue.
5. Centrifuge the tubes at 4 °C, 600 × *g* for 30 s, and remove the PBS using the aspirator.
6. Move the tissue into the gentleMACS tubes, no more than 5 mL of tissue per tube.
7. Fill the tubes with PBS, and run them on the gentleMACS Dissociator until the tissue and PBS are dispersed into a homogenous pulp.
8. Filter the pulp with PBS successively through a 100 μm metal net, 70 μm cell strainer, and lastly a 40 μm cell strainer.
9. Collect the pulp and repeat the **steps 6–8**, then discard of the tissue.
10. Continue with the rest of the decidua parietalis and basalis tissue.
11. After all tissues have been processed, pool the cell suspension from the same tissue in 50 mL tubes, and centrifuge the tubes at 4 °C, 600 × *g* for 8 min.
12. Re-suspend the pellets in a small volume of RPMI supplemented with EDTA, and pool the cell suspension in one 50 mL tube.
13. Count the cells using a hemocytometer or by using a cell counter. Divide the cell suspension into 50 mL tubes with no more than 50 × 10^6 cells/tube.
14. Dilute the cell suspension with RPMI with EDTA to a total volume of 35 mL.
15. Using reverse pipetting, add 10 mL of Ficoll below the cell suspension.
16. Centrifuge the tubes at room temperature, 400 × *g* for 30 min, 1 acceleration, 0 brake.
17. Transfer the white cell layer to new tubes, and fill them up with PBS. Wash the cells by three centrifugations at 4 °C, 600 × *g*

for 8 min. After the third wash step, re-suspend the cells in complete medium.

18. Count the cells in a hemocytometer or by using a cell counter.
19. The cells can either be used fresh or frozen for later analysis. Subheading 3.2.6 describes how cells can be frozen.

3.2.6 Freezing of Isolated Placental Mononuclear Cells

Freezing of placental cells is optional.

1. Re-suspend the cells to a desired concentration in complete medium.
2. Add 500 μL of cell suspension per cryotube.
3. Add 500 μL of cold freezing medium to each cryotube, resulting in a final DMSO concentration of 10%.
4. Place the cryotubes in a freezing container, and immediately transfer the container to a −80 °C freezer.
5. ≥24 h later, transfer the cryotubes to a liquid nitrogen for long-term storage.

3.3 Activation of Placental and Peripheral MAIT Cells

3.3.1 Microbial Stimulation

Decide on your desired microbe for stimulation. *E. coli* has been most commonly used to stimulate MAIT cells. We have used a resident strain of *E. coli* isolated from a fecal sample from a child [3, 15]. Others have used the *E. coli* strains DH5α [16, 17] or D21 [12] or other bacteria such as *Mycobacterium tuberculosis* [18], *Pseudomonas aeroginosa*, *Klebsiella pneumoniae*, *Lactobacillus acidophilus*, *Staphylococcus aureus*, or *Staphylococcus epidermidis* [19]. The use of fungus species, such as *Candia albicans*, has also been reported [12, 19]. The bacteria can be fixed by incubating with 1% PFA for 15 min.

1. Use freshly isolated mononuclear cells from decidua, intervillous blood, peripheral blood, or cord blood, or thaw frozen cells isolated previously. Dilute in complete medium to a concentration of 3×10^6 cells/mL.
2. Add 250 μL of cell suspension to a flat-bottomed 96-well plate, resulting in 750,000 mononuclear cells/well. Add cells to at least two wells per condition, one unstimulated and the other stimulated with PMA/Ionomycin (*see* **Note 5**).
3. Add inactivated bacteria to the wells marked for stimulation at an appropriate multiplicity of infection (MOI) (*see* **Note 6**). We used an MOI of 40. Add the same volume of PBS to the wells marked to be unstimulated.
4. Add anti-CD28, 1.25 μg/mL, to the wells marked for stimulation.
5. Mix the contents of each well, and incubate for 12 h at 37 °C and 5% CO_2.

6. After 12 h, add 10 μg/mL BFA to all wells, and incubate for another 4 h.
7. Harvest the cells, and centrifuge at 4 °C, 600 × *g* for 8 min
8. Wash the cells and stain for flow cytometric analysis as described in Subheading 3.4.

3.3.2 PMA/Ionomycin Stimulation

An unspecific, yet standardized and well-studied mode of T cell activation is the use of PMA together with ionomycin. PMA activates protein kinase C, and ionomycin raises intracellular levels of calcium, together resulting in activation of NF-κB and NFAT, thus mimicking a physiological T cell activation [20].

1. Use freshly isolated mononuclear cells, or thaw frozen cells previously isolated. Dilute in complete medium to a concentration of 3 × 10^6 cells/mL.
2. Add 250 μL of cell suspension to a flat-bottomed 96-well plate, resulting in 750,000 mononuclear cells/well. Add cells to at least two wells per condition, one unstimulated and the other stimulated with PMA/Ionomycin (*see* **Note 5**).
3. Add PMA (25 ng/mL) and ionomycin (1 μg/mL) to the wells marked for stimulation and BFA (10 μg/mL) to all wells.
4. Mix the contents of each well, and incubate for 5 h at 37 °C and 5% CO_2.
5. Harvest the cells, and centrifuge at 4 °C, 600 × *g* for 8 min.
6. Wash again by centrifugation at 4 °C, 600 × *g* for 8 min and stain for flow cytometric analysis.

3.4 Phenotypic and Functional Analysis of Placental MAIT Cells by Flow Cytometry

3.4.1 Extra- and Intracellular Staining for Flow Cytometry

1. Plate cells in a V-bottomed 96-well plate, ≤1 × 10^6 cells/well (*see* **Notes 7–9**).
2. Wash once in FACS buffer by centrifugation at 4 °C, 600 × *g* for 4 min, and add the extracellular antibodies for the base staining, and any additional extracellular antibodies you plan to use (*see* Table 1 and **Note 10**) in FACS buffer so that the total volume including antibodies is 100 uL/well.
3. Mix the contents of each well. Incubate for 30 min in 4 °C, protected from light.
4. Wash twice in FACS buffer by centrifugation at 4 °C, 600 × *g* for 4 min.
5. If you only plan to use extracellular antibodies, add the viability cell marker 7AAD, in a total volume of 50 μL of FACS buffer in each well, and mix the contents. Incubate for 6 min in room temperature, protected from light. Add a desired amount of FACS buffer, and transfer to FACS tubes.
6. If you plan to use intracellular staining, do not add 7AAD.

7. Use the BD Cytofix/Cytoperm™ kit. Re-suspend in 100 μL of CytoFix/Perm. Incubate for 20 min in 4 °C, protected from light.
8. Add 100 μL of Perm/Wash, and wash by centrifugation at 4 °C, 600 × *g* for 4 min.
9. Wash again in 200 μL Perm/Wash by centrifugation at 4 °C, 600 × *g* for 4 min.
10. Add the intracellular antibodies, and add Perm/Wash so that the total volume in each well, including antibodies, is 100 μL, and mix the contents of each well. Incubate for 40 min in 4 °C, protected from light.
11. Wash twice in 200 μL Perm/Wash by centrifugation at 4 °C, 600 × *g* for 4 min.
12. Re-suspend in a desired amount of FACS buffer, and transfer to FACS tubes.

3.4.2 Data Acquisition by Flow Cytometry

1. Using single-stained control beads and unstained cells, set up the flow cytometer so that positive and negative peaks are clearly separated, and that spectral overlap between fluorochromes is minimized. If 7AAD is in your panels, use single-stained cells instead of beads. Lastly, run one of your stained sample tubes (for a few seconds) to verify that positively and negatively stained cells will be captured.
2. Run the samples, the fluorescence minus one controls (FMOs), and the single-stained beads and cells using the settings above.

3.4.3 Data Analysis

1. After exporting your data, load it into FlowJo.
2. Sort your data into the compensation folder (single-stained beads and single-stained cells where applicable), and groups sorted by the staining panels (samples).
3. Use the *compensation wizard* tool in FlowJo to create a compensation matrix based on your single-stained beads and cells. Double-check the results, and adjust the settings so that the positive and negative peaks are properly gated. Create a separate compensation matrix for each panel so that the included control samples match the included antibodies.
4. The gating strategy used by us can be seen in Fig. 2. Start by plotting SSC-A (*y*) against FSC-A (*x*) and gate a broad lymphocyte gate. Then plot FSC-A (*y*) against FSC-H (*x*) and gate the single cells. If you have used 7AAD, plot it on the *x*-axis, SSC-A on the *y*-axis, and gate the 7AAD-negative, live cells. Then plot SSC-A (*y*) against CD3 (*x*) and gate the CD3-positive cells. Then plot CD4 (*y*) against CD8 (*x*) and gate the double-positive cells in the upper right quadrant, but change the settings of the gate by unclicking the "Events

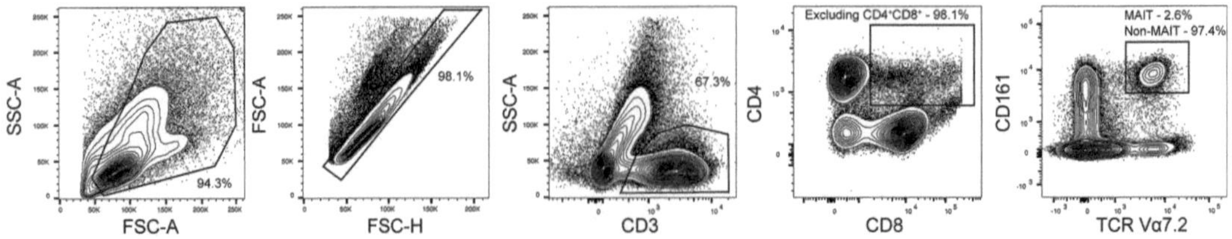

Fig. 2 Gating strategy for identification of MAIT and non-MAIT T cells. Graphs showing the extracellular part of the staining of a representative sample of intervillous blood. We recommend first gating lymphocytes, then single cells, and then CD3$^+$ cells. After this, do an inverted gate, excluding cells positive for both CD4 and CD8. Then gate MAIT cells, and duplicate the gate and change the settings to exclude, rather than include the MAIT cells. Thereafter continue to subgate on MAIT cells and make comparisons with the non-MAIT T cells from the same sample

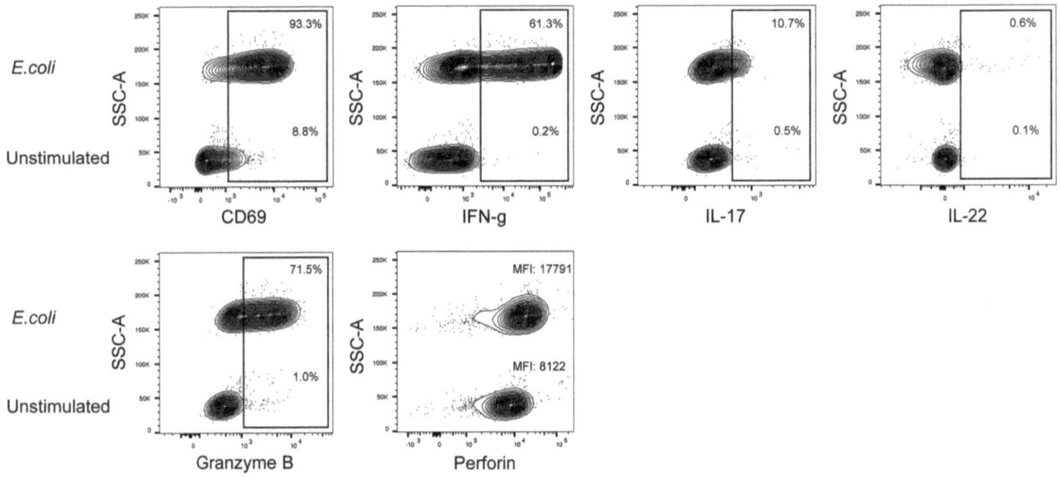

Fig. 3 Staining and gating strategies of MAIT cells stimulated with *E. coli* or left unstimulated. Graphs showing extra- and intracellular staining and gating strategies of a representative sample of intervillous blood (same sample as in Fig. 2), comparing paired samples either stimulated with *E. coli* or left unstimulated. The numbers in the graphs depicting the perforin staining represent the median fluorescent intensity (MFI)

Inside," thus selecting the cells not in the gate. Then plot CD161 (y) against TCR Vα7.2 (x), and gate the MAIT cells. Right-click the MAIT cell gate, and choose "Duplicate Population." Select the duplicated gate, and unclick the "Events Inside," thus resulting in MAIT cells, and non-MAIT cells, one population excluding the other, and both excluding cells double positive for the markers CD4 and CD8.

5. Then gate the MAIT cells based on their expression of markers of interest. A representative staining of the intracellular markers used by us can be seen in Fig. 3. We recommend the use of unstimulated cells both for gating and for comparisons with the *E. coli*-stimulated cells, although FMO controls can also be helpful in the gating process.

4 Notes

1. This is based on the maximal plasma concentration of misoprostol which is reached within 30 min. The half-life of misoprostol in plasma is 20–40 min. Thus, if samples of peripheral blood are taken between 30 and 60 min after the administration of the drug, a potential effect on immune cells should be in effect at the time of sample collection.
2. In many cases, blood will continue to seep out from the maternal side of the placenta, so after sufficient washing has been performed, continue directly to the collection of intervillous blood.
3. If a placenta has tissue damages after delivery, the integrity between the maternal and fetal parts of the placenta must be assumed to be broken. Thus, there is a risk that the intervillous blood will be contaminated by fetal cells, and the placenta should be discarded. However, the placenta could still be used for other purposes, such as collection of tissue for immunohistochemistry or tissue cultures.
4. To prevent the blood to coagulate, the Petri dish can be covered with a thin layer of heparin. If the blood coagulates in spite of the use of heparin, use a cell strainer, 70 μm pore size, to wash away the blood clots with PBS. However, using heparin or washing out blood clots will dilute the intervillous blood to an unknown concentration, thereby making it difficult to get an accurate result from a cell count analysis. This could be solved by collecting a representative sample of intervillous blood straight into a heparin tube, which is then used for analysis.
5. This is dependent on the number of cells that you have, and the proportion of MAIT cells in your sample. If you have enough cells, it is recommended to increase the number of wells to attain enough cells for later analysis by flow cytometry.
6. Different microbes as well as different strains of the same microbe will not activate MAIT cells to the same extent. If you use too few microbes, you will get a weak or absent response. If the stimulation is too strong, you might not be able to detect actual differences in responses when comparing different MAIT cells, although they might exist. If the stimulation is too strong, it can also result in methodological difficulties to evaluate your experiments with flow cytometry. When T cells are activated, the expression of CD3 and the T cell receptor (TCR) is downregulated on the cell surface. As CD3 and the MAIT cell TCR Vα7.2 are commonly used for the identification of MAIT cells by flow cytometry, a strong stimulation can render MAIT cells difficult to identify by this method. It is

therefore highly recommended that you titrate the MOI of the microbial stimulus you plan to use to get an optimal strength. The use of anti-CD28 will increase the stimulation of MAIT cells in your assay. CD69 is a good marker to evaluate MAIT cell activation when optimizing your assay. The activation of MAIT cells in your assay will also depend on the incubation time, thus time titrations are also recommended. We used bacteria from the same batch, frozen in aliquots. These aliquots were then thawed and used directly for titrations, and then for the experiments. This gave us a stimulation strength where we were able to observe differences between the different conditions. Others have also used freshly fixated bacteria for each experiment. It is recommended to decide whether to use freshly fixated cells or aliquots of fixated cells prepared in advance before the start of experiments. BFA is toxic when added to cell cultures, and therefore only added during the last 4 h of the 16-h incubation period. However, this means that you will only be able to investigate the effect of the activation of these last 4 h when you evaluate your experiment. If you want to evaluate MAIT cell activation during the whole incubation period, BFA should be added at the start of the incubation. However, we recommend that you investigate whether this affects the general performance of the assay.

7. In cases where you know that the number of MAIT cells is very few, but you have the requisite number of mononuclear cells following the cell culture stimulation, the staining for flow cytometry can be set up in duplicates or even triplicates. This will provide a higher number of MAIT cells available for analysis, and thus improve the sensibility as well as the credibility of your results.
8. We recommend that you add extra wells with stimulated cells for the extra- and intracellular markers you plan to use, and for which the staining does not create a clear separation between positive and negative (FMOs). In our experience, the results from the FMOs do not differ between cells isolated from different tissues and blood. However, this cannot always be assumed, so we recommend that you test whether you need separate FMOs for each kind of sample, or if one FMO can be used for the same marker on cells isolated from two or more sites.
9. If using the dye 7AAD, or non-mouse antibodies, add cells to an extra well per staining dye/antibody for later use in the compensation matrix.
10. We recommend that you titrate your antibodies, as well as the combination of antibodies and fluorochromes (your panels) you plan to use, before using them in experiments.

Acknowledgments

H.K. was supported by the Swedish Research Council, the Swedish Childhood Cancer Foundation, the Cancer Society in Stockholm, the Swedish Cancer Foundation, Stockholm County Council, and Karolinska Institutet. MS was supported by Karolinska Institutet and Stockholm County Council.

References

1. Kay HH, Nelson DM, Wang Y (2011) The placenta from development to disease. Chichester, West Sussex
2. Bartmann C, Segerer SE, Rieger L, Kapp M, Sutterlin M, Kammerer U (2014) Quantification of the predominant immune cell populations in decidua throughout human pregnancy. Am J Reprod Immunol (New York, NY : 1989) 71(2):109–119. https://doi.org/10.1111/aji.12185
3. Solders M, Gorchs L, Gidlof S, Tiblad E, Lundell AC, Kaipe H (2017) Maternal adaptive immune cells in decidua parietalis display a more activated and coinhibitory phenotype compared to decidua basalis. Stem Cells Int 2017:8010961. https://doi.org/10.1155/2017/8010961
4. Solders M, Gorchs L, Erkers T, Lundell AC, Nava S, Gidlof S, Tiblad E, Magalhaes I, Kaipe H (2017) MAIT cells accumulate in placental intervillous space and display a highly cytotoxic phenotype upon bacterial stimulation. Sci Rep 7(1):6123. https://doi.org/10.1038/s41598-017-06430-6
5. Lashley LE, van der Hoorn ML, van der Mast BJ, Tilburgs T, van der Lee N, van der Keur C, van Beelen E, Roelen DL, Claas FH, Scherjon SA (2011) Changes in cytokine production and composition of peripheral blood leukocytes during pregnancy are not associated with a difference in the proliferative immune response to the fetus. Hum Immunol 72(10):805–811. https://doi.org/10.1016/j.humimm.2011.05.027
6. Tilburgs T, Scherjon SA, van der Mast BJ, Haasnoot GW, Versteeg VDV-MM, Roelen DL, van Rood JJ, Claas FH (2009) Fetal-maternal HLA-C mismatch is associated with decidual T cell activation and induction of functional T regulatory cells. J Reprod Immunol 82(2):148–157. https://doi.org/10.1016/j.jri.2009.05.003
7. Apps R, Murphy SP, Fernando R, Gardner L, Ahad T, Moffett A (2009) Human leucocyte antigen (HLA) expression of primary trophoblast cells and placental cell lines, determined using single antigen beads to characterize allotype specificities of anti-HLA antibodies. Immunology 127(1):26–39. https://doi.org/10.1111/j.1365-2567.2008.03019.x
8. Persson G, Melsted WN, Nilsson LL, Hviid TVF (2017) HLA class Ib in pregnancy and pregnancy-related disorders. Immunogenetics 69(8–9):581–595. https://doi.org/10.1007/s00251-017-0988-4
9. Kjer-Nielsen L, Patel O, Corbett AJ, Le Nours J, Meehan B, Liu L, Bhati M, Chen Z, Kostenko L, Reantragoon R, Williamson NA, Purcell AW, Dudek NL, McConville MJ, O'Hair RA, Khairallah GN, Godfrey DI, Fairlie DP, Rossjohn J, McCluskey J (2012) MR1 presents microbial vitamin B metabolites to MAIT cells. Nature 491(7426):717–723. https://doi.org/10.1038/nature11605
10. Treiner E, Duban L, Bahram S, Radosavljevic M, Wanner V, Tilloy F, Affaticati P, Gilfillan S, Lantz O (2003) Selection of evolutionarily conserved mucosal-associated invariant T cells by MR1. Nature 422(6928):164–169. https://doi.org/10.1038/nature01433
11. Lepore M, Kalinichenko A, Colone A, Paleja B, Singhal A, Tschumi A, Lee B, Poidinger M, Zolezzi F, Quagliata L, Sander P, Newell E, Bertoletti A, Terracciano L, De Libero G, Mori L (2014) Parallel T-cell cloning and deep sequencing of human MAIT cells reveal stable oligoclonal TCRbeta repertoire. Nat Commun 5:3866. https://doi.org/10.1038/ncomms4866
12. Dias J, Leeansyah E, Sandberg JK (2017) Multiple layers of heterogeneity and subset diversity in human MAIT cell responses to distinct microorganisms and to innate cytokines. Proc Natl Acad Sci U S A. https://doi.org/10.1073/pnas.1705759114
13. Solders M, Gorchs L, Tiblad E, Gidlof S, Leeansyah E, Dias J, Sandberg JK, Magalhaes I, Lundell AC, Kaipe H (2019)

Recruitment of MAIT cells to the intervillous space of the placenta by placenta-derived chemokines. Front Immunol 10:1300. https://doi.org/10.3389/fimmu.2019.01300

14. Gold MC, Eid T, Smyk-Pearson S, Eberling Y, Swarbrick GM, Langley SM, Streeter PR, Lewinsohn DA, Lewinsohn DM (2013) Human thymic MR1-restricted MAIT cells are innate pathogen-reactive effectors that adapt following thymic egress. Mucosal Immunol 6(1):35–44. https://doi.org/10.1038/mi.2012.45
15. Ostblom A, Adlerberth I, Wold AE, Nowrouzian FL (2011) Pathogenicity island markers, virulence determinants malX and usp, and the capacity of Escherichia coli to persist in infants' commensal microbiotas. Appl Environ Microbiol 77(7):2303–2308. https://doi.org/10.1128/AEM.02405-10
16. Kurioka A, Ussher JE, Cosgrove C, Clough C, Fergusson JR, Smith K, Kang YH, Walker LJ, Hansen TH, Willberg CB, Klenerman P (2015) MAIT cells are licensed through granzyme exchange to kill bacterially sensitized targets. Mucosal Immunol 8(2):429–440. https://doi.org/10.1038/mi.2014.81
17. Le Bourhis L, Dusseaux M, Bohineust A, Bessoles S, Martin E, Premel V, Core M, Sleurs D, Serriari NE, Treiner E, Hivroz C, Sansonetti P, Gougeon ML, Soudais C, Lantz O (2013) MAIT cells detect and efficiently lyse bacterially-infected epithelial cells. PLoS Pathog 9(10):e1003681. https://doi.org/10.1371/journal.ppat.1003681
18. Gold MC, Cerri S, Smyk-Pearson S, Cansler ME, Vogt TM, Delepine J, Winata E, Swarbrick GM, Chua WJ, Yu YY, Lantz O, Cook MS, Null MD, Jacoby DB, Harriff MJ, Lewinsohn DA, Hansen TH, Lewinsohn DM (2010) Human mucosal associated invariant T cells detect bacterially infected cells. PLoS Biol 8(6):e1000407. https://doi.org/10.1371/journal.pbio.1000407
19. Le Bourhis L, Martin E, Peguillet I, Guihot A, Froux N, Core M, Levy E, Dusseaux M, Meyssonnier V, Premel V, Ngo C, Riteau B, Duban L, Robert D, Huang S, Rottman M, Soudais C, Lantz O (2010) Antimicrobial activity of mucosal-associated invariant T cells. Nat Immunol 11(8):701–708. https://doi.org/10.1038/ni.1890
20. Chatila T, Silverman L, Miller R, Geha R (1989) Mechanisms of T cell activation by the calcium ionophore ionomycin. J Immunol 143(4):1283–1289

Chapter 5

Methods for High-Dimensional Flow Cytometry Analysis of Human MAIT Cells in Tissues and Peripheral Blood

Benedikt Strunz, Christine L. Zimmer, Jonna Bister, Martin A. Ivarsson, and Niklas K. Björkström

Abstract

Mucosal-associated invariant T (MAIT) cells can be found throughout the human body, in peripheral blood, at mucosal sites, and, among other organs, in the liver. As unconventional T cells, MAIT cells have the capacity to readily respond to bacterial infections and are also engaged during anti-viral responses. To thoroughly investigate the MAIT cell phenotype and function in such conditions, multi-color flow cytometry is an appropriate and powerful tool. Yet, the recent rapid technological development within this methodology, with generation of highly complex data, has increased the need for downstream dimensionality reducing methods to fully interpret obtained results. Among such methods, stochastic neighbor embedding (SNE) analysis stands out as it provides intuitive low-dimensional representations of complex data. Here, we describe techniques and workflow for high-dimensional state-of-the-art investigation and analysis of human MAIT cells from blood and peripheral tissues.

Key words Human, MAIT cells, Immunophenotyping, Multi-color flow cytometry, SNE

1 Introduction

Flow cytometry has become indispensable to phenotype and assess the function of immune cells. As the number of flow cytometric channels and respective fluorochromes increases, it is now readily feasible for most immunology research labs to perform 18-color staining experiments which can be further expanded with the latest technology. These advances enable investigation of rare subsets of immune cells at high resolution. On the other hand, as the obtained data become more complex, advanced downstream analysis pipelines are necessary to analyze the resulting high-dimensional data. In particular, conventional 2D-gating becomes challenging, and instead algorithms that can reduce data dimensionality and that automatically cluster cells based on phenotypes are useful and powerful for such downstream analysis. One of the most commonly used algorithm is *stochastic neighbor embedding* (SNE) [1]. Here,

Helen Kaipe and Isabelle Magalhaes (eds.), *MAIT Cells: Methods and Protocols*, Methods in Molecular Biology, vol. 2098, https://doi.org/10.1007/978-1-0716-0207-2_5,

the cells are clustered in a two-dimensional space according to their phenotypic similarity, resulting in an easy-to-grasp depiction of multivariate relationships [2].

Mucosal-associated invariant T (MAIT) cells represent a subgroup of T cells that is specialized in recognition of bacterial metabolites presented on the MHC class 1-like molecule MR1 [3]. As their name implies, they reside in mucosal tissues and are also, among other organs, enriched in the liver where they represent up to 30% of all T cells [4]. Of note, MAIT cells are present early during fetal development and already at this stage display a heterogeneous phenotype that further evolves as an individual matures [5–8].

Because of this high level of heterogeneity, a workflow containing high-parameter flow cytometry assessment and downstream SNE analysis is well suited for the study of MAIT cells. Except for the multiple subpopulations of MAIT cells that are present during development and at steady state in the adult, the cells also respond to infection and inflammation in distinct ways depending on the organ and the specific condition. For example, their frequency in peripheral blood is reduced in a number of chronic conditions such as HCV, HIV, and mycobacterial infections with ensuing phenotypic and functional alterations [9–12]. Recently, reduction in MAIT cell frequency and dysfunction of the cells were also shown in the context of primary sclerosing cholangitis [12, 13]. The mechanism behind this reduction in chronic infectious and inflammatory conditions is yet to be defined. However, it is likely that high-dimensional cytometry and proper assessment of phenotypes will aid in unraveling this phenomenon.

In summary, a standardized way to study MAIT cells will be useful in order to determine their role in health and disease. Here, we describe a guideline on how to isolate and stain for human MAIT cells from both peripheral blood and tissues. Finally, we outline a workflow for downstream analysis using SNE.

2 Materials

2.1 Culture Media and Solutions

1. Incomplete medium for tissue digestion: RPMI containing 2 mM L-glutamine and 100 mM penicillin/streptomycin.
2. Enzyme for tissue digestion: 0.25 mg/mL collagenase II and 0.2 mg/mL DNase.
3. Complete culture medium: RPMI containing 2 mM L-glutamine, 100 mM penicillin/streptomycin, and 10% heat-inactivated (30 min at 56 °C) fetal calf serum (FCS).
4. FACS buffer: 1× PBS containing 2 mM EDTA and 2% heat-inactivated FCS.
5. Density gradient separation liquid such as Lymphoprep or Ficoll.

2.2 Flow Cytometer Hardware, Analysis Software

1. Flow cytometer (e.g., BD Fortessa) equipped with five lasers (388, 405, 488, 561, and 640 nm) and 18 filters for acquisition of stained cells.
2. FlowJo version 9 or 10 for data analysis, most recent version of FlowJo 10 for SNE and UMAP analysis.
3. R version 3.3.1 for SNE analysis.

2.3 Other Material

1. For antibody specification, *see* Tables 1 and 2 for the antibodies used.
2. Fixation and permeabilization buffer: FOXP3/Transcription factor staining buffer set.
3. LIVE/DEAD Fixable Aqua Dead Cell Stain Kit (DCM).

Table 1
Antibody list

Antibodies	Fluorochrome	Clone	Company
CD103	BUV395	Ber-ACT8	BD Biosciences
CD56	BUV737	NCAM16.2	BD Biosciences
CD4	BB515	RPA-T4	BD Biosciences
CD49a	AF647	TS2/7	Biolegend
CD45	AF700	HI30	Biolegend
CD57	APC-Vio770	TB03	Miltenyi
PD-1	BV421	EH12.2H7	Biolegend
CD14	V500	M5E2	BD Biosciences
CD15	V500	HI98	BD Biosciences
CD19	V500	SJ24C1	BD Biosciences
CD161-biotin		191B8	Miltenyi
CD8	Qdot605	3B5	Invitrogen
CD38	BV650	HB-7	Biolegend
CD25	BV711	BC96	Biolegend
HLA-DR	BV785	L234	Biolegend
TCRVa7.2	PE	3C10	Biolegend
MR-1 tetramer	PE	–	NIH
CD69	PE-Cf594	FN50	BD Biosciences
CD3	PE-Cy5	UCHT1	Biolegend
CD27	PE-Cy5.5	1A4CD27	Beckman Coulter
CD127	PE-Cy7	R.34.34	Beckman Coulter

Table 2
Flow cytometer settings

Laser (nm)	Detector	Filter (nm)	Panel
388	FL1	379/28	CD103
388	FL2	740/35	CD56
488	FL3	530/40	CD4
639	FL4	670/30	CD49a
639	FL5	730/45	CD45
639	FL6	780/60	CD57
405	FL7	450/50	PD-1
405	FL8	525/50	CD14 CD19 DCM
405	FL9	585/42	CD161
405	FL10	610/20	CD8
405	FL11	670/30	CD38
405	FL12	710/50	CD25
405	FL13	780/60	HLA-DR
561	FL14	586/15	TCRVa7.2
561	FL15	610/20	CD69
561	FL16	661/20	CD3
561	FL17	710/50	CD27
561	FL18	780/60	CD127

4. Qdot585 Streptavidin Conjugate is used in a second staining step to label biotin-conjugated CD161 (see details below).
5. Compensation beads.
6. Magnets.
7. 70 μm filters.
8. Scalpels.
9. Petri dishes.
10. FACS tubes.
11. 5 and 50 mL tubes.
12. Aspiration pipette (7.5 mL).
13. Water bath with magnetic stir.
14. Syringe plunger of 5 or 10 mL syringe.
15. 96-well V-bottom plate.

3 Methods

3.1 Isolation of Immune Cells from Peripheral Blood

1. Overlay 25 mL of heparinized whole blood on 15 mL lymphoprep/Ficoll solution in a 50 mL tube.
2. Centrifuge at 800 × *g* for 20 min at room temperature without break.
3. Recover the interphase with an aspiration pipette and transfer the cells to a new 50 mL tube.
4. Add PBS to the 50 mL tube until 40 mL.
5. Centrifuge for 5 min at 500 × *g* at room temperature.
6. Discard supernatant and repeat **steps 4** and **5**.

3.2 Preparation of Immune Cells from Tissue

Isolation of cells from endometrial tissue is used as an example. Other tissues may require other methods.

1. Transfer the tissue to a Petri dish of appropriate size and add 200 μL of incomplete medium (*see* **Note 1**).
2. Mechanically dissect the tissue using two scalpels to break up the tissue into small pieces.
3. Transfer all pieces into a 5 mL tube.
4. Wash the Petri dish with an additional 300 μL of incomplete medium to collect remaining cells and transfer this to the tube containing the tissue pieces.
5. Add collagenase II and DNase in a final concentration of 0.25 mg/mL and 0.2 mg/mL, respectively.
6. Add a magnet of appropriate size to the tube containing the tissue.
7. Digest the tissue in a water bath that is set to 37 °C for 30 min with the magnetic stirrer turned on (*see* **Note 2**).
8. Stop the tissue digestion by adding 500 μL of complete culture media and push cells through a filter with the pore size of 70 μm using a syringe plunger into a 50 mL tube.
9. Add an additional 5 mL of PBS or complete medium through the 70 μm filter to improve the yield of extracted cells.
10. Centrifuge at 500 × *g* for 5 min at room temperature.
11. Remove the supernatant and resuspend the cells in PBS.
12. Repeat **step 10**.

3.3 Multi-Color Flow Cytometry

1. Prepare the antibody mix in FACS buffer. Tables 1 and 2 provide a list of suggested antibodies (mAb) (*see also* **Notes 3** and **4**). At this point it should be considered which staining controls are necessary to include in the experiment (*see* **Note 5**). Always make excess antibody cocktail.

2. Plate 1–2 × 10^6 cells/well in a 96-well V-bottom plate. Since the frequency of MAIT cells in certain tissues might be low, several wells (duplicate/triplicate that are stained in parallel and pooled before acquisition) per sample might be necessary (*see* **Note 6**).
3. Centrifuge the plate at 500 × *g* for 3 min at room temperature and discard the supernatant.
4. Add 50 μL of the antibody mix and resuspend well.
5. Incubate for 20 min at room temperature in the dark (*see* **Note 7**).
6. Add 150 μL of FACS buffer to the wells and resuspend.
7. Centrifuge the plate at 500 × *g* for 3 min at room temperature and discard the supernatant.
8. Add 200 μL of FACS buffer to the wells and resuspend.
9. Centrifuge the plate at 500 × *g* for 3 min at room temperature and discard the supernatant.
10. If a biotinylated antibody is used, a secondary staining step containing the streptavidin conjugate and DCM is necessary (*see* **Note 8**). Add 50 μL of the secondary staining mix and resuspend well. In case only directly conjugated antibodies are used, proceed with **step 13**.
11. Incubate for 15 min at room temperature.
12. Repeat **steps 6–9** to wash.
13. Fix the cells by resuspending them in 200 μL fixation buffer. If intracellular staining is performed, fix cells for 45 min at room temperature and continue with **steps 14–19** below. Otherwise, stop the fixation after 10–15 min and continue with washing away the fixative according to **steps 6–7**, and then store the cells for flow cytometry according to **step 22** and onwards.
14. The intracellular mAb mix can be prepared during the primary stain and fixation.
15. After fixation centrifuge the plate at 500 × *g* for 3 min at room temperature and discard the supernatant containing fixing agent (dispose of accordingly).
16. Add 200 μL of permeabilization buffer to the wells and resuspend.
17. Centrifuge the plate at 500 × *g* for 3 min at room temperature and discard the supernatant.
18. For intracellular staining, add 50 μL of intracellular staining mix to each well. Incubate at room temperature in the dark for 40 min.
19. Add 150 μl of permeabilization buffer to the wells and resuspend.

20. Centrifuge the plate at 500 × *g* for 3 min at room temperature and discard the supernatant.
21. Repeat **steps 6** and 7 to wash away any permeabilization buffer.
22. Resuspend cells in an appropriate amount of FACS buffer (depending on cell density, 100–200 μL final volume) and transfer to FACS tubes. Additional dilution can be made in the FACS tube.
23. Acquire compensation beads, samples, and controls on the flow cytometry instrument. In order to perform a multiparametric analysis on the relatively rare MAIT cells, we recommend to acquire at least 1–4 × 10^6 events per sample.

3.4 Analysis of Flow Cytometric Data and SNE Analysis

1. Analyze the acquired data files from flow cytometry with FlowJo. Use the compensation beads to generate a compensation matrix (*see* **Note 9**). Note that cells stained in tissue and peripheral blood samples can differ in their fluorescence and the need for compensation can therefore also be different, even when the samples are acquired at the same time.
2. Next, identify MAIT cells in the hierarchical gating tree (*see* Fig. 1). After removal of cell debris and exclusion of other cell types, gate on $CD3^+$ T cells and among those identify MAIT cells via co-expression of CD161 and Vα7.2. As $CD4^+CD161^+V\alpha7.2^+$ cells contain a substantial percentage of non-MAIT cells [14], exclude these cells in order to achieve a purer MAIT cell population. Alternatively, use the MR1 tetramer on $CD161^+$ $CD3^+$ cells to identify MAIT cells (*see* **Note 10**).
3. For SNE analysis, either the built-in version in FlowJo 10 or publicly available R packages can be used. It is possible to perform SNE analysis on single samples or on combined samples. If the purpose is to perform SNE analysis of compiled samples with the most recent version of FlowJo 10 (10.6), we recommend to first export a similar number of events from each file together with compensated flow cytometry parameters into new fcs-file for each sample. Next these files are

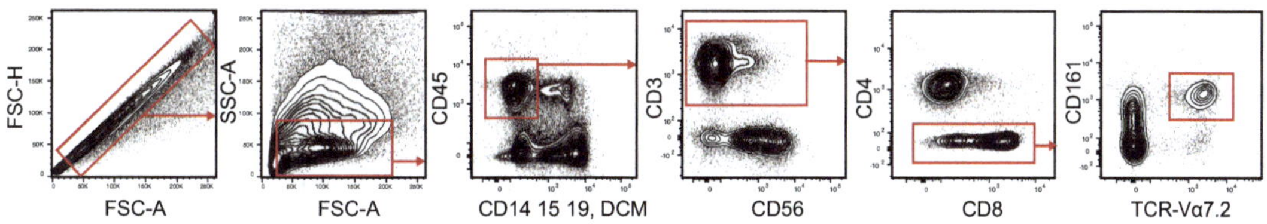

Fig. 1 Gating strategy for MAIT cells. After removal of doublets, lymphocytes were selected before gating on live and lineage negative cells. Next, $CD3^+$ T cells were gated and $CD4^+$ T cells were removed. Finally, MAIT cells were identified via CD161 and TCRVα7.2 co-expression

concatenated and finally SNE analysis can be performed on this concatenated file within FlowJo. In order to deconvolute the file afterwards to allow for comparison of groups, keywords identifying the respective groups should be added before samples are exported and concatenated. If necessary, normalize parameters in-between samples before concatenating the samples and running the SNE algorithm, for this for instance FlowCore and FlowStats can be used. This is of particular importance if samples and obtained data are derived from distinct experiments since this might yield inter-experimental variation and possibly resulting in artifacts. Thus, internal controls and other measures to reduce these confounders should be considered when performing SNE analysis on samples from independent experiments.

4. SNE analysis with publicly available packages needs exporting of either .csv or .fcs-files from the compensated and gated MAIT cell populations containing similar numbers of cells. Next, data is imported into R, and SNE analysis is performed on the data frame and then displayed (packages that can be used for this are, for example, flowCore, plyr, Rtsne, ggplot2, viridis, viridisLite, and grid) [15]. An exemplary SNE analysis of MAIT cells derived from peripheral blood and tissue is attached in Fig. 2.

3.5 Prospects to Move Beyond 18-Color Flow Cytometry

SNE and similar analysis methods can naturally be applied to any cytometry data with even more than 18 parameters. Conventional flow cytometry now allows up to 30 parameter analysis. While this promises to deliver more information for each individual sample analyzed, the more complex compensation matrices will require even more rigorous control of the data to avoid false signals. Cytometry time of flight (CyTOF) uses antibodies conjugated to metal particles instead of fluorophores and offers yet even more parameters, and is less dependent on compensation since the signals are more discrete than in flow cytometry. However, even with CyTOF, a small degree of spillover exists between channels close to each other and to channels 16 units apart due to oxidization of the metal particles [16]. Moreover, signal intensity is in general lower with CyTOF compared to flow cytometry, which means that more controls and optimization of the staining panel is needed. Ideally, all markers used in CyTOF should therefore be verified with conventional flow cytometry on the very sample type (e.g., tissue or blood cells) to be used. Also, if the intensity of a key marker is very low, this will influence downstream data analysis such as SNE.

Another recent addition to the flow cytometry arena includes techniques using oligo-tagged antibodies (CITE-seq) [17]. In combination with RNAseq and transcriptomic analyses, this approach may prove very powerful, and SNE analysis will be

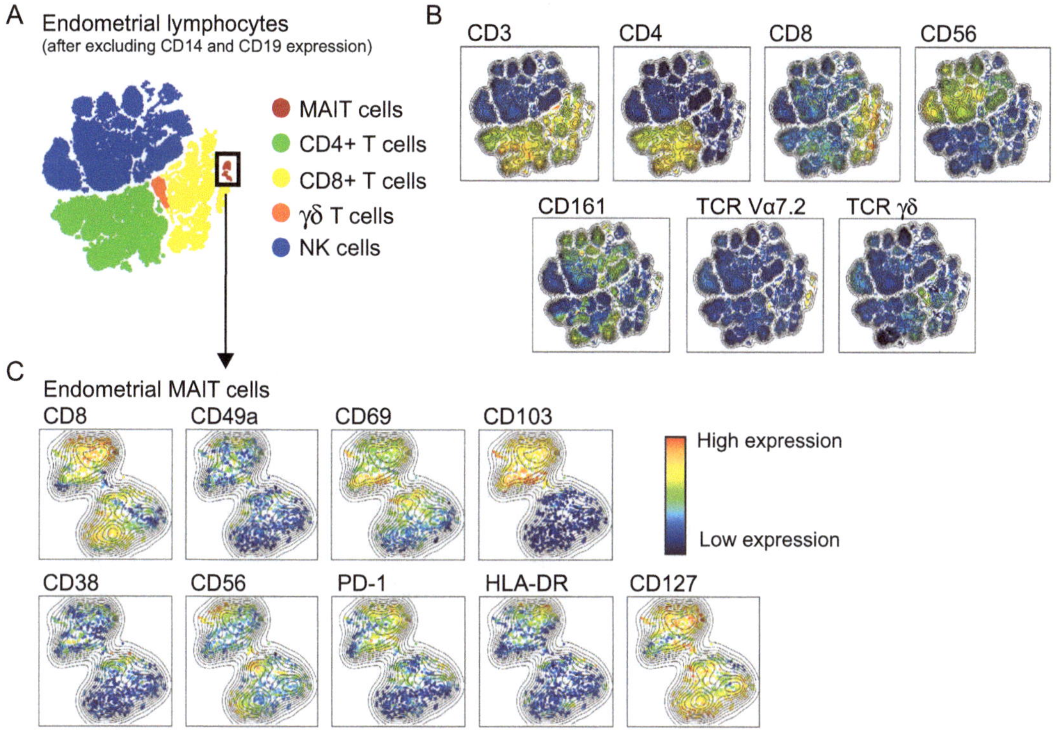

Fig. 2 SNE analysis of MAIT cells in endometrial tissue. After importing the MAIT cell data into R, SNE analysis was performed and the markers of interest plotted. (**a**) SNE map with T cell subsets and NK cells color-coded. (**b**) Relative expression of indicated markers overlaid on SNE map in A. (**c**) New SNE map on MAIT cells from the dataset in (**a**). $CD69^+$ MAIT cells are pre-gated to ensure tissue-residency phenotype, CD69 still forms a gradient of relative expression, $n = 2$

applicable. The development of new bioinformatic tools to better cluster and visualize high-dimensional datasets is still ongoing. The Newell group recently published the Uniform Manifold Approximation and Projection (UMAP) method that is faster than SNE analysis and thus may be more applicable if datasets get larger [18]. For this, a FlowJo plugin is available, and UMAP analysis can be performed in FlowJo along the same lines as described above for SNE analysis.

Taken together, the recent advances in flow cytometry make it possible to study MAIT cells also in tissues in great detail including improved downstream analysis grasping the full complexity of these cells.

4 Notes

1. Pieces of tissue vary in size. For this reason, it is important to add enough medium to not let the cells dry out before the tissue is dissected. Due to differences in tissue structure, the

time for tissue dissection may vary and should be taken into consideration. For tissue digestion, the tube size should be chosen according the volume of tissue suspension (capacity of 5–50 mL).

2. For optimal tissue digestion, a correct water temperature is of importance. Hence, for optimal collagenase activity the water bath must have a temperature of exact 37 °C when the tube is placed into the water. Of note, enzymatic digestion might damage or lead to downregulation of surface proteins. It is recommended to test different incubation times and the effect on surface receptors beforehand. Furthermore, all tissue samples should be enzymatically digested for a similar amount of time to avoid differences in staining quality due to different ways of sample treatment.
3. Flow cytometry panel design is not covered in this methods chapter but needs to be carefully considered for 18-color panels. There are useful online tools that can help with panel design. Important to consider is the amount of spillover and excitation by each laser and the relative brightness of fluorochromes to be able to pair antigens of interest with appropriate fluorochrome.
4. All antibodies in the panel need to be titrated in order to optimize the staining. Excess antibody can bind at low affinity and create background staining, which will reduce the resolution and complicate the interpretation of the results.
5. In order to facilitate analysis, unstained controls, fluorescence minus one (FMO) controls, and controls stained with only DCM might be appropriate. FMO controls are samples that contain all antibodies in the panel except one, allowing accurate discrimination of positive versus negative signals. This is especially useful for channels with difficult compensation and when determining the level of expression above background for low expressed markers. Additionally, isotype controls should be considered, especially when testing for markers of unknown expression.
6. It is expected to find roughly 1–10% MAIT cells within the T cell compartment in both peripheral blood and oral mucosa [19], up to 30% in the liver [4], but only about 1% in endometrium [20]. Thus, especially in tissue samples the MAIT cell yield might be low.
7. Alternatively, the cells can be incubated at 4 or 37 °C. The purpose of staining at 4 °C is to slow down cellular processes in order to prevent receptor internalization and to reduce unspecific binding. Staining at 37 °C might be beneficial when staining for chemokine receptors as the subunits gather at the surface at higher temperatures [21].

8. If a biotinylated antibody is used in the primary staining step, a secondary staining with the fluorochrome–streptavidin conjugate is necessary. As DCM staining should be done shortly before fixation to get a true live/dead distinction, we recommend to add the dead cell staining dye in this second step together with the streptavidin conjugate. Naturally, without biotinylated antibodies, DCM can be added to the main antibody mix and **steps 10–12**, as presented above, can be left out.
9. To compensate spectral overlap when performing multi-color flow cytometry, it is necessary to have single-stained controls, which can be compensation beads (*see* Subheading 2.3, other materials) or cells. When using several fluorochromes, some portions of the emission spectra of a certain fluorochrome may fall within the detector of another channel. Therefore, the fluorochrome of the single-color bead that is detected in other channels than the intended is measured and compensated for in order to remove the overlapping signal. Compensation beads should be run in conjunction to each experiment. Compensation can also be done on the FACS instrument.
10. For more stringent identification of MAIT cells, the MR1-5OP-RU tetramer can be used since recent research has demonstrated that MAIT cells cannot be precisely identified using Vα7.2 in combination with CD161 [12, 22]. CD161 is highly expressed by other non-conventional T cells in tissues and might introduce erroneous results. Instead, the tetramer helps to identify MAIT cells more stringently. Alternatively, if Vα7.2 in combination with CD161 is used, the $CD4^{+}V\alpha7.2^{+}CD161^{+}$ cell population is the largest confounder to falsely identified MAIT cells (tetramer negative) and should therefore be excluded from the analysis. For optimal results, the tetramer staining should be performed at room temperature for 30 min as the first staining step without other antibodies being present. The MR1-5OP-RU tetramer can be obtained from the NIH Tetramer Core Facility at Emory University, Atlanta, NE.

References

1. Amir E-AD, Davis KL, Tadmor MD, Simonds EF, Levine JH, Bendall SC, Shenfeld DK, Krishnaswamy S, Nolan GP, Pe'er D (2013) visNe enables visualization of high dimensional single-cell data and reveals phenotypic heterogeneity of leukemia. Nat Biotechnol 31:545–552
2. Saeys Y, Gassen SV, Lambrecht BN (2016) Computational flow cytometry: helping to make sense of high-dimensional immunology data. Nat Rev Immunol 16:449–462
3. Kjer-Nielsen L, Patel O, Corbett AJ et al (2012) MR1 presents microbial vitamin B metabolites to MAIT cells. Nature 491:717–723
4. Godfrey DI, Uldrich AP, McCluskey J, Rossjohn J, Moody DB (2015) The burgeoning family of unconventional T cells. Nat Immunol 16:1114–1123
5. Koay H-F, Gherardin NA, Enders A et al (2016) A three-stage intrathymic development

pathway for the mucosal-associated invariant T cell lineage. Nat Immunol 17:1300–1311
6. Leeansyah E, Loh L, Nixon DF, Sandberg JK (2014) Acquisition of innate-like microbial reactivity in mucosal tissues during human fetal MAIT-cell development. Nat Commun 5:3143
7. Rahimpour A, Koay H-F, Enders A et al (2015) Identification of phenotypically and functionally heterogeneous mouse mucosal-associated invariant T cells using MR1 tetramers. J Exp Med 212:1095–1108
8. Dias J, Leeansyah E, Sandberg JK (2017) Multiple layers of heterogeneity and subset diversity in human MAIT cell responses to distinct microorganisms and to innate cytokines. Proc Natl Acad Sci U S A 114:E5434–E5443
9. Hengst J, Strunz B, Deterding K, Ljunggren H-G, Leeansyah E, Manns MP, Cornberg M, Sandberg JK, Wedemeyer H, Björkström NK (2016) Nonreversible MAIT cell-dysfunction in chronic hepatitis C virus infection despite successful interferon-free therapy. Eur J Immunol 46:2204–2210
10. Leeansyah E, Ganesh A, Quigley MF et al (2013) Activation, exhaustion, and persistent decline of the antimicrobial MR1-restricted MAIT-cell population in chronic HIV-1 infection. Blood 121:1124–1135
11. Kwon Y-S, Cho Y-N, Kim M-J et al (2015) Mucosal-associated invariant T cells are numerically and functionally deficient in patients with mycobacterial infection and reflect disease activity. Tuberculosis (Edinb) 95:267–274
12. Seth v E, Zimmer CL, Reuterwall-Hansson M, Barakat A, Arnelo U, Bergquist A, Ivarsson MA, Björkström NK (2018) Primary sclerosing cholangitis leads to dysfunction and loss of MAIT cells. Eur J Immunol 48:1997–2004
13. Böttcher K, Rombouts K, Saffioti F, Roccarina D, Rosselli M, Hall A, Luong T, Tsochatzis EA, Thorburn D, Pinzani M (2018) MAIT cells are chronically activated in patients with autoimmune liver disease and promote profibrogenic hepatic stellate cell activation. Hepatology 68:172–186
14. Dias J, Boulouis C, Gorin J-B et al (2018) The CD4−CD8− MAIT cell subpopulation is a functionally distinct subset developmentally related to the main CD8+ MAIT cell pool. Proc Natl Acad Sci U S A 115: E11513–E11522
15. Hengst J, Theorell J, Deterding K, Potthoff A, Dettmer A, Ljunggren H-G, Wedemeyer H, Björkström NK (2015) High-resolution determination of human immune cell signatures from fine-needle liver aspirates. Eur J Immunol 45:2154–2157
16. Han G, Spitzer MH, Bendall SC, Fantl WJ, Nolan GP (2018) Metal-isotope-tagged monoclonal antibodies for high-dimensional mass cytometry. Nat Protoc 13:2121–2148
17. Stoeckius M, Hafemeister C, Stephenson W, Houck-Loomis B, Chattopadhyay PK, Swerdlow H, Satija R, Smibert P (2017) Simultaneous epitope and transcriptome measurement in single cells. Nat Methods 9:2579
18. Becht E, McInnes L, Healy J, Dutertre C-A, Kwok IWH, Ng LG, Ginhoux F, Newell EW (2018) Dimensionality reduction for visualizing single-cell data using UMAP. Nat Biotechnol. https://doi.org/10.1038/nbt.4314
19. Sobkowiak MJ, Davanian H, Heymann R et al (2019) Tissue-resident MAIT cell populations in human oral mucosa exhibit an activated profile and produce IL-17. Eur J Immunol 49:133–143
20. Gibbs A, Leeansyah E, Introini A, Paquin-Proulx D, Hasselrot K, Andersson E, Broliden K, Sandberg JK, Tjernlund A (2016) MAIT cells reside in the female genital mucosa and are biased towards IL-17 and IL-22 production in response to bacterial stimulation. Mucosal Immunol. https://doi.org/10.1038/mi.2016.30
21. Berhanu D, Mortari F, De Rosa SC, Roederer M (2003) Optimized lymphocyte isolation methods for analysis of chemokine receptor expression. J Immunol Methods 279:199–207
22. Reantragoon R, Corbett AJ, Sakala IG et al (2013) Antigen-loaded MR1 tetramers define T cell receptor heterogeneity in mucosal-associated invariant T cells. J Exp Med 210:2305–2320

Chapter 6

In Situ Detection of MAIT Cells and MR1-Expressing Cells in Tissue Biopsies Utilizing Immunohistochemistry

Anna Gibbs, Michal J. Sobkowiak, Johan K. Sandberg, and Annelie Tjernlund

Abstract

The mucosa-associated invariant T (MAIT) cells are innate-like T cells that recognize microbial vitamin B2 metabolites presented via MR1, a MHC-I-related protein. MAIT cells are abundant in blood and mucosa, where they display a broad range of functions. Spatial distribution of cells and their proximity to other cells, including infected cells and antigen presenting cells, are crucial components of cell-mediated immunity. Here we describe techniques to detect MAIT cells and MR1-expressing cells in situ, which enable the visualization, distribution, and localization of these cells within their histological context. We provide specific protocols and describe potential advantages and limitations for each of the presented methodologies for studying MAIT cells in human tissues.

Key words MAIT cells, MR1, In situ detection, Immunohistochemistry, Immunofluorescence, Fixation, Antibodies and microscopy

1 Introduction

The mucosa-associated invariant T (MAIT) cells are a relatively new subset of unconventional T cells, which is abundant in blood, liver, and mucosal tissues [1, 2]. Human MAIT cells display extensive antimicrobial capacity against numerous pathogens, including direct killing of infected cells [3, 4] and production of pro-inflammatory cytokines [5, 6].

MAIT cells express a semi-invariant T cell receptor (TCR), including the Vα7.2 segment coupled with restricted Jα and Vβ repertoires, which together with the surface marker C-type lectin CD161 and/or interleukin (IL)-18 receptor α-subunit (IL-18Rα) can be used to identify human MAIT cells [6, 7]. The MAIT cell TCR recognizes microbial vitamin B2 metabolites presented via the evolutionarily conserved major histocompatibility complex (MHC) class I-related (MR) 1 molecule [8–10]. MR1 is ubiquitously expressed, hence various antigen harboring cells, including

Helen Kaipe and Isabelle Magalhaes (eds.), *MAIT Cells: Methods and Protocols*, Methods in Molecular Biology, vol. 2098, https://doi.org/10.1007/978-1-0716-0207-2_6,

professional antigen presenting cells and epithelial cells, have the capacity to activate MAIT cells [4, 11]. The recently developed MR1-antigen-loaded tetramer can be used to identify human MAIT cells in single cell suspension [9]. However, the knowledge of its application for MAIT cell detection in situ is currently limited.

In this chapter, we present in situ staining approaches that enable visualization and enumeration of MAIT cells and of MR1-expressing cells. Specifically, we provide detailed protocols for detection of these cells using a polymer detection system and immunofluorescence staining in several tissues, including mucosa of the female reproductive tract (cervix and endometrium) and buccal mucosa [12, 13]. The staining protocols can also be combined with detection of cytokines and cytolytic molecules as proxies for functional phenotypes of the cells of interest within the intact tissue section [14]. Microscopy methods have great advantage as a visualization tool compared to other cell-based techniques, and provide the fundamental approach to study single cells in small tissue sample. However, microscopy-based methods are also associated with certain limitations. Limited antibody availability and narrow application of certain reagents, including the MR1 tetramer as discussed above, restrict the assessment of several markers at a time, resulting in the insufficient phenotypical and functional cell analysis. Thus, combination of flow cytometry and in situ staining is very informative as these two methods complement each other, and thus advantageous to perform concurrently to maximize the data obtained from limited amounts of tissue.

To visualize the spatial distribution of cells and structures in tissue is important in order to understand local host immune defenses. Digitalized image analysis of tissues is rapidly improving [15, 16], and these techniques allow for objective enumeration, characterization, and spatial visualization of cells/structures in intact tissue, and can handle a large set of samples in an objective and high-throughput manner. This is especially interesting for the clinical trial setting. Furthermore, an increasing amount of data indicates that the tissue compartments have their own immunological niches, which are not necessarily reflected in blood [17, 18]. It is thus of major importance to study the antimicrobial immune responses of MAIT cells in the tissues with the help of microscopy and image analysis tools.

2 Materials

2.1 Tissue Preparation

1. Cryopreserved biopsy imbedded in OCT (optimal cutting temperature) compound.
2. Superfrost Plus Gold Microscope Slides.
3. Phosphate-buffered saline (PBS).

4. 2% formaldehyde in PBS.
5. Coplin jar.

2.2 Polymer Detection System

1. TBS-H-Sap wash buffer: Tris-buffered saline (TBS) with 1% HEPES (*4-(2-Hydroxyethyl) piperazine-1-ethanesulfonic acid*) and 0.1% Saponin.
2. Antibody Diluent.
3. 2% hydrogen peroxide in TBS-H-Sap wash buffer.
4. Background Punisher and Denaturation buffer.
5. MACH3 Mouse HRP (horseradish peroxidase) kit and MACH3 Rabbit HRP kit.
6. MilliQ water.
7. Betazoid diaminobenzidine tetrahydrochloride (DAB) and Vina Green chromogens.
8. Hematoxylin and Pertex mounting medium.
9. Primary and secondary antibodies (Table 1).
10. Cover slips.

2.3 Immunofluorescence Detection System

1. PBS-H-Sap wash buffer: PBS with 1% HEPES and 0.1% Saponin.
2. Avidin/Biotin Blocking Kit.
3. Primary blocking buffer: 0.1% BSA-c (*bovine serum albumin*) in PBS-H-Sap wash buffer.
4. Secondary blocking buffer: 2% normal donkey serum + 0.1% BSA-c in PBS-H-Sap wash buffer.
5. MilliQ water.
6. DAPI (4′,6-diamidino-2-phenylindole).
7. Fluorescent mounting medium.
8. Primary and secondary antibodies (Table 1).
9. Cover slips

2.4 In Situ Detection and Imaging Analysis

1. Light or fluorescence microscope.
2. Microscope slide scanner: the stained tissue sections described here were scanned into digital images using a Panoramic 250 Flash Slide Scanner.
3. Imaging analysis software: the stained tissue sections described here were previewed and analyzed using the Panoramic Viewer imaging software.

Table 1
Antibodies used for in situ staining

Primary antibodies	Clone	Antibody source
Mouse anti-human Vα7.2	3C10	BioLegend
Goat anti-human IL-18Rα	AF840	R&D Systems
Mouse anti-human MR1	8F2.F9	Kindly provided by Dr. Ted Hansen School of Medicine, Washington University, St Louis, MO
Biotinylated mouse anti-human HLA-DR	L243	BD Biosciences
FITC-conjugated mouse anti-human CD11c	B-ly6	BD Biosciences
Rat anti-human Langerin	929F3.01	Dendritics
Secondary antibodies		*Antibody source*
Biotinylated rabbit anti-goat		Beckman Coulter
Alexa Fluor 488-conjugated Streptavidin		Thermo Fisher Scientific
Alexa Fluor 594-conjugated donkey anti-mouse IgG		Thermo Fisher Scientific
Alexa Fluor 488-conjugated donkey anti-rat IgG		Thermo Fisher Scientific

3 Methods

3.1 Tissue Sample Preparation

1. Cut the cryopreserved biopsies in 8 μm thickness using a cryostat and mount the tissue section on a SuperFrost Plus Gold slide.
2. Fix the section slide in 2% formaldehyde solution for 10 min at room temperature (RT) (*see* **Note 1**).
3. Wash the section slide with PBS for 10 min at RT.
4. Air-dry the section slides for 30 min.
5. Proceed further with either polymer detection protocol or immunofluorescence staining protocol.

3.2 In Situ Detection of MAIT Cells with Polymer Detection System

The section slides should be washed between each of the following incubation steps marked with ∗, using TBS-H-Sap as wash buffer. A representative image of human buccal mucosa stained for Vα7.2 in combination with IL-18Rα, using the following protocol, is shown in Fig. 1.

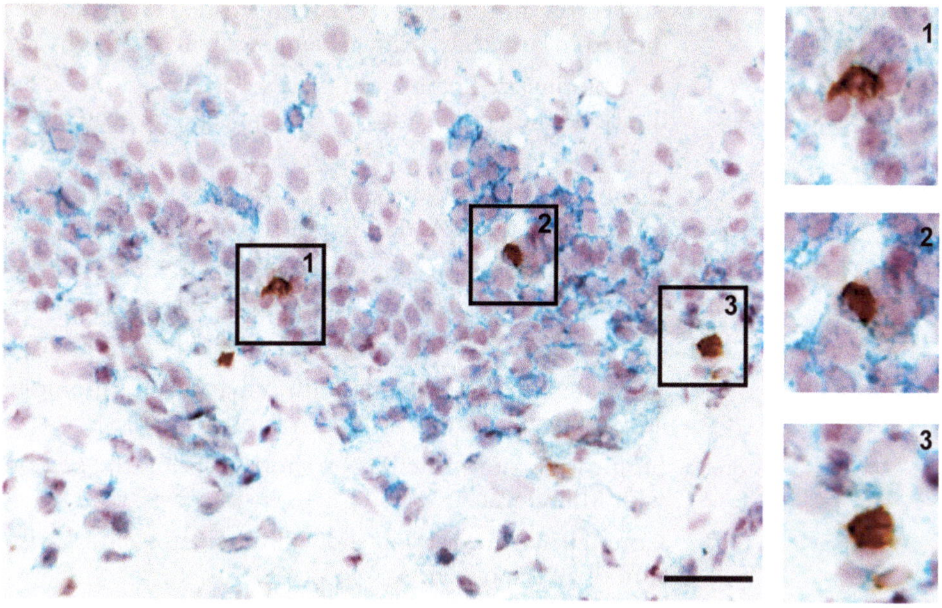

Fig. 1 Immunohistochemical staining of MAIT cells in the human buccal mucosa. Representative bright field image of a buccal mucosal tissue section stained for Vα7.2 in brown (DAB) and IL-18Rα in green (Vina Green) using MACH3 polymer detection system. Hematoxylin (blue) was used as a counterstain for visualization of cell nuclei. The image on the left was collected using a 40× objective and the scale bar represents 60 μm. The images on the right are magnified view of the regions as indicated by the boxes on the image to the left

1. Wash the section slide.
2. Block endogenous peroxidase by incubation of the section slide with 2% hydrogen peroxide for approximately 10 min at RT (*see* **Note 2**)∗.
3. Add Background Punisher solution for 5 min at RT (*see* **Note 3**)∗.
4. Incubate the section slide with mouse anti-human Vα7.2 antibody (1:12 dilution, diluted in Antibody Diluent) overnight at RT∗.
5. Incubate the section slide with MACH3 Mouse Probe (included in the MACH3 Mouse HRP kit) for 10 min at RT∗.
6. Incubate the section slide with MACH3 Mouse HRP Polymer (included in the MACH3 Mouse HRP kit) for 10 min at RT∗.
7. Incubate the section slide with DAB chromogen for 2–5 min at RT (*see* **Note 4**).
8. Stop the staining reaction by rinsing the section slide with MilliQ water.
9. Incubate the section slide with denaturation buffer for 5 min at RT∗.

10. Add goat anti-human IL-18Rα antibody (1:200 dilution, diluted in Antibody Diluent) and incubate overnight at RT∗.
11. Add biotinylated rabbit anti-goat secondary antibodies (1:500 dilution, diluted in Antibody Diluent) and incubate for 30 min at RT∗.
12. Incubate the section slide with MACH3 Rabbit Probe (included in the MACH3 Rabbit HRP kit) for 10 min∗.
13. Incubate the section slide with MACH3 Rabbit HRP Polymer (included in the MACH3 Rabbit HRP kit) for 10 min at RT.
14. Wash with TBS-H-Sap buffer and thereafter with MilliQ water.
15. Incubate the section slide with Vina Green chromogen for approximately 10 min at RT (*see* **Notes 4** and **5**).
16. Stop the staining reaction by shortly rinsing the section slide with MilliQ water.
17. Counterstain the section slide with hematoxylin for visualization of all cell nuclei.
18. Leave the section slide to air-dry overnight (*see* **Note 5**) and mount it with the Pertex mounting medium.
19. Store the slide at +4 °C.

3.3 In Situ Detection of MAIT Cells with Immunofluorescence Staining

The section slides should be washed between each of the following incubation steps, marked with ∗, using PBS-H-Sap as wash buffer. Representative images of human female genital mucosa stained for Vα7.2 in combination with IL-18Rα, using the following protocol, are shown in Fig. 2.

1. Wash the section slide.
2. Incubate the section slide with the primary blocking buffer for 30 min at RT (*see* **Note 6**).
3. Flip the section slide and add mouse anti-human Vα7.2 antibody (diluted 1:12 in primary blocking buffer) and incubate overnight at RT∗.
4. Incubate the section slide with the secondary blocking buffer for 30 min at RT (*see* **Note 6**).
5. Flip the section slide and add the secondary antibody Alexa Fluor 594-conjugated donkey anti-mouse (1:200 dilution, diluted in the secondary blocking buffer) (*see* **Note 7**). Incubate for 1 h at RT in the dark∗.
6. Add the goat anti-human IL-18Rα antibody (1:200 dilution, diluted in primary blocking buffer), and incubate overnight at RT∗.
7. Add the secondary antibody Alexa Fluor 488-conjugated donkey anti-goat (1:100 dilution, diluted in the secondary blocking buffer). Incubate for 1 h at RT in the dark∗

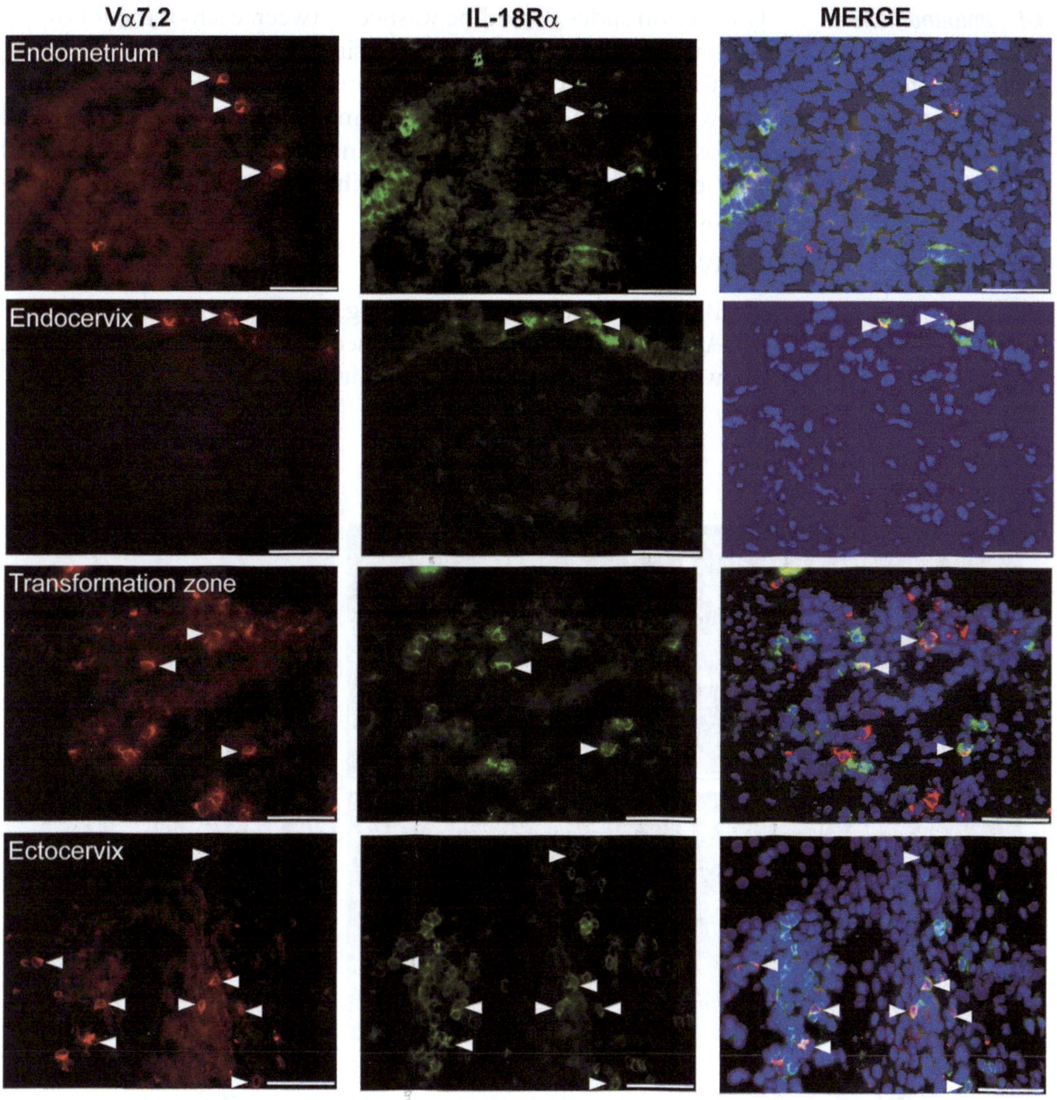

Fig. 2 Immunofluorescence staining of MAIT cells in the human genital mucosa. Representative fluorescent images of endometrial (first row), endocervical (second row), transformation zone (third row), and ectocervical (fourth row) tissue stained for Vα7.2 (red) and IL-18Rα (green) cells. DAPI (blue) was used as a counterstain for visualization of cell nuclei. Vα7.2$^+$IL-18Rα$^+$ (MAIT) cells are shown in yellow and are indicated by the white arrows. The images were collected using a 40× objective and the scale bars represent 60 μm (Reproduced from Ref. [12], with permission from Mucosal Immunology)

8. Wash in MilliQ water.
9. Counterstain the section slide with DAPI for visualization of all cell nuclei.
10. Wash in MilliQ water.
11. Mount with Fluorescent Mounting Medium.
12. Store the slide in dark at +4 °C.

3.4 Immunofluorescence Detection of Potential Antigen Presenting Cells Expressing MR1

The section slides should be washed between each of the following incubation steps, marked with ∗, using PBS-H-Sap as wash buffer. Use the protocol described above under Subheading 3.3, with the following changes. Representative images of human female genital mucosa stained for $MR1^+$ in combination with markers associated with detection of antigen presenting cells, using the following protocol, are shown in Fig. 3.

1. Wash the section slide.
2. Block endogenous biotin using Avidin/Biotin Blocking Kit. Add Avidin and incubate the section slide for 15 min at RT, wash, and then add Biotin and incubate for 15 min∗.

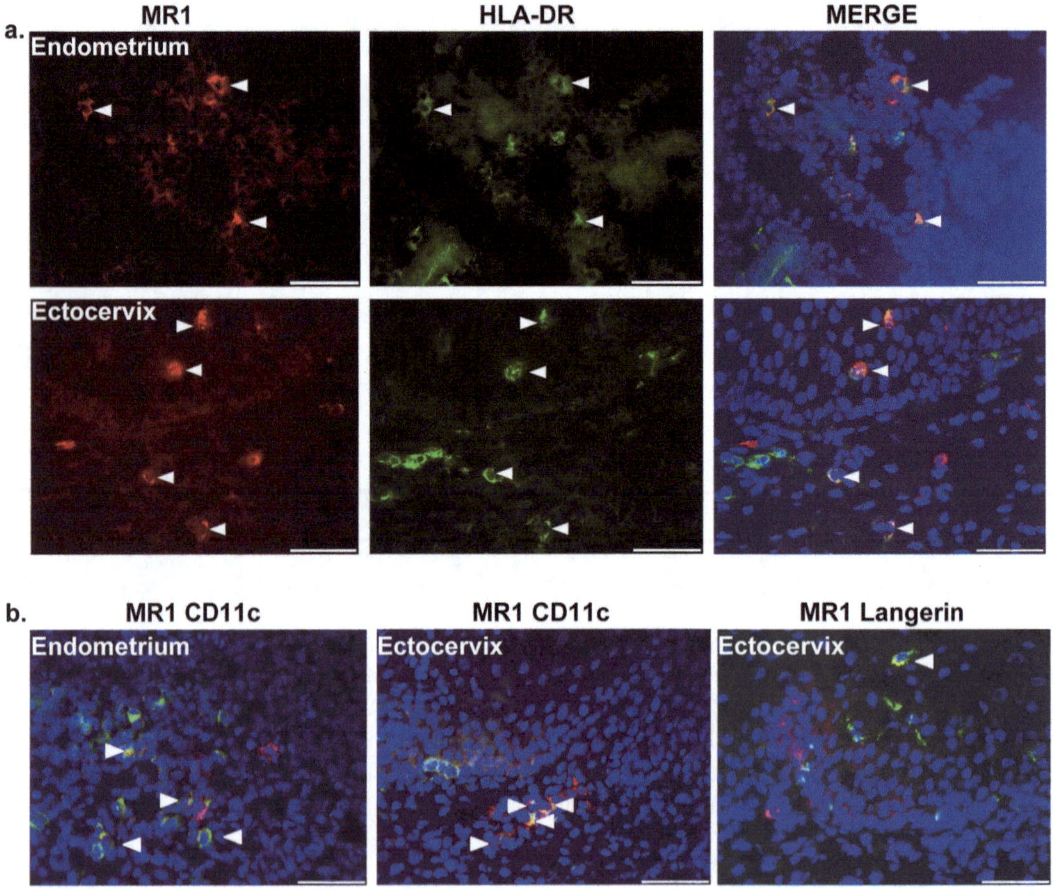

Fig. 3 Immunofluorescence staining of $MR1^+$ cells in the human genital mucosa. Representative fluorescent images from endometrial and ectocervical tissue stained for (**a**) MR1 (red) and HLA-DR (green), and for (**b**) MR1 (red) together with either of the following surrogate marker for antigen-presenting cells: CD11c (green) or Langerin (green). DAPI (blue) was used as a counterstain for visualization of cell nuclei. Double-positive cells are shown in yellow and are indicated by the white arrows. The images were collected using a 40× objective and the scale bars represent 60 μm (Reproduced from Ref. [12], with permission from Mucosal Immunology)

3. Incubate the section slide with the primary blocking buffer for 30 min at RT (*see* **Note 6**).
4. Flip the section slide and add primary mouse anti-human MR1 antibody (diluted 1:25 in primary blocking buffer) and incubate overnight at RT*.
5. Incubate the section slide with the secondary blocking buffer for 30 min at RT (*see* **Note 6**).
6. Flip the section slide and add the secondary antibody Alexa Fluor 594-conjugated donkey anti-mouse (1:200 dilution, diluted in the secondary blocking buffer) (*see* **Note** 7). Incubate for 1 h at RT in the dark.
7. Add biotinylated mouse anti-human HLA-DR antibody, or FITC-conjugated mouse anti-human CD11c, or rat anti-human Langerin (1:500, 1: 100, and 1: 200 dilution, respectively, diluted in primary blocking buffer) (*see* **Note 8**), incubate overnight at RT*.
8. Add streptavidin-conjugated Alexa Fluor 488 (1:800 dilution in primary blocking buffer) to the HLA-DR staining (*see* **Note** 7). Add the secondary Alexa Fluor 488-conjugated donkey anti-rat IgG antibody (1:100 dilution, diluted in the secondary blocking buffer) to the Langerin staining. Incubate for 1 h at RT in the dark*.
9. Wash with MilliQ water.
10. Counterstain the section slide with DAPI for visualization of all cell nuclei.
11. Wash in MilliQ.
12. Mount with Fluorescent Mounting Medium.
13. Store the slide in dark at +4 °C.

3.5 Visualization of Stained Cells

The immunohistochemical results gained from performing the polymer staining can be visualized using normal light microscopy, whereas the immunofluorescence staining is evaluated using a fluorescence microscope. Immunohistochemical staining usually generates strong positive staining, and DAB is a superior chromogen that is easily visible and thus good to use if the markers of interest are expressed in low abundance. Furthermore, there is usually little problem with background due to non-specific staining if correct blocking has been performed (*see* **Notes 2**, **3**, and **6**). However, when it comes to performing double or multiple staining, immunofluorescence staining has the advantage of visualizing each stained marker in its own channel, as well as together with the other stained markers in the color-combined mode. However, for the immunohistochemical method, all stains are visible in the bright field, and it can thus be hard to distinguish the different colors superimposed. Being able to toggle between the different channels makes it easy to evaluate if the markers of interest co-localize to the same cell or area within the tissue section (*see* **Note 9**). A drawback with immunofluorescence staining is autofluorescence, which is present to a different degree in most tissues, and thus the intensity signal from the markers of interest need to be clearly distinct from the autofluorescent intensity (i.e., good signal-to-noise ratio needs to be achieved).

An advantage of using a digital slide scanner is that the whole tissue section can easily be scanned at high resolution in relatively short time. Saved digital images make up a great data source that can be visualized, analyzed, and re-analyzed when needed.

4 Notes

1. 2% formaldehyde solution should be freshly prepared from 37% formaldehyde, dilute in PBS.
2. Some tissues or cells contain endogenous peroxidase, which needs to be quenched or blocked in order to prevent non-specific background when using HRP-conjugated reagents such as the MACH3 mouse or Rabbit HRP Polymers. If the tissue contains endogenous peroxidase, a reaction will occur when applying 2% hydrogen peroxide to the tissue section. This reaction is visible; small bubbles can be observed. If there is a strong reaction, wash, add some more of the 2% hydrogen peroxide solution, and repeat this step until no new bubbles appear.
3. Background Punisher can be used if there is a problem with non-specific background, which may vary depending on the source of tissue used.
4. The incubation time with the chromogens (DAB or Vina Green) needs to be adjusted for each staining condition depending on the quality of primary antibody/staining intensity in the analyzed tissue sample.
5. The Vina Green chromogen is sensitive to denaturation, and prolonged exposure to TBS/PBS will fade the Vina Green staining. Thus, Vina Green must always be used as the last chromogen, avoid long washes, and make sure that the tissue is dry before mounting the slides with a solvent-based mounting medium such as Pertex.
6. Blocking with sera or a protein such as BSA may prevent non-specific binding of antibodies to tissue or to Fc receptors. Serum is a common blocking agent as it contains antibodies that bind to non-specific sites. For secondary blocking buffer, it is recommended to use approximately 2% of a serum that matches the species in which the secondary antibody is made.

7. Alexa-conjugated antibodies as well as Alexa-conjugated streptavidin need to be titrated for each staining condition, depending on which primary antibody the Alexa-conjugated product should detect as well as the source of tissue.
8. HLA-DR, CD11c, and Langerin were here used as surrogate markers for antigen-presenting cells.
9. Always include negative controls consisting of either irrelevant isotype control antibody or incubations in the presence of secondary antibody alone. For double staining always include single staining for each marker of interest to control for cross reactivity between the different antibodies used.

Acknowledgments

This work was supported by grants to A.G. from Swedish Physicians against AIDS Foundation, the Swedish Society of Medicine, Tore Nilsons Research Foundation for Medical Research, and the Sigurd and Elsa Goljes Memorial Fund. J.K.S. was supported by grants from the Swedish Research Council (2016-03052), the Swedish Cancer Society (CAN 2017/777), the Swedish Heart-Lung Foundation (20180675), and the US National Institutes of Health (R01DK108350). A.T. was supported by grants from the Clas Groschinsky Minnesfond, Magnus Bergvalls Stiftelse, and the Swedish Society of Medicine.

References

1. Le Bourhis L, Mburu YK, Lantz O (2013) MAIT cells, surveyors of a new class of antigen: development and functions. Curr Opin Immunol 25:174–180
2. Garner LC, Klenerman P, Provine NM (2018) Insights into mucosal-associated invariant T cell biology from studies of invariant natural killer T cells. Front Immunol 9:1478
3. Kurioka A, Ussher JE, Cosgrove C, Clough C, Fergusson JR, Smith K, Kang YH, Walker LJ, Hansen TH, Willberg CB, Klenerman P (2015) MAIT cells are licensed through granzyme exchange to kill bacterially sensitized targets. Mucosal Immunol 8:429–440
4. Le Bourhis L, Dusseaux M, Bohineust A, Bessoles S, Martin E, Premel V, Core M, Sleurs D, Serriari NE, Treiner E, Hivroz C, Sansonetti P, Gougeon ML, Soudais C, Lantz O (2013) MAIT cells detect and efficiently lyse bacterially-infected epithelial cells. PLoS Pathog 9:e1003681
5. Dias J, Leeansyah E, Sandberg JK (2017) Multiple layers of heterogeneity and subset diversity in human MAIT cell responses to distinct microorganisms and to innate cytokines. Proc Natl Acad Sci U S A 114:E5434–E5443
6. Le Bourhis L, Martin E, Peguillet I, Guihot A, Froux N, Core M, Levy E, Dusseaux M, Meyssonnier V, Premel V, Ngo C, Riteau B, Duban L, Robert D, Huang S, Rottman M, Soudais C, Lantz O (2010) Antimicrobial activity of mucosal-associated invariant T cells. Nat Immunol 11:701–708
7. Martin E, Treiner E, Duban L, Guerri L, Laude H, Toly C, Premel V, Devys A, Moura IC, Tilloy F, Cherif S, Vera G, Latour S, Soudais C, Lantz O (2009) Stepwise development of MAIT cells in mouse and human. PLoS Biol 7:e54
8. Huang S, Martin E, Kim S, Yu L, Soudais C, Fremont DH, Lantz O, Hansen TH (2009) MR1 antigen presentation to mucosal-associated invariant T cells was highly conserved in evolution. Proc Natl Acad Sci U S A 106:8290–8295

9. Reantragoon R, Corbett AJ, Sakala IG, Gherardin NA, Furness JB, Chen Z, Eckle SB, Uldrich AP, Birkinshaw RW, Patel O, Kostenko L, Meehan B, Kedzierska K, Liu L, Fairlie DP, Hansen TH, Godfrey DI, Rossjohn J, McCluskey J, Kjer-Nielsen L (2013) Antigen-loaded MR1 tetramers define T cell receptor heterogeneity in mucosal-associated invariant T cells. J Exp Med 210:2305–2320
10. Treiner E, Duban L, Bahram S, Radosavljevic M, Wanner V, Tilloy F, Affaticati P, Gilfillan S, Lantz O (2003) Selection of evolutionarily conserved mucosal-associated invariant T cells by MR1. Nature 422:164–169
11. McWilliam HE, Villadangos JA (2018) MR1 antigen presentation to MAIT cells: new ligands, diverse pathways? Curr Opin Immunol 52:108–113
12. Gibbs A, Leeansyah E, Introini A, Paquin-Proulx D, Hasselrot K, Andersson E, Broliden K, Sandberg JK, Tjernlund A (2017) MAIT cells reside in the female genital mucosa and are biased towards IL-17 and IL-22 production in response to bacterial stimulation. Mucosal Immunol 10:35–45
13. Sobkowiak MJ, Davanian H, Heymann R, Gibbs A, Emgard J, Dias J, Aleman S, Kruger-Weiner C, Moll M, Tjernlund A, Leeansyah E, Sallberg Chen M, Sandberg JK (2019) Tissue-resident MAIT cell populations in human oral mucosa exhibit an activated profile and produce IL-17. Eur J Immunol 49:133–143
14. Gibbs A, Buggert M, Edfeldt G, Ranefall P, Introini A, Cheuk S, Martini E, Eidsmo L, Ball TB, Kimani J, Kaul R, Karlsson AC, Wahlby C, Broliden K, Tjernlund A (2018) Human immunodeficiency virus-infected women have high numbers of CD103−CD8+ T cells residing close to the basal membrane of the ectocervical epithelium. J Infect Dis 218:453–465
15. https://medicalfuturist.com/digital-future-pathology.
16. Koos B, Kamali-Moghaddam M, David L, Sobrinho-Simoes M, Dimberg A, Nilsson M, Wahlby C, Soderberg O (2015) Next-generation pathology—surveillance of tumor microecology. J Mol Biol 427:2013–2022
17. Steinbach K, Vincenti I, Merkler D (2018) Resident-memory T cells in tissue-restricted immune responses: for better or worse? Front Immunol 9:2827
18. Wilk MM, Mills KHG (2018) CD4 TRM cells following infection and immunization: implications for more effective vaccine design. Front Immunol 9:1860

Part II

Functional Analysis of MAIT Cells

Chapter 7

Human MAIT Cell Activation In Vitro

Joachim P. Hagel, Lucy C. Garner, Matthew Bilton, Hema Mehta, Tianqi Leng, Carl-Philipp Hackstein, Prabhjeet Phalora, Ali Amini, Hossain D. Akther, Nicholas M. Provine, Matthew Edmans, Christian B. Willberg, and Paul Klenerman

Abstract

Mucosal-associated invariant T (MAIT) cells are an abundant innate-like T cell subset in humans, enriched in mucosal tissues and the liver. MAIT cells express a semi-invariant T cell receptor (TCR) and recognize microbial-derived riboflavin metabolites presented on the MHC Class I-like molecule MR1. In addition to activation via the TCR, MAIT cells can also be activated in response to cytokines such as IL-12 and IL-18, in contrast to conventional T cells. Here we describe TCR-dependent and -independent methods for MAIT cell activation. The TCR-dependent approaches include stimulation with microbead- or plate-bound anti-CD3/anti-CD28 antibodies, and with 5-OP-RU or paraformaldehyde (PFA)-fixed *E. coli* in the presence of antigen-presenting cells (APCs). The latter method includes a combination of TCR- and cytokine-mediated stimulation. The TCR-independent methods include direct stimulation with the recombinant cytokines IL-12 and IL-18, and indirect stimulation with TLR-4/TLR-8 agonists or influenza A virus in the presence of APCs. Finally, we outline a protocol to analyze activated MAIT cells using flow cytometry.

Key words MAIT cells, Cytokines, MR1, Invariant T cell receptor, Vα7.2, Innate, Interleukin-12, Interleukin-18, *E. coli*, CD3, CD28, Beads, Plate-bound, 5-OP-RU, Toll-like receptor

1 Introduction

Innate-like subsets of T cells, including Natural Killer T (NKT), Mucosal-Associated Invariant T (MAIT), and γδ-T cells, operate at the interface of the innate and adaptive immune responses. In contrast to T cell receptors (TCRs) of conventional T cells, the TCRs of innate-like T cells show more limited diversity, and function in a similar manner to pattern recognition receptors through recognition of common pathogen-associated molecular motifs [1–3]. A prominent example is the semi-invariant MAIT cell TCR (Vα7.2-Jα33/20/12 TCRα chain paired with a limited Vβ repertoire) that recognizes microbial riboflavin metabolites presented by the MHC class Ib molecule MR1 on antigen-presenting cells

Helen Kaipe and Isabelle Magalhaes (eds.), *MAIT Cells: Methods and Protocols*, Methods in Molecular Biology, vol. 2098, https://doi.org/10.1007/978-1-0716-0207-2_7,

(APCs) [4–6]. This allows MAIT cells to respond rapidly to public antigens and thus bridge the temporal gap between the responses of innate and adaptive immunity [2]. Compared with the TCR expressed by conventional T cells, the MAIT cell TCR is hyporesponsive. This may be necessary to ensure tolerance to commensal microorganisms, given the promiscuity of the MAIT cell TCR in regard to pathogen recognition [7, 8]. The majority of human MAIT cells (about 80%) are CD8+ and almost all MAIT cells express the marker CD161 [9–12], which is highly expressed by innate-like T cell subsets compared with conventional T cells [3, 12–16]. An important characteristic of MAIT cells is their rapid activation in response to cytokines such as IL-12 and IL-18, which they share with other CD161-expressing T cells [11, 17]. Such cytokines are produced by activated APCs in the context of bacterial or viral infections and induce MAIT cell responses either on their own or in combination with TCR signals [11, 18].

Here we describe several TCR-dependent and -independent experimental approaches for the activation of MAIT cells. Protocols to activate MAIT cells via their TCR include using anti-CD3 and anti-CD28 antibodies coated on microbead particles or immobilized on a plate. Moreover, we describe an activation protocol using the riboflavin intermediate 5-OP-RU that can be bound and presented on MR1 by APCs to induce specific MAIT cell activation. We also outline how MAIT cells can be activated by combined TCR and cytokine stimulation using an in vitro experimental model in which APCs (within Peripheral Blood Mononuclear Cells (PBMCs) or THP-1 cells) are challenged with PFA-fixed *E. coli*. MAIT cells are activated by the riboflavin metabolites presented on MR1, and cytokines produced by activated APCs. Furthermore, we explain how to stimulate MAIT cells in a TCR-independent manner using cytokines such as IL-12 and IL-18, either directly by adding recombinant cytokines or indirectly by stimulating APCs to produce IL-12 and IL-18 with Toll-like receptor (TLR) agonists. Finally, we extend this MR1-independent activation to an in vitro viral infection model in which virus-activated APCs produce cytokines that activate MAIT cells. As one possible way to analyze TCR-dependent or -independent MAIT cell activation, we provide a protocol to assess expression of IFNγ, TNFα, GzmB, and CD69 by flow cytometry.

2 Materials

2.1 TCR-Dependent MAIT Cell Activation

1. R10 medium: RPMI 1640 Medium, 10% heat-inactivated fetal calf serum (FCS), 1% L-glutamine, and 1% penicillin/streptomycin.

2. Phosphate Buffered Saline (PBS), pH 7.4, sterile.
3. MACS Buffer: PBS, pH 7.2, 0.5% BSA (or alternatively 2% FCS), 2 mM EDTA.
4. PBMCs (peripheral blood mononuclear cells).
5. CD8 MicroBeads (Miltenyi Biotec).
6. THP-1 cells (THP-1 ECACC 88081201), European Collection of Authenticated Cell Cultures.
7. T Cell Activation/Expansion Kit, human (contains anti-Biotin MACSiBead Particles and anti-CD3 and anti-CD28 antibodies to generate immobilized anti-CD3 and anti-CD28 on MACSiBead Particles, Miltenyi Biotec).
8. 96-well U-bottom cell culture plates.
9. 96-well flat-bottom Nunc™ MaxiSorp™ ELISA Plate.
10. PMA (Phorbol 12-myristate 13-acetate).
11. Ionomycin 1mM in DMSO.
12. Brefeldin A Solution, 1000×.
13. Anti-human CD3 antibody, clone UCHT1.
14. Anti-human CD28 antibody, clone CD28.2.
15. Isotype control antibodies for anti-human CD3 and CD28.
16. 5-OP-RU (5-(2-oxopropylideneamino)-6-dribitylaminouracil) obtained from Prof. David P. Fairlie, University of Queensland, Australia.
17. DMSO.
18. LB Broth with agar (Lennox) tablet or powder microbial growth medium.
19. LB Broth (Lennox) powder microbial growth medium.
20. Petri dishes (100 × 15 mm).
21. DH5α or TOP10 *E. coli* cells.
22. 0.25 L Erlenmeyer flasks.
23. Rattler Plating Beads, 4.5 mm.
24. Paraformaldehyde solution 2% in PBS.
25. Tube rotator.

2.2 TCR-Independent MAIT Cell Activation

1. R10 medium: RPMI 1640 Medium, 10% heat-inactivated fetal calf serum (FCS), 1% L-glutamine, and 1% penicillin/streptomycin.
2. PBMCs.
3. CD8 MicroBeads (Miltenyi Biotec).
4. THP-1 cells (THP-1 ECACC 88081201), European Collection of Authenticated Cell Cultures.

5. 96-well U-bottom cell culture plates.
6. Recombinant human IL-12.
7. Recombinant human IL-18.
8. PMA (Phorbol 12-myristate 13-acetate).
9. Ionomycin, 1 mM in DMSO.
10. Brefeldin A Solution, 1000×.
11. TLR8 agonist ssRNA40/Lyovec.
12. TLR4 agonist *E. coli* K12 LPS.
13. Influenza virus strain A/WSN/1933(H1N1).

2.3 Analysis of MAIT Cell Activation Via Flow Cytometry

1. 96-well U-bottom or 96-well V-bottom cell culture plates.
2. PBS, pH 7.4.
3. PBS with 5% FCS.
4. LIVE/DEAD™ Fixable Near-IR Dead Cell Stain (Invitrogen).
5. Antibodies: refer to Tables 1, 2, and 3
6. Formaldehyde solution 2% in PBS.
7. Permeabilization Buffer 10× (Thermo Fisher Scientific).

3 Methods

3.1 TCR-Dependent MAIT Cell Activation

3.1.1 MAIT Cell Activation Using Anti-CD3/Anti-CD28: MACSiBead Particles

Here we describe how to activate MAIT cells in a non-specific way by exposing MAIT cells within PBMCs or isolated CD8+ T cells to anti-CD3 and anti-CD28 antibodies immobilized on MACSiBead Particles. The anti-CD3 and anti-CD28 antibodies bind CD3 and CD28, respectively, on T cells including MAIT cells, inducing CD3 and CD28 clustering and T cell activation [19–22] (Figs. 1a and 2).

1. Anti-CD3/anti-CD28—MACSiBead Particles (T Cell Activation/Expansion Kit, human) are prepared according to manufacturer's instructions with a few alterations.
   (a) Add 100 μL anti-CD3-Biotin, 100 μL anti-CD28-Biotin, and 100 μL MACS Buffer to a sterile and sealable 2 mL tube under aseptic conditions (*see* **Note 1**).
   (b) Vortex anti-Biotin MACSiBead Particles thoroughly for 30 s as they have a tendency to sediment quickly.
   (c) Add 500 μL anti-Biotin MACSiBead Particles to the antibody mix.
   (d) Add 200 μL MACS Buffer to obtain a final volume of 1 mL.
   (e) To allow the beads to be loaded with the antibodies, incubate at 2–8 °C for at least 2 h under constant, gentle rotation using a tube rotator.

Table 1
Example antibody panel 1 for analysis with MACSQuant Analyzer 10, Miltenyi Biotec

Antibody	Clone	Color	Manufacturer	Dilution
Antibodies for surface antigens				
CD3	OKT3	eFluor 450	Invitrogen	1:100
CD8	BW135/80	VioGreen	Miltenyi Biotec	1:25
Vα7.2 TCR	3C10	FITC	BioLegend	1:50
CD161	191B8	PE	Miltenyi Biotec	1:50
Antibodies for intracellular antigens				
TNFα	MAb11	PerCP-Cy5.5	BioLegend	1:25
IFNγ	4S.B3	PE-Cy7	BioLegend	1:100
Granzyme B	GB12	APC	Invitrogen	1:100

Table 2
Example antibody panel 2 for analysis with MACSQuant Analyzer 10, Miltenyi Biotec

Antibody	Clone	Color	Manufacturer	Dilution
Antibodies for surface antigens				
CD3	OKT3	eFluor 450	Invitrogen	1:100
CD8	BW135/80	VioGreen	Miltenyi Biotec	1:25
CD69	FN50	FITC	eBioscience	1:50
CD161	191B8	PE	Miltenyi Biotec	1:50
Vα7.2 TCR	3C10	PerCP-Cy5.5	BioLegend	1:50
Antibodies for intracellular antigens				
IFNγ	4S.B3	PE-Cy7	BioLegend	1:100
Granzyme B	GB12	APC	Invitrogen	1:100

(f) Do not remove the antibody mix from the loaded beads. Loaded beads can be stored for up to 4 months at 2–8 °C.

2. Count PBMCs or isolated CD8+ T cells (For CD8+ T cell isolation we recommend using CD8 MicroBeads and following manufacturer's instructions).
3. Resuspend cells in R10 medium at a concentration of 10 million PBMCs or 2 million CD8+ T cells per 1 mL.
4. Seed 100 μL of cells per well into a 96-well U-bottom cell culture plate. This results in a final cell density of 1 million PBMCs or 0.2 million CD8+ T cells per well. Cell numbers can be adapted to your needs (*see* **Note 2**).

Table 3
Example antibody panel 3 for analysis with BD LSR II, BD Biosciences

Antibody	Clone	Color	Manufacturer	Dilution
Antibodies for surface antigens				
CD69	FN50	eFluor 450	eBioscience	1:50
CD8	BW135/80	VioGreen	Miltenyi Biotec	1:25
CD3	UCHT1	BV605	BioLegend	1:100
CD161	191B8	PE	Miltenyi Biotec	1:50
Vα7.2 TCR	3C10	PE-Cy7	BioLegend	1:50
CD4	SK3	APC	BioLegend	1:100
Antibodies for intracellular antigens				
IFNγ	45-15	FITC	Miltenyi Biotec	1:50
Granzyme B	GB11	AF700	BioLegend	1:100
TNFα	MAb11	PerCP-Cy5.5	BioLegend	1:50

The panel was optimized using a BD LSR II setup including violet (405 nm), blue (488 nm), and red (633 nm) lasers with the following filter configuration: Violet array filters: 450/50, 525/50, 560/20, 585/15, 605/12, 655/8. Blue array filters: 530/30, 575/26, 610/20, 670/14, 695/40, 780/60. Red array filters: 670/14, 730/45, 780/60

5. To prepare the loaded MACSiBead particles, resuspend or vortex beads thoroughly, as beads tend to sediment quickly to the bottom of the container.
6. Take an aliquot of loaded MACSiBead particles (5 μL per 1 million cells, which equals 0.5 million beads; bead-to-cell ratio 1:2 for PBMCs and isolated CD8+ T cells) and transfer into a new sterile Eppendorf tube.
7. Add at least 100–200 μL R10 medium and centrifuge at 500 × *g* for 5 min to separate the beads from any unbound antibody.
8. Aspirate the supernatant completely and resuspend in 100 μL R10 medium per 0.5 million beads if PBMCs are used or resuspend in 100 μL R10 medium per 0.1 million beads if CD8+ T cells are used.
9. Add 100 μL of the bead mix to the wells containing the cells and mix well. This results in a total culture medium volume of 200 μL (*see* **Note 3**).
10. As a negative control, add 100 μL R10 medium instead of beads. Further controls can include conditions with unloaded or isotype-loaded MACSiBead particles.
11. As a positive control, add 100 μL 2× PMA + Ionomycin mix, diluted in R10 medium, to the cells and mix well (final concentration: PMA = 5–50 ng/mL, we recommend 10 ng/mL;

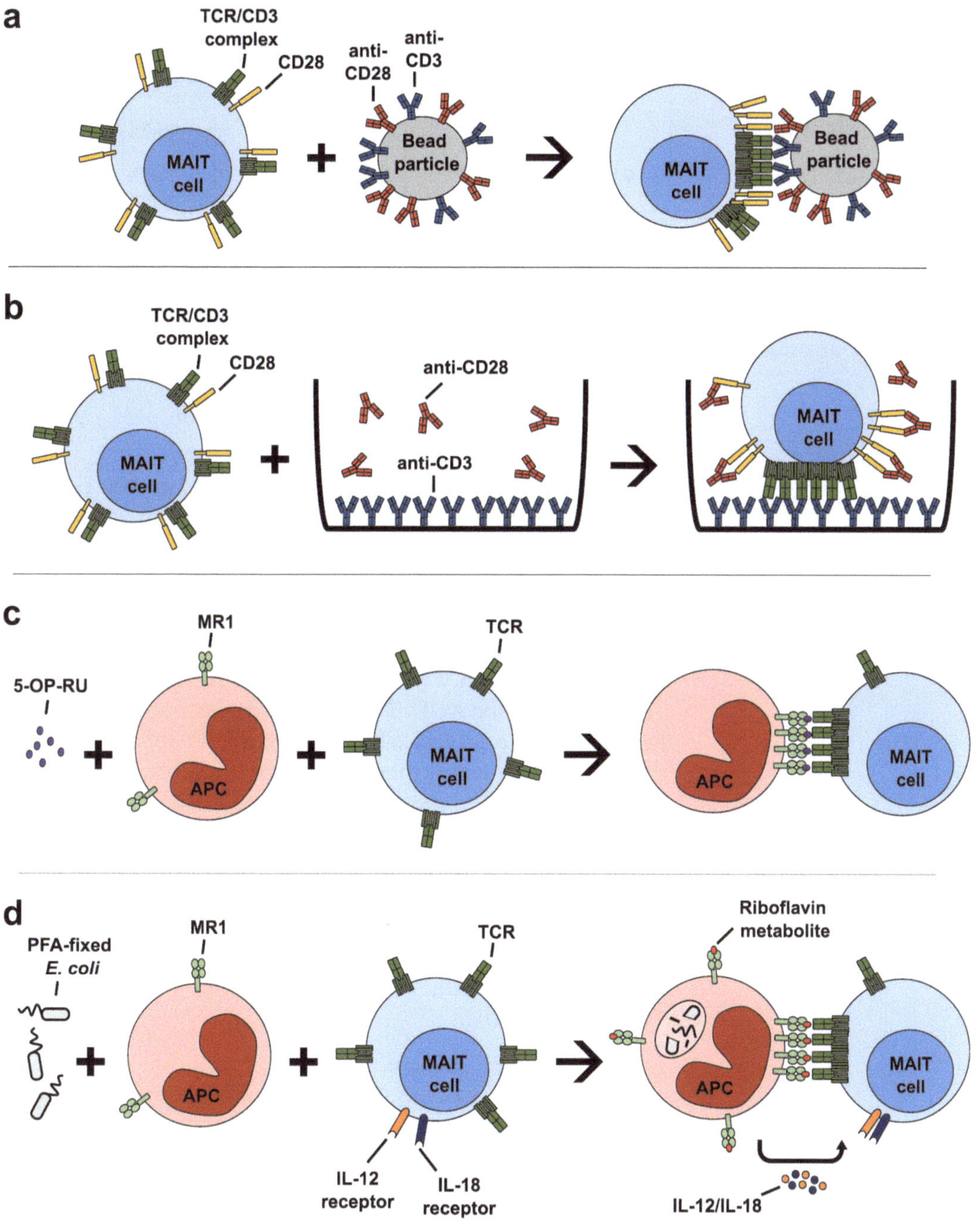

Fig. 1 Methods to activate MAIT cells in a TCR-dependent manner. (**a**) MAIT cells can be activated by anti-CD3 and anti-CD28 antibodies that are coupled to beads. The beads bind CD3 and CD28 on MAIT cells, which results in clustering of those molecules leading to MAIT cell activation. (**b**) MAIT cells can be activated by anti-CD3 antibody that is immobilized on a plate in combination with soluble anti-CD28 antibody. The plate-immobilized anti-CD3 antibodies bind and cluster CD3, and the soluble anti-CD28 antibodies bind and crosslink CD28 on MAIT cells. This results in MAIT cell activation. (**c**) MAIT cells can be activated in the presence of APCs by the administration of the riboflavin intermediate 5-OP-RU. 5-OP-RU can be bound by MR1

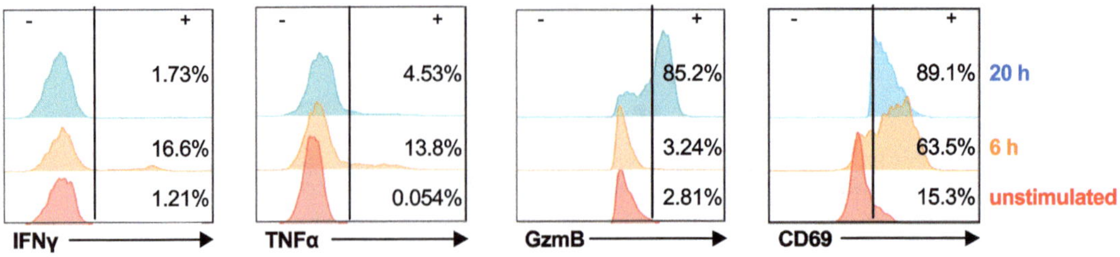

Fig. 2 MAIT cell activation using anti-CD3/anti-CD28 beads. Isolated CD8+ T cells were stimulated using anti-CD3/anti-CD28 beads and subsequently analyzed by flow cytometry. Example histograms show the kinetics of IFNγ, TNFα, GzmB, and CD69 expression by MAIT cells at 6 and 20 h post-stimulation. Half-offset histograms show modal *y*-axis scaling (% max. count per condition)

Ionomycin: 0.66–1 μM, we recommend 0.66 μM) (*see* **Note 4**).

12. Incubate at 37 °C, 5% CO_2 for 24 h (the incubation time can be varied depending on the experimental question, *see* **Note 5**).
13. If the experiment is to be analyzed by flow cytometry including staining for intracellular cytokines, add Brefeldin A to all conditions 4 h before termination of the assay. Prepare a 1:50 dilution of Brefeldin A 1000× stock solution in R10 medium. Add 10 μL to each well containing 200 μL cultured cell suspension and mix well. The final working concentration of Brefeldin A is 3 μg/mL. Incubate at 37 °C and 5% CO_2 for the remaining 4 h and then continue with Subheading 3.3.

3.1.2 MAIT Cell Activation Using Plate-Bound Anti-CD3/Anti-CD28

This protocol describes how to activate MAIT cells within PBMCs or isolated CD8+ T cells by exposing them to anti-CD3 antibodies, which are immobilized on a plate, and soluble anti-CD28 antibodies. The anti-CD3 antibodies bind CD3 on T cells, resulting in CD3 clustering and the soluble anti-CD28 antibodies crosslink CD28. This results in the activation of T cells, including MAIT cells [20, 23] (Fig. 1b).

1. To coat the plate, prepare a 2.5 μg/mL solution of anti-CD3 antibody or IgG isotype control (negative control) in sterile PBS (*see* **Note 6**).
2. Add 50 μL anti-CD3 or IgG isotype control antibody solution to the wells of a 96-well flat-bottom Nunc™ MaxiSorp™

Fig. 1 (continued) on the APCs and presented to MAIT cells. The MAIT cell TCR recognizes and binds 5-OP-RU in the context of MR1, which leads to MAIT cell activation. (**d**) MAIT cells can be activated in the presence of APCs by administration of PFA-fixed *E. coli*. The bacteria activate the APCs, which take up and process the *E. coli*. Bacterial-derived riboflavin metabolites are then presented on MR1 to MAIT cells. Additionally, the activated APCs express cytokines including IL-12 and IL-18, which can be recognized by the MAIT cells via IL-12 and IL-18 receptors. The combination of TCR and cytokine signal leads to MAIT cell activation

ELISA Plate. Use aseptic techniques when handling the plate and cover the plate with the lid of a sterile standard 96-well cell culture plate.

3. Seal the plate tightly with Parafilm to avoid evaporation and incubate at 4 °C overnight or at 37 °C for 2 h.
4. To wash the wells, add 150 μL sterile PBS or R10 medium under aseptic conditions, then aspirate and discard all the liquid.
5. Wash two more times with 200 μL sterile PBS or R10 medium.
6. Add 200 μL R10 medium containing 10% FCS and incubate for at least 2 h at 37 °C to block the plate.
7. In the meantime, prepare and count PBMCs or isolated CD8+ T cells (For CD8+ T cell isolation we recommend using CD8 MicroBeads and following manufacturer's instructions).
8. Resuspend PBMCs or CD8+ T cells in R10 medium at a concentration of 2 million cells per 1 mL.
9. Aspirate and discard the medium of the blocked Nunc™ MaxiSorp™ ELISA Plate and seed 100 μL cells per well, resulting in a cell density of 0.2 million cells per well (*see* **Note 7**). Work fast to avoid that the coated wells get dried out.
10. Prepare a 2× anti-CD28 antibody or IgG isotype antibody mix (2 μg/mL) diluted in R10.
11. Add 100 μL antibody mix to 100 μL cell suspension and mix well, resulting in a final anti-CD28 antibody concentration of 1 μg/mL (*see* **Note 6**) and a total culture medium volume of 200 μL (*see* **Note 3**).
12. Incubate at 37 °C, 5% CO_2 for 24 h (the incubation time can be varied depending on the experimental question, *see* **Note 5**).
13. If the experiment is to be analyzed by flow cytometry including staining for intracellular cytokines, add Brefeldin A to all conditions 4 h before termination of the assay. Prepare a 1:50 dilution of Brefeldin A 1000× stock solution in R10 medium. Add 10 μL to each well containing 200 μL cultured cell suspension and mix well. The final working concentration of Brefeldin A is 3 μg/mL. Incubate at 37 °C and 5% CO_2 for the remaining 4 h and then continue with Subheading 3.3.

3.1.3 MAIT Cell Activation by 5-OP-RU

Here we outline a protocol to activate MAIT cells using 5-OP-RU, a riboflavin intermediate. This metabolite can be bound by MR1 on APCs and presented to MAIT cells. Recognition of 5-OP-RU in the context of MR1 by the MAIT cell TCR induces MAIT cell activation (Figs. 1c and 3). This experiment can be performed using MAIT cells and APCs within PBMCs or using MAIT cells within isolated CD8+ T cells in combination with THP-1 cells as APCs [24–26].

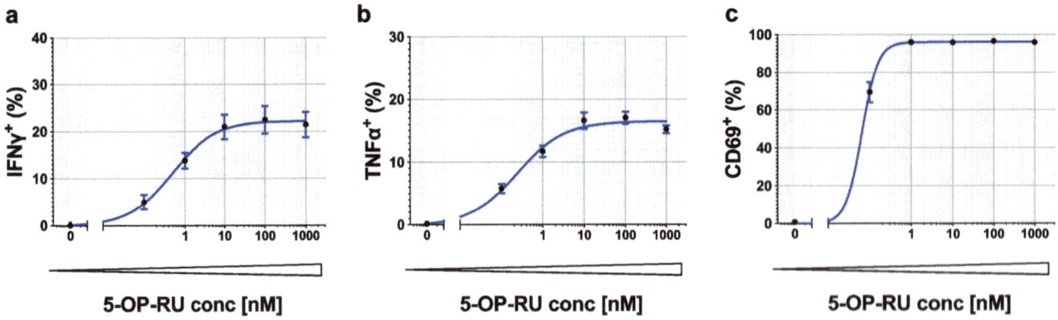

Fig. 3 Stimulation of MAIT cells with 5-OP-RU in the presence of APCs. Increasing concentrations of 5-OP-RU were added to isolated CD8+ T cells and THP-1 cells as indicated. Cells were incubated for 24 h and subsequently analyzed by flow cytometry. (**a**) Frequency of IFNγ+ MAIT cells. (**b**) Frequency of TNFα+ MAIT cells. (**c**) Frequency of CD69+ MAIT cells. Data points represent mean ± SEM of five donors

1. Prepare and count PBMCs or THP-1 cells and isolated CD8+ T cells (For CD8+ T cell isolation we recommend using CD8 MicroBeads and following manufacturer's instructions).
2. Resuspend cells in R10 medium at a concentration of 10 million PBMCs per 1 mL or 4 million CD8+ T cells per 1 mL and 2 million THP-1 cells per 1 mL.
3. Seed 100 μL of PBMCs or alternatively 50 μL of CD8+ T cells per well into a 96-well U-bottom cell culture plate. When using isolated CD8+ T cells, add 50 μL THP-1 cells into each well. This results in a final cell density of 1 million PBMCs or 0.2 million CD8+ T cells and 0.1 million THP-1 cells per well. Cell numbers can be adapted to your needs (*see* **Notes 2** and **8**).
4. Dilute 5-OP-RU in R10 medium to obtain a 2× mix (20 nM).
5. Add 100 μL of the 2× 5-OP-RU mix to the wells containing 100 μL cell suspension and mix well. This results in a final 5-OP-RU concentration of 10 nM. The concentration of 5-OP-RU can be adapted according to your experimental needs (*see* **Note 9**).
6. As a negative control, prepare a 2× mix of DMSO diluted in R10 medium (because the 5-OP-RU stock is diluted in DMSO), and add 100 μL to the cells instead of 5-OP-RU.
7. Incubate at 37 °C, 5% CO_2 for 24 h (the incubation time can be varied depending on the experimental question, *see* **Note 5**).
8. If the experiment is to be analyzed by flow cytometry including staining for intracellular cytokines, add Brefeldin A to all conditions 4 h before termination of the assay. Prepare a 1:50 dilution of Brefeldin A 1000× stock solution in R10 medium. Add 10 μL to each well containing 200 μL cultured cell suspension and mix well. The final working concentration of Brefeldin A is 3 μg/mL. Incubate at 37 °C and 5% CO_2 for the remaining 4 h and then continue with Subheading 3.3.

3.1.4 MAIT Cell Activation by PFA-Fixed E. coli

In this experimental model, MAIT cells are activated within PBMCs (model 1) or within CD8+ T cells in the presence of THP-1 cells (model 2). APCs within PBMCs or THP-1 cells are activated by PFA-fixed *E. coli*. This results in the processing of the bacteria within the APCs, upregulation of MR1 expression, presentation of bacterial riboflavin metabolites on MR1, and expression of IL-12 and IL-18 by the APCs. MAIT cells are specifically activated by recognizing the riboflavin metabolites in the context of MR1 in combination with binding IL-12 and IL-18 to respective receptors on their surface [5, 6, 11, 27, 28] (Figs. 1d and 4). TCR-mediated MAIT cell activation dominates at early time points (6 h), while both TCR and cytokine signals contribute to later MAIT cell activation (24 h) [11].

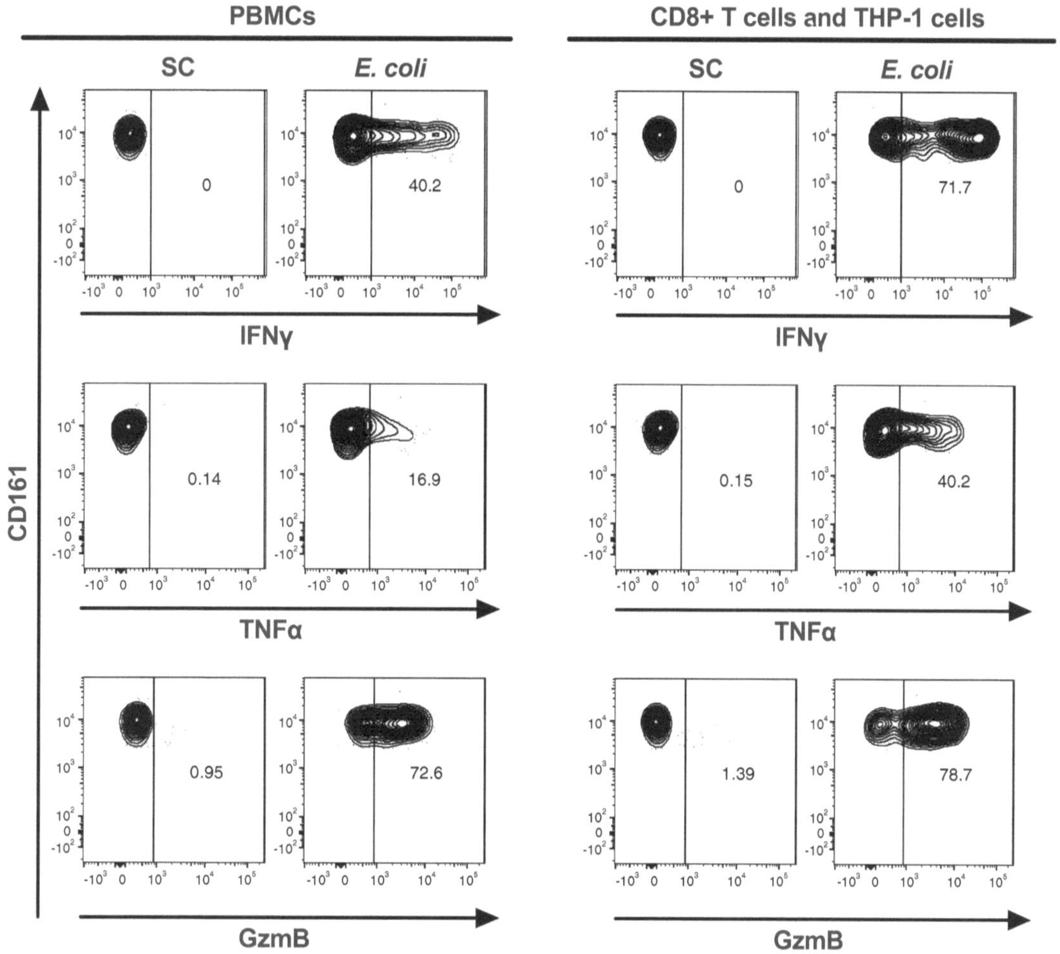

Fig. 4 Activation of MAIT cells with *E. coli*. PFA-fixed *E. coli* or sterility control (SC) was added to PBMCs or CD8+ T cells and THP-1 cells. Cells were incubated for 24 h and subsequently analyzed by flow cytometry. Graph shows an example of flow cytometry plots of IFNγ, TNFα, and GzmB production by CD8+Va7.2+CD161++ MAIT cells

Preparation of PFA-Fixed *E. coli* [27–30]

1. Preparation of LB agar plates
   - (a) Add LB Broth with agar powder or tablets to deionized water following manufacturer's instructions and make up the quantity required for the desired number of plates (approximately 10 mL LB agar per Petri dish) within a glass bottle.
   - (b) Close the bottle loosely with the cap and label it with autoclave tape.
   - (c) Autoclave the LB agar medium at 121 °C to sterilize.
   - (d) When the LB agar medium has cooled enough to handle safely, but before it begins to solidify, pour it into sterile Petri dishes under aseptic conditions to form a layer of approximately 5 mm. Avoid bubbles.
   - (e) In order to prevent condensation within the plates, leave them to solidify and dry at room temperature. When the agar plates are solidified and free of moisture, store them upside down at 2–8 °C.
2. Obtaining *E. coli* colonies on LB agar plates
   - (a) Warm LB agar plates to 37 °C.
   - (b) From now on, work near a Bunsen burner flame or within a laminar flow cabinet.
   - (c) Take the *E. coli* stock vial (strain DH5α or TOP10) from the −80 °C freezer and transfer it onto ice. Scrape the surface of the frozen bacterial stock with an inoculation loop.
   - (d) Streak the bacteria onto a plate with an inoculation loop as shown in Fig. 5. This procedure is necessary to obtain isolated colonies.
   - (e) Incubate plates in a microbiological incubator overnight (approx.16 h) at 37 °C.
   - (f) Remove the plates containing the *E. coli* cultures. Single colonies should be visible. Store the plates at 2–8 °C for up to 2 months.
3. Growing up *E. coli* cultures
   - (a) To prepare LB Broth medium, add LB Broth powder to deionized water within a glass bottle following manufacturer's instructions. Swirl to mix. Powder will not completely dissolve. Loosen the bottle cap for sterilization. Autoclave at 121 °C to sterilize. Allow medium to cool before use.
   - (b) Add 50 mL of LB Broth medium into a 0.25 L Erlenmeyer flask for the bacterial culture (BC). Prepare a second flask for the sterility control (SC).

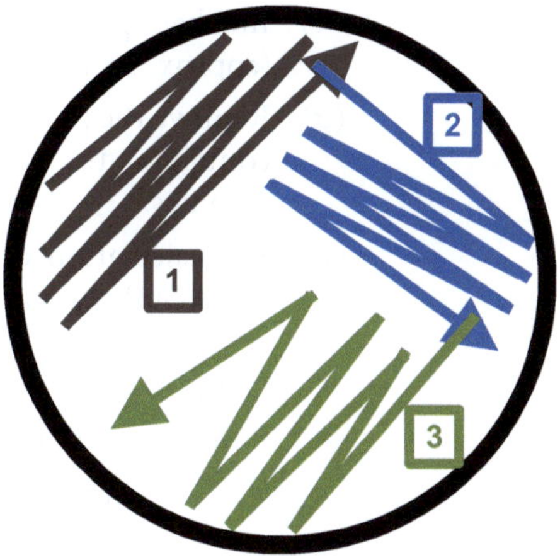

Fig. 5 Streaking technique to obtain single *E. coli* colonies on LB agar plates. Spread bacteria with an inoculation loop on the LB agar plate to obtain streak 1. To further separate the bacteria, use a fresh sterile inoculation loop and streak through streak 1 to create streak 2. Repeat using a fresh sterile inoculation loop, crossing through streak 2 to create streak 3

(c) Use a pipette with a sterile 200 μL tip and pick five bacterial colonies from the agar plate and eject the tip into the BC flask (*see* **Note 10**).

(d) Take a fresh tip and eject it into the SC flask. This will be the no bacteria control.

(e) Cover the flasks with aluminum foil. NB: cover should not be too tight, as this will prevent gas exchange.

(f) Incubate the flasks in a shaking incubator (225 rpm) at 37 °C for 16 h.

4. Quantification of bacteria: Determination of CFU [31]

(a) Prepare a 1/10 serial dilution from BC and SC ranging from 1/10 to $1/10^{10}$. For the initial 1/10 dilution, take 50 μL of BC or SC and add it to 450 μL sterile PBS. Work under aseptic conditions.

(b) To plate each of the dilutions, add 50 μL to the center of an agar plate.

(c) Pour 4–5 Rattler Plating spherical glass beads into the plate and close the plate lid.

(d) Shake the plate back and forth and from side to side to spread the bacteria over the plate. Ensure that the beads cover the entire plate so that there is an even distribution of bacteria.

(e) To remove the beads, invert the plate. Use fresh beads for each dilution and plate.

(f) Incubate plates in a microbiological incubator overnight (approx. 16 h) at 37 °C.

(g) Remove the plates and count bacterial colonies. The plates containing the dilutions of the SC should not contain any colonies.

(h) For calculation of CFU, use the plate containing from 30 to 300 colonies.

$$\frac{\mathrm{CFU}}{\mathrm{mL}} = \frac{\text{number of colonies} \times \text{dilution factor}}{\text{volume of bacterial culture plated}} = \frac{\text{number of colonies} \times \text{dilution factor}}{0.05\ \mathrm{mL}}$$

$$\text{total CFU} = \frac{\mathrm{CFU}}{\mathrm{mL}} \times \text{total volume of bacterial culture} = \frac{\mathrm{CFU}}{\mathrm{mL}} \times 50\ \mathrm{mL}$$

5. PFA fixation of *E. coli*

(a) Transfer the BC and SC into 50 mL Falcon tubes. Treat the SC in exactly the same way you treat the BC throughout the whole protocol.

(b) Pellet bacteria through centrifugation (10 min at 4000–6000 × *g*). Discard supernatant.

(c) Resuspend bacteria in 50 mL sterile PBS. Make sure the bacterial pellet is thoroughly resuspended by pipetting up and down (this is necessary for all the washing steps and for the fixation).

(d) Centrifuge (10 min at 4000–6000 × *g*) and discard supernatant.

(e) Resuspend the bacterial pellet in 25 mL 2% PFA, diluted in PBS, and fix at room temperature for 20 min.

(f) Add 25 mL sterile and filtered (0.22 μm) PBS and centrifuge for 10 min at 4000–6000 × *g*. Discard supernatant.

(g) Wash twice with 50 mL sterile filtered PBS (10 min at 4000–6000 × *g*).

(h) Resuspend bacteria in sterile filtered PBS at a concentration of 1 million CFU per 0. 1 μL. Add an equal volume of sterile filtered PBS to the SC tube.

(i) Aliquots should be prepared and stored at 2–8 °C for up to 2 months.

Experimental Setup for MAIT Cell Activation by PFA-Fixed *E. coli*

1. Prepare and count PBMCs or THP-1 cells and isolated CD8+ T cells (For CD8+ T cell isolation we recommend using CD8 MicroBeads and following manufacturer's instructions) (*see* **Note 11**).
2. Resuspend cells in R10 medium at a concentration of 10 million PBMCs per 1 mL or 4 million CD8+ T cells per 1 mL and 2 million THP-1 cells per 1 mL.
3. Seed 100 μL of PBMCs or alternatively 50 μL of CD8+ T cells per well into a 96-well U-bottom cell culture plate. When using isolated CD8+ T cells, add 50 μL THP-1 cells into each well. This results in a final cell density of 1 million PBMCs or 0.2 million CD8+ T cells and 0.1 million THP-1 cells per well. Cell numbers can be adapted to your needs (*see* **Notes 2** and **8**).
4. Dilute PFA-fixed *E. coli* in R10 medium to obtain a 2× mix (we recommend 20 million CFU per 100 μL for the PBMC assay or 6 million CFU per 100 μL for the THP-1 assay) (*see* **Note 12**).
5. Add 100 μL of the 2× *E. coli* mix to the wells containing 100 μL cell suspension and mix well. This results in a final concentration of 10 CFU per PBMC or 30 CFU per THP-1 cell.
6. As a negative control, prepare a 2× mix of the sterility control (SC). Add 100 μL of the 2× SC mix to the control wells containing 100 μL cell suspension and mix well.
7. Incubate at 37 °C, 5% CO_2 for 6 or 24 h (the incubation time can be varied depending on the experimental question, *see* **Note 5**).
8. If the experiment is to be analyzed by flow cytometry including staining for intracellular cytokines, add Brefeldin A to all conditions 4 h before termination of the assay. Prepare a 1:50 dilution of Brefeldin A 1000× stock solution in R10 medium. Add 10 μL to each well containing 200 μL cultured cell suspension and mix well. The final working concentration of Brefeldin A is 3 μg/mL. Incubate at 37 °C and 5% CO_2 for the remaining 4 h and then continue with Subheading 3.3.

3.2 TCR-Independent MAIT Cell Activation

Here we describe how to activate MAIT cells within PBMCs or isolated CD8+ T cells by recombinant IL-12 and IL-18 [11] (Figs. 6a and 7).

3.2.1 MAIT Cell Activation by Cytokines IL-12 + IL-18

1. Prepare and count PBMCs or isolated CD8+ T cells (For CD8+ T cell isolation we recommend using CD8 MicroBeads and following manufacturer's instructions).
2. Resuspend cells in R10 medium at a concentration of 10 million PBMCs or 2 million CD8+ T cells per 1 mL

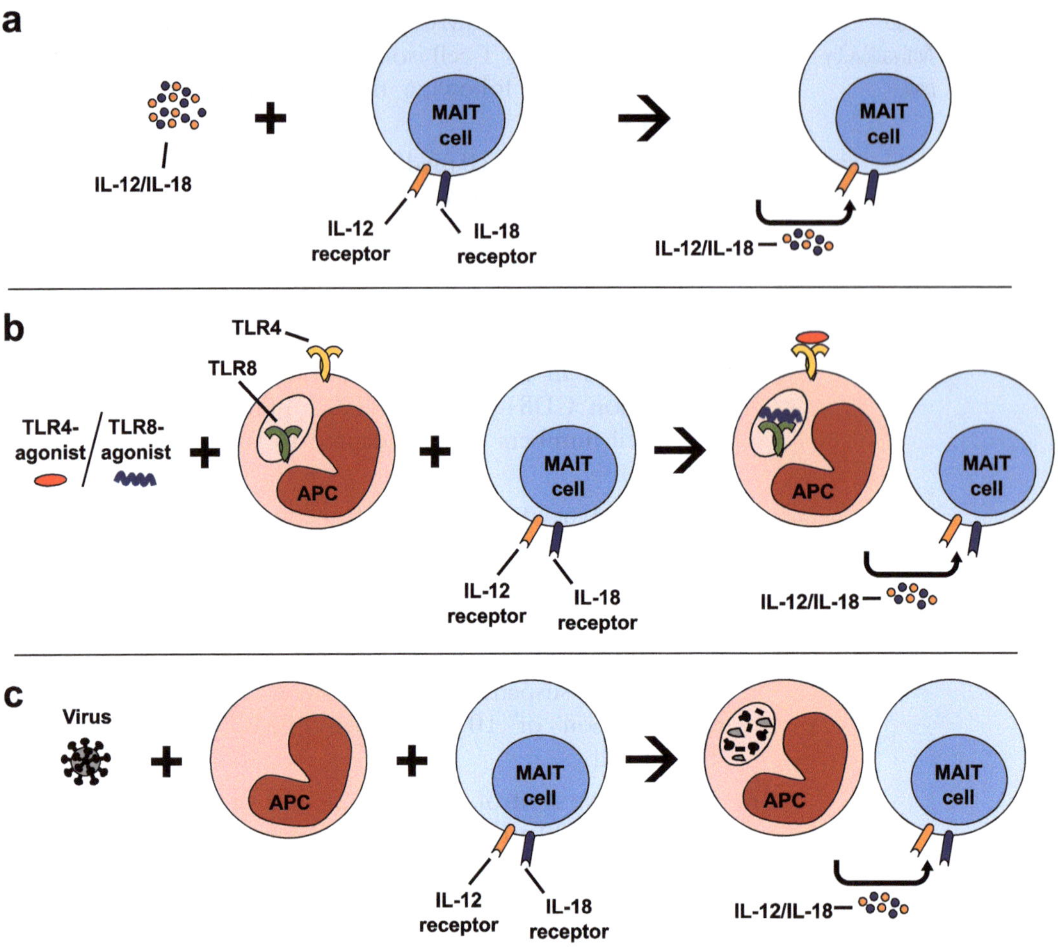

Fig. 6 Methods to activate MAIT cells in a TCR-independent manner. (**a**) MAIT cells can be activated by cytokines such as the combination of IL-12 and IL-18. The cytokines signal through IL-12 and IL-18 receptors on MAIT cells, inducing activation. (**b**) MAIT cells can be indirectly activated by TLR agonists in the presence of APCs. The administration of TLR8-agonist to TLR8-expressing APCs within PBMCs or TLR4-agonist to TLR4-expressing THP-1 cells induces APC activation. Activated APCs produce cytokines including IL-12 and IL-18 that can activate MAIT cells. (**c**) MAIT cells can be activated indirectly by the administration of certain viruses in the presence of APCs. The APCs become activated by sensing pathogen-associated molecular patterns of the viruses resulting in the expression of cytokines such as IL-12 and IL-18 and hence MAIT cell activation

3. Seed 100 μL cells per well into a 96-well U-bottom cell culture plate. This results in a final cell density of 1 million PBMCs or 0.2 million CD8+ T cells per well. Cell numbers can be adapted to your needs (*see* **Note 2**).
4. To generate a 2× IL-12 + IL-18 mix, add 1 μL of each cytokine (stock concentration: 100 μg/mL) per 1 mL R10 medium.
5. Add 100 μL of the 2× IL-12 + IL-18 mix to 100 μL cell suspension. This results in a total culture medium volume of

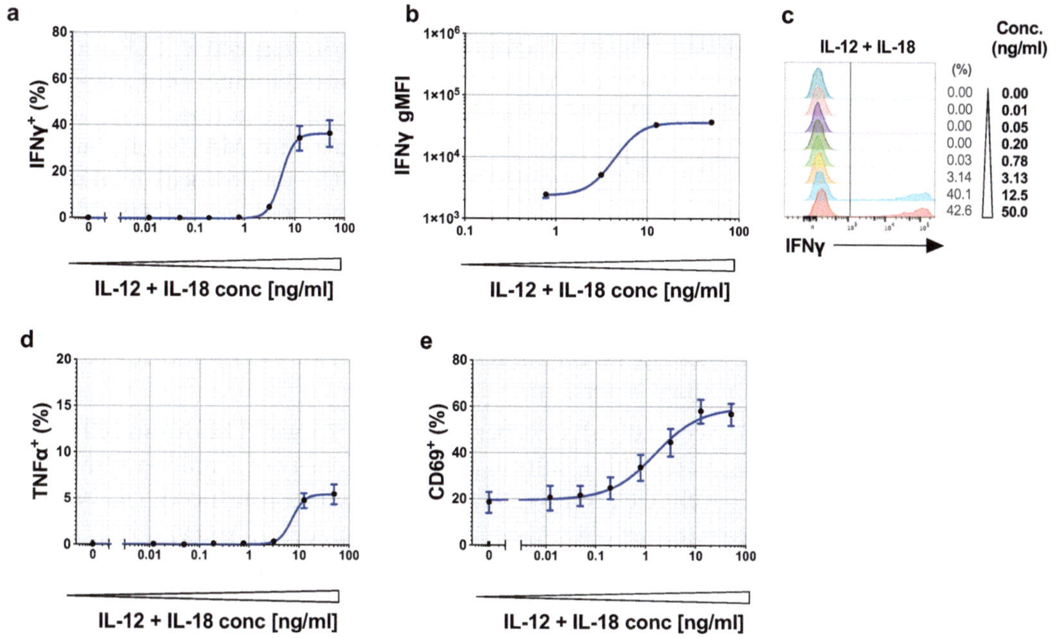

Fig. 7 MAIT cell stimulation with IL-12 + IL-18. CD8+ T cells were stimulated with increasing concentrations of IL-12 + IL-18 as indicated. Cells were incubated for 24 h and subsequently analyzed by flow cytometry. Responses of CD161++ T cells, including CD161++Va7.2+ MAIT cells and CD161++Va7.2- T cells, are shown. (**a**) Frequency of IFNγ+ cells. (**b**) IFNγ gMFI of IFNγ+ cells. (**c**) Representative flow cytometry plot of IFNγ expression by CD161++ T cells. Half-offset histogram shows modal *y*-axis scaling (% max. count per condition). (**d**) Frequency of TNFα+ cells. (**e**) Frequency of CD69+ cells. Graphs show mean ± SEM of four donors

200 μL (*see* **Note 3**) and a final cytokine concentration of 50 ng/mL each. The cytokine concentration can be adapted to your experimental needs (*see* **Note 13**).

6. Prepare a negative control by adding 100 μL R10 medium instead of cytokines to the cells and mix well.
7. Prepare a positive control by adding 100 μL 2× PMA + Ionomycin mix, diluted in R10 medium, to the cells and mix well (final concentration: PMA = 5–50 ng/mL, we recommend 10 ng/mL; Ionomycin: 0.66–1 μM, we recommend 0.66 μM) (*see* **Note 4**).
8. Incubate at 37 °C, 5% CO_2 for 24 h (the incubation time can be varied depending on the experimental question, *see* **Note 5**).
9. If the experiment is to be analyzed by flow cytometry including staining for intracellular cytokines, add Brefeldin A to all conditions 4 h before termination of the assay. Prepare a 1:50 dilution of Brefeldin A 1000× stock solution in R10 medium. Add 10 μL to each well containing 200 μL cultured cell suspension and mix well. The final working concentration of Brefeldin A is 3 μg/mL. Incubate at 37 °C and 5% CO_2 for the remaining 4 h and then continue with Subheading 3.3.

3.2.2 Indirect MAIT Cell Activation Using TLR4 or TLR8 Agonists

In this experimental model APCs are activated by TLR4 or TLR8 agonists. The TLR agonists bind to the respective TLR on APCs, stimulating the production of relevant cytokines (e.g., IL-12 and IL-18) for MAIT cell activation [11, 32] (Fig. 6b).

Follow protocol A in order to activate MAIT cells indirectly through TLR8 agonists using PBMCs, or protocol B to activate MAIT cells indirectly through TLR4 agonists using CD8+ T cells in combination with THP-1 cells.

Protocol A: Activate MAIT Cells Indirectly Through TLR8 Agonists Within PBMCs

1. Prepare and count PBMCs.
2. Resuspend cells in R10 medium at a concentration of 10 million PBMCs per 1 mL.
3. Seed 100 μL cells per well into a 96-well U-bottom cell culture plate. This results in a final cell density of 1 million PBMCs per well. Cell numbers can be adapted to your needs (*see* **Note 2**).
4. Prepare a 2× mix of TLR8 agonist (ssRNA40/Lyovec) by adding 2 μL TLR8 agonist (stock: 100 μg/mL) per 100 μL R10 medium.
5. Add 100 μL of the 2× TLR8 agonist mix to 100 μL cell suspension. This results in a total culture medium volume of 200 μL (*see* **Note 3**) and a final TLR agonist concentration of 1 μg/mL each. The final TLR agonist concentration can be adapted to your experimental needs (*see* **Note 14**).
6. Prepare a negative control by adding 100 μL R10 medium instead of TLR agonist to the cells and mix well.
7. Prepare a positive control by adding 100 μL 2× PMA + Ionomycin mix, diluted in R10 medium, to the cells and mix well (final concentration: PMA = 5–50 ng/mL, we recommend 10 ng/mL; Ionomycin: 0.66–1 μM, we recommend 0.66 μM) (*see* **Note 4**).
8. Incubate at 37 °C, 5% CO_2 for 24 h (the incubation time can be varied depending on the experimental question, *see* **Note 5**).
9. If the experiment is to be analyzed by flow cytometry including staining for intracellular cytokines, add Brefeldin A to all conditions 4 h before termination of the assay. Prepare a 1:50 dilution of Brefeldin A 1000× stock solution in R10 medium. Add 10 μL to each well containing 200 μL cultured cell suspension and mix well. The final working concentration of Brefeldin A is 3 μg/mL. Incubate at 37 °C and 5% CO_2 for the remaining 4 h and then continue with Subheading 3.3.

Protocol B: Activate MAIT Cells Indirectly Through TLR4 Agonists Using CD8+ T Cells in the Presence of THP-1 Cells

1. Prepare and count THP-1 cells.
2. Resuspend cells in R10 medium at a concentration of 1 million THP-1 cells per 1 mL.

3. Seed 100 μL THP-1 cell suspension per well into a 96-well U-bottom cell culture plate. This results in a cell density of 0.1 million THP-1 cells per well (*see* **Note 8**).
4. Prepare a 2× mix of TLR4 agonist (*E. coli* K12 LPS) by adding 2 μL TLR4 agonist (stock: 100 μg/mL) per 100 μL R10 medium.
5. Add 100 μL of the 2× TLR4 agonist mix to 100 μL cell suspension. This results in a total culture medium volume of 200 μL (*see* **Note 3**) and a final TLR agonist concentration of 1 μg/mL each. The final TLR agonist concentration can be adapted to your experimental needs (*see* **Note 14**).
6. Prepare a negative control by adding 100 μL R10 medium instead of TLR agonist to the THP-1 cells and mix well.
7. Incubate at 37 °C, 5% CO_2 for 24 h.
8. Before continuing with the plate, prepare and count isolated CD8+ T cells. (For CD8+ T cell isolation we recommend using CD8 MicroBeads and following manufacturer's instructions).
9. Resuspend CD8+ T cells in R10 medium at a concentration of 2 million cells per 1 mL.
10. Centrifuge the plate containing THP-1 cells for 5 min at 350 × *g*. Aspirate and discard supernatant.
11. Wash THP-1 cells by resuspending cells in 200 μL fresh medium. Centrifuge again, and aspirate and discard the supernatant.
12. Resuspend THP-1 cells in 100 μL R10 medium.
13. Add 100 μL of CD8+ T cell suspension per well to the stimulated THP-1 cells. This results in a final cell density of 0.2 million CD8+ T cells and 0.1 million THP-1 cells per well. Cell numbers can be adapted to your needs (*see* **Note 8**).
14. Incubate at 37 °C, 5% CO_2 for 24 h (the incubation time can be varied depending on the experimental question, *see* **Note 5**).
15. If the experiment is to be analyzed by flow cytometry including staining for intracellular cytokines, add Brefeldin A to all conditions 4 h before termination of the assay. Prepare a 1:50 dilution of Brefeldin A 1000× stock solution in R10 medium. Add 10 μL to each well containing 200 μL cultured cell suspension and mix well. The final working concentration of Brefeldin A is 3 μg/mL. Incubate at 37 °C and 5% CO_2 for the remaining 4 h and then continue with Subheading 3.3.

3.2.3 MAIT Cell Activation Using Viruses

MAIT cells can be indirectly activated by certain viruses in the presence of APCs. The viruses stimulate cytokine production by the APCs, for example, IL-12, IL-18, and/or IL-15, which in turn can activate MAIT cells [18] (Fig. 6c). Here we describe a protocol for the activation of MAIT cells within PBMCs using influenza strain A/WSN/1933 (H1N1), which activates MAIT cells mainly through the induction of IL-18 secretion by APCs. Please refer to reference [18] for information about using other viruses such as dengue or hepatitis C virus.

1. Prepare and count PBMCs.
2. Resuspend cells in R10 medium at a concentration of 10 million PBMCs per 1 mL.
3. Seed 100 μL cells per well into a 96-well U-bottom cell culture plate. This results in a final cell number of 1 million PBMCs per well. Cell numbers can be adapted to your needs (*see* **Note 2**).
4. Prepare a 2× mix of influenza virus strain A/WSN/1933 (H1N1) in R10 medium (final MOI of PBMC: 0.05–2.5; we recommend a final MOI of 1 for optimal stimulation). Refer to reference [18] for any information about influenza virus preparation.
5. Add 100 μL of the 2× influenza virus mix to 100 μL cell suspension and mix well. This results in a total culture medium volume of 200 μL (*see* **Note 3**).
6. As a negative control, add 100 μL R10 medium instead of influenza virus to the cells and mix well.
7. Prepare a positive control by adding 100 μL 2× PMA + Ionomycin mix, diluted in R10 medium, to the cells and mix well (final concentration: PMA = 5–50 ng/mL, we recommend 10 ng/mL; Ionomycin: 0.66–1 μM, we recommend 0.66 μM) (*see* **Note 4**).
8. Incubate at 37 °C, 5% CO_2 for 24 h (the incubation time can be varied depending on the experimental question, *see* **Note 5**).
9. If the experiment is to be analyzed by flow cytometry including staining for intracellular cytokines, add Brefeldin A to all conditions 4 h before termination of the assay. Prepare a 1:50 dilution of Brefeldin A 1000× stock solution in R10 medium. Add 10 μL to each well containing 200 μL cultured cell suspension and mix well. The final working concentration of Brefeldin A is 3 μg/mL. Incubate at 37 °C and 5% CO_2 for the remaining 4 h and then continue with Subheading 3.3.

3.3 Analysis of MAIT Cell Activation Via Flow Cytometry

3.3.1 Staining for Flow Cytometry

1. Centrifuge the 96-well U-bottom plate containing the cells for 5 min at 350 × *g*. If cells were cultured in a flat-bottom plate, e.g., for a plate-bound anti-CD3/anti-CD28 stimulation assay, transfer cells first to a 96-well U-bottom plate to ensure proper cell pellet formation during centrifugation. For very low cell numbers, V-bottom plates can be used for staining.
2. Discard supernatant.

3. To wash cells, resuspend the pellet in 200 μL PBS and centrifuge (5 min at 350 × *g*). Discard the supernatant.
4. Resuspend cells in 50 μL PBS-diluted viability dye (1:1000 dilution of LIVE/DEAD™ Fixable Near-IR Dead Cell Stain). Incubate for 20–30 min at 2–8 °C.
5. Add 150 μL PBS, centrifuge (5 min at 350 × *g*) and discard the supernatant.
6. To wash cells, resuspend the pellet in 200 μL PBS and centrifuge (5 min at 350 × *g*). Discard the supernatant.
7. Resuspend cells in 50 μL antibody mix containing the antibodies for surface antigens (prepare the antibody mix in advance by diluting antibodies in PBS containing 5% FCS. Tables 1, 2, and 3 lists recommended dilutions of three different panels for use with two different flow cytometers. For optimal performance, we recommend titrating the antibodies before use. This is especially important when using a flow cytometer different from those listed. Refer to **Notes 15** and **16** for further considerations regarding panel design). Incubate for 20–30 min at 2–8 °C.
8. Add 150 μL PBS and centrifuge for 5 min at 350 × *g*. Discard the supernatant.
9. Wash cells in 200 μL PBS, centrifuge (5 min at 350 × *g*) and discard supernatant.
10. Resuspend cells in 100 μL 2% formaldehyde solution (diluted in PBS) and incubate for 10 min at 2–8 °C.
11. Add 100 μL PBS and centrifuge at 500 × *g* for 5 min.
12. Wash cells in 200 μL PBS, centrifuge (500 × *g* for 5 min), and discard supernatant.
13. Resuspend cells in 200 μL 1× Permeabilization Buffer (diluted in distilled water) and incubate for 10 min.
14. Centrifuge (500 × *g* for 5 min) and discard supernatant.
15. Resuspend cells in 50 μL antibody mix containing the antibodies for intracellular antigens (prepare the antibody mix in advance by diluting antibodies in 1× Permeabilization Buffer. Tables 1, 2, and 3 lists recommended dilutions of three different panels for use with two different flow cytometers. For optimal performance, we recommend titrating the antibodies before use. This is especially important when using a flow cytometer different from those listed. Refer to **Note 16** for further considerations regarding panel design). Incubate for 30 min at 2–8 °C.
16. Add 150 μL 1× Permeabilization Buffer and centrifuge for 5 min at 350 × *g*. Discard the supernatant.

17. Wash cells in 200 μL 1× Permeabilization Buffer, centrifuge (500 × *g* for 5 min), and discard supernatant.
18. Resuspend cells in 100 μL PBS containing 5% FCS.
19. Analyze samples on the flow cytometer.

3.3.2 Gating Strategy to Identify MAIT Cells and Analyze Their Activation

1. Exclude doublets by gating on singlets using the parameters FSC-A and FSC-H (Fig. 8).
2. Identify lymphocytes by size (FSC-A) and granularity (SSC-A).
3. Exclude dead cells by gating on live/dead marker-negative cells (*see* **Note 15**).

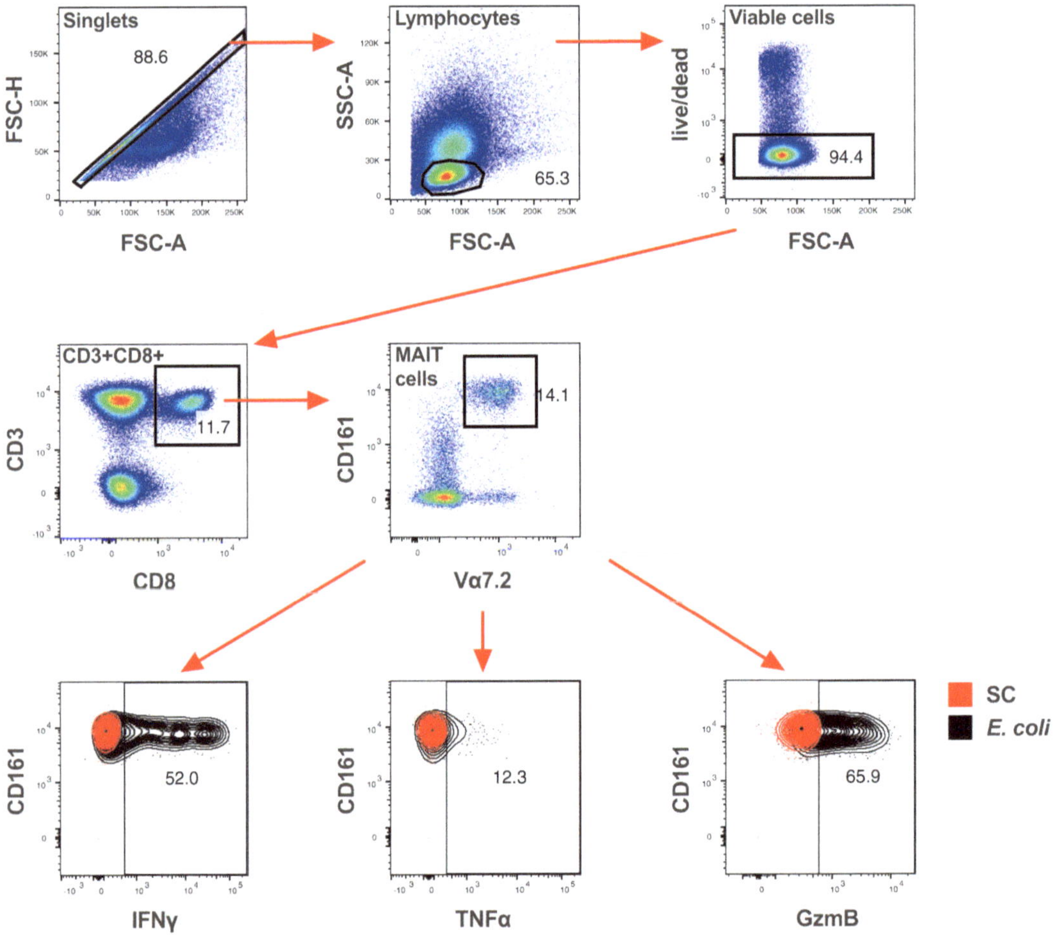

Fig. 8 Gating strategy to identify MAIT cells with their activation profile. The gating strategy is demonstrated on unstimulated (SC, sterility control) and *E. coli*-stimulated PBMCs (24 h). Doublets are first excluded, then lymphocytes are defined by size (FSC-A) and granularity (SSC-A), and viable cells identified by gating on live/dead marker-negative cells. Next, CD3+CD8+ T cells are selected, and MAIT cells are identified as CD161+ +Va7.2+ cells. Finally, the activation profile of MAIT cells is determined by analyzing the expression of IFNγ, TNFα, and GzmB

4. Gate on CD3+CD8+ T cells. Optional: If the staining panel includes CD4, CD4+ T cells can be excluded to identify CD3+CD4-CD8+ T cells for further analysis. Depending on the experimental question, additional cell populations can be identified and analyzed, including CD3+CD4+CD8−, CD3+CD4−CD8−, and CD3+CD4+CD8+ T cells (*see* **Note 17**).
5. Plot CD161 vs. Vα7.2 TCR. In the CD3+CD8+ T cell gate, two CD161+ populations can be distinguished, a CD161 intermediate population (CD161+) and a CD161 bright population (CD161++). MAIT cells are identified as CD161++Vα7.2+ T cells (*see* **Note 18**).
6. Within the MAIT cell gate, the expression of cytokines, activation markers, or other readouts can be analyzed, for example, IFNγ, TNFα, GzmB, or CD69 expression, depending on the antibody staining panel.

4 Notes

1. The manufacturer recommends using 100 μL of anti-CD3-Biotin, 100 μL anti-CD28-Biotin, and 100 μL of anti-CD2-Biotin to load the Anti-Biotin MACSiBead Particles. For better comparability with other assays and work from other groups, we load the beads only with anti-CD3 and anti-CD28 antibodies and substitute the anti-CD2 antibody with 100 μL MACS Buffer. This approach successfully activates T cells including MAIT cells. If your needs and experimental setup require, 100 μL MACS Buffer can be replaced with 100 μL anti-CD2-Biotin.
2. Cell numbers can be adapted depending on the needs and experimental setting. We recommend culturing 0.25–2 million (PBMCs) or 0.1–1 million (CD8+ T cells) per well of a 96-well U-bottom cell culture plate, depending on required cell numbers for analysis and on assay duration. For stimulations of >5 days, we recommend not to exceed 0.25 million cells per well. For an overnight stimulation, cell numbers of up to 2 million cells per well can be used. If the experimental design allows, high cell densities can be accommodated for a longer time period by replacing the media and any stimuli periodically.
3. We usually culture cells in a final volume of 200 μL per well of a 96-well U-bottom cell culture plate. If any additional reagents are added, make sure to adapt the cited volumes of cell suspension and reagents to obtain a final culture medium volume of 200 μL.
4. Increasing concentrations of PMA and Ionomycin correlates with increased cell death of MAIT cells, most likely due to activation-induced cell death.

5. Incubation time can be varied depending on the experimental question. Time courses from 0 h up to 120 h can be performed. Nutrient deprivation and accumulation of harmful metabolites can compromise long-term experiments of >5 days. If the experimental question allows, the media (and if necessary, any stimuli) should be replaced periodically from day 5. As mentioned in **Notes 2**, 7, and **8**, it is important to carefully consider cell numbers.
6. In our standard protocol we use anti-CD3 at a concentration of 2.5 μg/mL in combination with soluble anti-CD28 at a concentration of 1 μg/mL. Depending on the experimental question, the efficacy of other antibody concentrations should be explored. For example, co-stimulation with other antibodies, cytokines, or other stimuli could require suboptimal stimulation with antibody concentrations as low as 0.1–0.5 μg/mL. Stronger responses could possibly be reached using higher antibody concentrations with up to 10 μg/mL. Please note that overstimulation can lead to increased cell death most likely due to activation-induced cell death. In addition, high anti-CD3/anti-CD28 antibody concentrations can lead to downregulation of the TCR which can make subsequent analysis via flow cytometry difficult when staining for TCR-associated antigens.
7. When using a plate-bound approach of anti-CD3 antibodies with soluble CD28 antibodies to stimulate MAIT cells, we recommend a cell density of 0.1–0.5 million PBMCs or CD8 + T cells per well of a 96-well flat-bottom cell culture plate. Efficacy of lower and higher cell densities should be titrated for particular experimental questions. Very low cell densities could compromise experiments due to insufficient cell-to-cell contact. Higher cell densities could compromise experiments due to insufficient contact of the cells with anti-CD3 adsorbed to the well surface.
8. When culturing CD8+ T cells in the presence of THP-1cells, we recommend cell densities of 0.2 million CD8+ T cells and 0.1 million THP-1 cells per well (CD8+ T cell to THP-1 ratio of 2:1). Other cell densities and ratios can be tested for particular experimental needs. Depending on the duration of the experiment, we recommend not to use less than 0.1 million CD8+ T cells and 0.05 million THP-1 cells per well or exceed 2 million total cells per well in overnight assays.
9. When stimulating MAIT cells by using 5-OP-RU in the presence of APCs, we recommend using a 5-OP-RU concentration of 10 nM for optimal stimulation. Certain experimental questions can require other 5-OP-RU concentrations, for example, for suboptimal stimulation. Therefore, a range of concentrations (e.g., 1 pM–1 μM) should be tested and titrated for their efficacy before use (Fig. 3).

10. We recommend adding one bacterial colony per 10 mL LB Broth medium. The total volume of bacterial culture can be scaled up or down depending on the final number of bacteria required. Remember to adjust Erlenmeyer flask size to the culture volume to allow optimal oxygen supply to the bacteria. We recommend an Erlenmeyer flask capacity to LB Broth volume ratio of 5:1.
11. MAIT cells can be stimulated in an MR1-dependent manner using PFA-fixed *E. coli* in a model with PBMCs (model 1) or with isolated CD8+ T cells in combination with THP-1 cells (model 2). In model 1, PFA-fixed *E. coli* are processed by APCs contained within the PBMCs. MAIT cells become activated by *E. coli*-derived riboflavin metabolites presented by MR1 on APCs, in combination with cytokines, such as IL-12 + IL-18, produced by the activated APCs. In model 2, PFA-fixed *E. coli* are processed by THP-1 cells that activate MAIT cells within isolated CD8+ T cells via MR1 and cytokines. THP-1 cells are more efficient APCs compared to those contained within PBMCs, and frequencies of activated MAIT cells are usually higher in model 2 compared with model 1.
12. We recommend performing a pilot experiment with each fresh batch of PFA-fixed *E. coli* to test their potency to activate MAIT cells. The dose–response should be investigated at a range of 0–100 CFU per cell in steps of 10 CFU and smaller steps of 5 CFU at <20 CFU. The CFU required for maximal activation (we recommend staining for intracellular IFNγ expression) should be fairly consistent and is in our hands around 10 CFU per PBMC and 30 CFU per THP-1 cell. As the PFA-fixed *E. coli* degrade over time, their potency to activate MAIT cells decreases. Therefore, we recommend producing new batches frequently. The fixed bacteria should be reasonably stable for up to 8 weeks at 2–8 °C.
13. When stimulating MAIT cells with the cytokines IL-12 and IL-18, we recommend a concentration of 50 ng/mL each for optimal stimulation. Certain experimental questions require simulation with suboptimal IL-12 + IL-18 concentrations as low as 1–10 ng/mL. Excessively high cytokine concentrations greater than 50 ng/mL can lead to increased cell death through overstimulation. The efficacy of other concentrations should be tested and titrated for their specific purpose before use (Fig. 7).
14. We recommend a TLR4 agonist (*E. coli* K12 LPS) or TLR8 agonist (ssRNA40/Lyovec) concentration of 1 μg/mL to stimulate MAIT cells indirectly within CD8+ T cells plus THP-1 cells or within PBMCs, respectively. Depending on your experimental needs, TLR agonist concentration can be

adapted and further titrated. The manufacturer InvivoGen recommends a range of 10 ng–10 μg/mL for *E. coli* K12 LPS, and 0.25–10 μg/mL for ssRNA40/Lyovec.

15. The antibody staining panel can be extended by using the channel of the viability dye as a dump channel. The dump channel includes cells which are not of interest and could contaminate downstream analysis, for example, pan-γδTCR or CD14. By adding additional markers of the same color as the viability marker to the antibody staining mix, those cell populations can be merged in one channel. During analysis, dead cells and unwanted cell populations can be excluded together by gating on dump channel fluorochrome-negative cells.
16. Strong MAIT cell activation can lead to downregulation of CD3 and Vα7.2 TCR expression, which makes subsequent analysis difficult. In such cases, intracellular staining of CD3 and Vα7.2 TCR can rescue some of the CD3 and Vα7.2 TCR staining intensity. To stain for those markers intracellularly, include the CD3 and Vα7.2 TCR antibodies in the antibody staining mix at **step 15** instead of **step 7**.
17. Most MAIT cells are CD3+CD4−CD8+, but subsets have been identified within other CD3+ T cell populations including CD4-CD8-, CD4+CD8−, and CD4+CD8+ [9].
18. In flow cytometric analysis, MAIT cells are usually defined by high expression of the surrogate marker CD161 in combination with Vα7.2 TCR. Alternative surrogate markers can be used for MAIT cell identification, including CD26 and IL18Rα, but are less reliable than CD161 [9]. MR1-Ag tetramers are now freely available from the NIH for accurate MAIT cell identification. Use of the MR1 tetramer is particularly important in populations where MAIT cell frequency is low or where staining with surrogate markers is not very reliable, e.g., CD4+ T cells [9].

Acknowledgments

This research was supported by the Wellcome Trust (WT109965MA), NIHR Senior Fellowship (PK), NIHR Biomedical Research Centre, Oxford, NIH U19 I082360, Chinese Scholarship Council (TL). The authors wish to acknowledge the BRC Oxford GI Biobank and BRC Oxford IBD Cohort in collecting and making the samples/data available that were used in the generation of this publication.

References

1. Chua W-J, Truscott SM, Eickhoff CS et al (2012) Polyclonal mucosa-associated invariant T cells have unique innate functions in bacterial infection. Infect Immun 80:3256–3267
2. Godfrey DI, Uldrich AP, Mccluskey J et al (2015) The burgeoning family of unconventional T cells. Nat Immunol 16:1114–1123
3. Gao Y, Williams AP (2015) Role of innate T cells in anti-bacterial immunity. Front Immunol 6:302
4. Porcelli S, Yockey CE, Brenner MB et al (1993) Analysis of T cell antigen receptor (TCR) expression by human peripheral blood CD4-8- alpha/beta T cells demonstrates preferential use of several V beta genes and an invariant TCR alpha chain. J Exp Med 178:1–16
5. Gold MC, Cerri S, Smyk-Pearson S et al (2010) Human mucosal associated invariant T cells detect bacterially infected cells. PLoS Biol 8:e1000407
6. Le Bourhis L, Martin E, Péguillet I et al (2010) Antimicrobial activity of mucosal-associated invariant T cells. Nat Immunol 11:701–708
7. Turtle CJ, Delrow J, Joslyn RC et al (2011) Innate signals overcome acquired TCR signaling pathway regulation and govern the fate of human CD161hi CD8 + semi-invariant T cells. Blood 118:2752–2762
8. Wencker M, Turchinovich G, Di Marco Barros R et al (2014) Innate-like T cells straddle innate and adaptive immunity by altering antigen-receptor responsiveness. Nat Immunol 15:80–87
9. Gherardin NA, Souter MN, Koay H-F et al (2018) Human blood MAIT cell subsets defined using MR1 tetramers. Immunol Cell Biol 96:507–525
10. Walker LJ, Kang YH, Smith MO et al (2012) Human MAIT and CD8αα cells develop from a pool of type-17 precommitted CD8+T cells. Blood 119:422–433
11. Ussher JE, Bilton M, Attwod E et al (2014) CD161++CD8+T cells, including the MAIT cell subset, are specifically activated by IL-12 +IL-18 in a TCR-independent manner. Eur J Immunol 44:195–203
12. Martin E, Treiner E, Duban L et al (2009) Stepwise development of MAIT cells in mouse and human. PLoS Biol 7:e1000054
13. Lanier LL, Chang C, Phillips JH (1994) Human NKR-P1A. A disulfide-linked homodimer of the C-type lectin superfamily expressed by a subset of NK and T lymphocytes. J Immunol 153:2417–2428
14. Battistini L, Borsellino G, Sawicki G et al (1997) Phenotypic and cytokine analysis of human peripheral blood gamma delta T cells expressing NK cell receptors. J Immunol 159:3723–3730
15. Takahashi T, Dejbakhsh-Jones S, Strober S (2006) Expression of CD161 (NKR-P1A) defines subsets of human CD4 and CD8 T cells with different functional activities. J Immunol 176:211–216
16. Exley M, Porcelli S, Furman M et al (1998) CD161 (NKR-P1A) costimulation of CD1d-dependent activation of human T cells expressing invariant V alpha 24 J alpha Q T cell receptor alpha chains. J Exp Med 188:867–876
17. Fergusson JRR, Smith KEE, Fleming VMM et al (2017) CD161 defines a transcriptional and functional phenotype across distinct human T cell lineages. Cell Rep 9:1075–1088
18. van Wilgenburg B, Scherwitzl I, Hutchinson EC et al (2016) MAIT cells are activated during human viral infections. Nat Commun 7:11653
19. Zinser ME, Highton AJ, Kurioka A et al (2018) Human MAIT cells show metabolic quiescence with rapid glucose-dependent upregulation of granzyme B upon stimulation. Immunol Cell Biol 96:666–674
20. Leng T, King T, Friedrich M, et al. (2018) TCR and inflammatory signals tune human MAIT cells to exert specific tissue repair and effector functions. bioRxiv 475913
21. Slichter CK, McDavid A, Miller HW et al (2016) Distinct activation thresholds of human conventional and innate-like memory T cells. JCI insight 1:e86292
22. Tang X-Z, Jo J, Tan AT et al (2013) IL-7 licenses activation of human liver intrasinusoidal mucosal-associated invariant T cells. J Immunol 190:3142–3152
23. Bennett MS, Trivedi S, Iyer AS et al (2017) Human mucosal-associated invariant T (MAIT) cells possess capacity for B-cell help. J Leukoc Biol 102:1261–1269
24. Corbett AJ, Eckle SBG, Birkinshaw RW et al (2014) T-cell activation by transitory neo-antigens derived from distinct microbial pathways. Nature 509:361–365
25. Eckle SBG, Birkinshaw RW, Kostenko L et al (2014) A molecular basis underpinning the T cell receptor heterogeneity of mucosal-associated invariant T cells. J Exp Med 211:1585–1600
26. Mak JYW, Xu W, Reid RC et al (2017) Stabilizing short-lived Schiff base derivatives of

5-aminouracils that activate mucosal-associated invariant T cells. Nat Commun 8:14599
27. Dias J, Sobkowiak MJ, Sandberg JK et al (2016) Human MAIT-cell responses to Escherichia coli: activation, cytokine production, proliferation, and cytotoxicity. J Leukoc Biol 100:233–240
28. Kurioka A, Ussher JE, Cosgrove C et al (2015) MAIT cells are licensed through granzyme exchange to kill bacterially sensitized targets. Mucosal Immunol 8:429–440
29. Bykowski T, Stevenson B (2008) Aseptic technique. Curr Protoc Essent Lab Tech 11: A.4D.1–A.4D.11
30. Sanders ER (2012) Aseptic laboratory techniques: plating methods. J Vis Exp:e3064
31. Prusokas A, Hawkins M, Nieduszynski C, et al (2018) The effectiveness of glass beads for **plating cell cultures. bioRxiv 241752
32. Jo J, Tan AT, Ussher JE et al (2014) Toll-like receptor 8 agonist and bacteria trigger potent activation of innate immune cells in human liver. PLoS Pathog 10:e1004210

Chapter 8

Interactions Between MAIT Cells and Dendritic Cells

Mariolina Salio and Vincenzo Cerundolo

Abstract

Mucosal-associated invariant T (MAIT) cells recognize intermediates of the vitamin B2 biosynthetic pathway present in a variety of bacteria, presented by the monomorphic MR1 molecules. Because of their central role in shaping adaptive immunity through interaction with dendritic cells (DCs) and B cells, their manipulation can be of translational relevance. We describe a method to routinely isolate and maintain MAIT cells from peripheral blood and to investigate their activity using DC as targets.

Key words MAIT, MR1 dendritic cells, CD40L

1 Introduction

In the past 25 years it has become clear that lymphocyte populations bearing TCR with limited diversity, recognizing non-peptide antigens presented by monomorphic molecules and with a memory-like phenotype, orchestrate adaptive immunity through dendritic cell (DC) maturation [1]. These lymphocyte populations are collectively referred to as unconventional T cells, and include γδ T cells with various restrictions and MHC class Ib-restricted T cells. Among these, the best characterized populations are MR1-restricted MAIT cells and CD1d-restricted iNKT cells.

MAIT cells are an abundant lymphocyte population, which represents up to 10% of circulating CD3 T cells, and are enriched at mucosal sites and in the liver. MAIT cells are identified by co-expression of the C-type lectin CD161 and a semi-invariant T cell receptor (TCR), composed of the Vα7.2 chain, jointed preferentially to Jα33 or Jα20 segments and paired with a limited number of β chains, Vβ2 and Vβ13 among the most frequent. MAIT cells recognize unstable adducts derived from intermediates in the riboflavin (vitamin B2) pathway, and present in a number of bacterial and fungal species but not made by mammalian cells [2]. Recent work has highlighted details of MR1-restricted antigen presentation [3], and the biological relevance of MAIT cells in microbial

Helen Kaipe and Isabelle Magalhaes (eds.), *MAIT Cells: Methods and Protocols*, Methods in Molecular Biology, vol. 2098, https://doi.org/10.1007/978-1-0716-0207-2_8,

infections has been extensively reviewed [4]. In addition, like other innate lymphocytes, MAIT cells can be activated by IL-12 and IL-18 independently of TCR engagement [5], and a role of this pathway during several viral infections has been demonstrated [6].

We have recently reported that human MAIT cells induce CD40L-dependent dendritic cell (DC) maturation, both ex vivo and upon in vitro culture [7]. This observation opens the avenue for therapeutic manipulation of MAIT cells to enhance antigen-specific adaptive immunity at mucosal sites.

2 Materials

The protocols described require laboratories equipped with a tissue culture facility and authorized to use blood and blood products. Good aseptic techniques are required for long-term maintenance of cells in culture.

General plasticware required: Tissue culture plates (96 wells, round- and flat-bottomed; 24, 12, and 6 wells; T75 flasks); tissue culture tubes (5, 15, and 50 mL, polypropylene); tissue culture, sterile 0.5 mL tubes.

Instruments: Flow cytometer and cell sorter, ELISA plate reader.

2.1 Lymphocyte Isolation

1. Human blood: can be purchased from designated blood banks or drawn from work colleagues. We prefer using the blood bank as supplier as the number of cells available for manipulation is far superior.
2. Washing medium: RPMI buffered with 10 mM Hepes.
3. Density gradient medium for mononuclear cell isolation.
4. Complete medium: RPMI supplemented with 10% fetal calf serum (FCS, tested batches), 1% sodium pyruvate, 1% non-essential amino acids, 10 mM Hepes, 2 mM glutamine, 1% pen/strep, and 50 μM 2-mercaptoethanol.
5. Magnetic cell separation: human CD14 magnetic MACS beads, LS columns, smart strainers (to remove cell aggregates), and magnet.
6. MACS buffer: PBS, 10% FCS, and 2 mM EDTA, cold (*see* **Note 1**).

2.2 Monocyte Differentiation

1. Hemocytometer.
2. DC medium: complete medium supplemented with 1000 U/mL recombinant human IL-4, 50 ng/mL recombinant human GM-CSF (*see* **Note 2**).
3. Freezing medium: 90% FCS, 10% DMSO, sterile filtered.

4. Magnetic cell separation: human CD2 magnetic MACS beads, LS columns, smart strainers (to remove cell aggregates), and magnet.

2.3 CD2 Purification and Cell Sorting

1. T cell medium: IMDM supplemented with 5% pooled human serum (from the blood bank, or AB male serum), 500 U/mL recombinant human IL-2, 1% sodium pyruvate, 1% non-essential amino acids, 10 mM Hepes, 2 mM glutamine, 1% pen/strep, and 50 μM 2-mercaptoethanol (*see* **Note 2**).
2. 5-A-RU: Toronto Research chemicals (A629245). Resuspend at 10 mg/mL in DMSO, protect from light, and store in small aliquots at −80 °C.
3. Methylglyoxal (MG) 40% in solution water.
4. Acetyl-6-formylpterin (Ac6FP): Schirks Laboratories (cat 11.418). Resuspend at 10 mg/mL in DMSO, protect from light, and store in small aliquots at −80 °C.
5. 10 mM Hepes-buffered RPMI.
6. Antibodies for cell sorting (*see* Table 1).

2.4 Purity Characterization

1. Antibodies for flow cytometry (*see* Table 1).
2. MACS buffer.

2.5 MAIT Cell Activation

1. Anti-MR1, CD40L, or IL-12p40/70 blocking antibodies (*see* Table 1 for clone numbers).
2. 10 mM Hepes-buffered RPMI.
3. RPMI 10% FCS.
4. 5-A-RU.
5. Methylglyoxal (MG) 40% in solution water.

2.6 MAIT Cell Activation with Bacterial Antigens

1. Bacteria producing vitamin B2 metabolites (i.e., *E. coli* DH5α).
2. Luria Broth (LB) for bacterial growth.
3. Gentamycin for killing extracellular bacteria after infection of antigen presenting cells.
4. Sterile filters (0.2 μm).
5. Anti-MR1, IL-12p40/70 blocking antibodies (*see* Table 1 for clone numbers).

2.7 ELISA

1. ELISA: antibody pairs (IFNγ and IL-12p40 and p70) for capture and detection, standards, resuspended in PBS–10% FCS and stored in small aliquots at −80 °C.
2. ELISA plates: 96-well half-area plates.
3. Coating buffer: 0.1 M $NaHCO_3$ pH ≅ 8.
4. Washing buffer: PBS, 0.1% Tween-20.

Table 1
List of antibodies

Specificity	Clone	Fluorophore	Company
CD3	UCHT1	APC ef780	eBioscience
CD3	UCHT1	BUV660	BD
CD4	OKT4	BV650	BioLegend
CD8	RPA-T8	BV711	BioLegend
CD161	HP-3G10	APC or BV510	BioLegend
Vα7.2	3C10	BV605 or PE	BioLegend
Vβ13.1	H131	PE	BioLegend
Vβ2	REA654	FITC	Miltenyi
CD14	M5E2	BV510	BioLegend
CD19	HIB19	BV510	BioLegend
CD40L	TRAP1	FITC	BD
CD80	2D10	BV421	BioLegend
Live/dead		Aqua, or near IR	BioLegend
MR1	26.5	Purified, PE, or APC	BioLegend
CD40L	24.31	Purified for blocking	BioLegend
IL12p70	C8.6	Purified for blocking	BD
Isotype controls		Purified	BioLegend
CD83	HB15e	FITC	BD
CD86	FUN1	APC or BB515	BD
CD80	L307.4	PE	BD
PDL1	29E.2AE		
CD25	M-A251	PE-CF594	BioLegend
Brilliant violet stain buffer			BD
HLA-DR	L243	BV421 or APC Cy7	BioLegend
CD2	RPA2.10	FITC	BioLegend
CD123	7G3	BV785	BioLegend
IFNγ	4S.B3	BUV737	BD
TNFα	Mab11	BUV395	BD
CD33	P67.6	PE or APC or BB630	BioLegend or BD
CD137	VI C-7	PE/Dazzle	BioLegend
IL-12p40	C11.5	PE	BioLegend

(continued)

Table 1
(continued)

Specificity	Clone	Fluorophore	Company
CD11b	M1/70	Alexa700	BioLegend
CD16	3G8	BV650	BioLegend
CD1c	L161	PECy7	BioLegend
BDCA2	201A	APC	BioLegend
CD40	5C3	BV711	BioLegend
CD141	M80	PerCPCy5.5	BioLegend
CD38	HIT2	BV785 or BB790	BioLegend or BD
Gammanorm (Fc block)			Octopharma
Foxp3 fix/perm buffer set			BioLegend

5. Blocking buffer: PBS, 10% FCS.
6. Avidin-peroxidase (1 mg/mL in water).
7. o-Phenylenediamine dihydrochloride tablets (5 or 10 mg).
8. ELISA developing solutions: 10 mM citric acid monohydrate, 250 mM Na_2HPO_4—$2H_2O$.
9. 30% H_2O_2.
10. 2N H_2SO_4.

2.8 Whole Blood Assay

1. Brefeldin or Golgistop for detection of intracellular cytokines by flow cytometry.
2. Red blood cell lysis solution.
3. Complete medium.

3 Methods

We routinely isolate MAIT cells and monocytes from healthy blood donors and maintain MAIT cells in culture over several weeks. We differentiate monocytes to DC over a 5-day culture period. Because of lack of allo-reactivity, MAIT cells can be assayed against autologous and allogeneic DC. This protocol can be easily adapted to large-scale screening of libraries of compounds with potential MAIT cell agonist activity.

3.1 Lymphocyte Isolation

1. Dilute blood with room temperature 10 mM Hepes-buffered RPMI (1:2 if donation up to 50 mL, 1:12 for leukocyte cones), in T75 flasks. We routinely buy ~10 mL leukocyte cones, residues from approximately 500 mL blood donations, and dilute them in 120 mL, then distribute in four Falcon tubes.

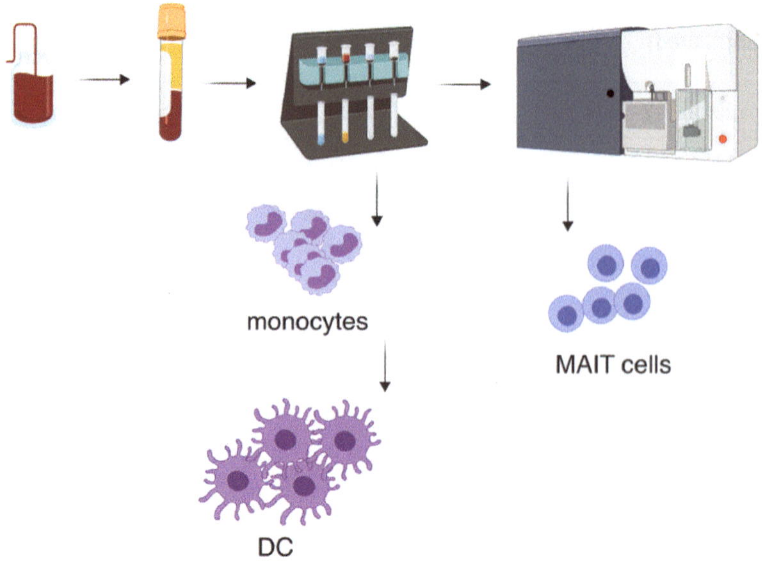

Fig. 1 Schematic diagram of the protocol. Blood is separated over lymphocyte separation medium, and monocytes are purified with magnetic beads and differentiated into DC; MAIT cells are sorted from the negative fraction collected after monocyte purification

2. In 50 mL Falcon tubes, carefully layer 30 mL diluted blood over 15 mL Lymphoprep™ or suitable density gradient medium for the isolation of mononuclear cells.
3. Centrifuge tubes 30 min at 400 g *without brake*, at room temperature.

   Carefully collect the mononuclear cells, at the interface between the plasma and the lymphocyte separation medium (Fig. 1). If starting with up to 50 mL blood, pool the mononuclear cells in one 50 mL Falcon tube; if starting from a leukocyte cone, distribute them over two Falcon tubes.
4. Wash two times with 10 mM Hepes-buffered RPMI, 100 g, room temperature, 15 min each spin.
5. Proceed immediately to monocyte separation:

   After the last spin, resuspend cells in 3 mL MACS buffer and add 200 μL CD14 MACS beads (if using leukocytes cones, i.e., 600–800 × 10^6 cells; otherwise scale proportionally the amounts of beads).
6. Incubate 16 min in the fridge, with gentle shaking after 8 min. Do *NOT* vortex.
7. Wash unbound beads with excess MACS buffer, and spin 10 min, 100 g at room temperature (*see* **Note 3**).
8. Resuspend cells in 3 mL MACS buffer and apply through a MACS strainer over one (if starting from 50 mL bloods) or two

(if starting from leukocyte cone) LS columns, prewashed with 3 mL MACS buffer. Do not leave the column unattended. Collect flow through ("negative fraction"), wash each column three times with 3 mL MACS buffer, and elute bound cells in 5 mL MACS buffer. Wash both fractions with excess 10 mM Hepes-buffered RPMI.

9. *Optional*: check eluted cells for purity. Pellet (100 g, RT, 3 min) 100 μL of cells and incubate on ice with anti-human CD14 antibody at the recommended concentration (*see* **Note 4**).

3.2 Monocytes Differentiation

1. Count cells eluted in **step 8** in Subheading 3.1 with hemocytometer or automatic cell counter (expect up to 120×10^6 cells, if starting from a leukocyte cone), resuspend in DC medium at $3\text{–}5 \times 10^5$/mL, distribute 3 mL/well of a 6-well plate.
2. Incubate 5 days in a 37°, 5% CO_2 incubator.
3. Monocytes can be frozen in freshly made freezing medium, in 10×10^6 aliquots (1 mL each) and used at different times. Frozen cells should be stored as soon as possible in liquid nitrogen to preserve viability. When preparing DC from a frozen aliquot of monocytes, thaw cells in a water bath and immediately wash in 10 mL DC medium (without cytokines). Resuspend and plate in DC medium containing IL-4 and GM-CSF as per material section. Expect 60–70% of the yield of DC as compared to fresh material.

3.3 CD2 Purification

1. Resuspend cell pellet from negative fraction in **step 8** in Subheading 3.1 in 3 mL MACS buffer and incubate with 200 μL CD2 MACS beads, following the same steps described above for monocytes.

   Expect around $200\text{–}400 \times 10^6$ cells after this step, if starting from a leukocyte cone (*see* **Note 5**).
2. CD2-purified cells can be immediately stained for cell sorting or stored overnight in the fridge in 20 mL complete medium and sorted the following day (this depends on the time of the day you receive the blood, we routinely do the sort the following day).
3. MAIT cell sorting:

   Pellet the CD2-purified cells at 100 g, 10 min, room temperature, and incubate 30 min on ice with antibody mix: Vα7.2 and CD161 antibodies are sufficient to isolate bona fide MAIT cells from adult blood samples (*see* **Note 6**). Depending on the experimental question, antibodies to CD4, CD8, and individual TCR β chains can be added and subsets of MAIT cell separated (*see* **Note 6**).

Depending on the frequency of MAIT cells in the donor, from 200 to 400 × 10^6 cells post CD2 separation expect to recover up to 6–8 × 10^6 MAIT cells. The sorting process requires up to 6 h, depending on the specifications of the machine.

4. At the end of the sort, wash cells two times in 10 mM Hepes-buffered RPMI and resuspend in T cell medium. Distribute cells in 48- or 24-well plates, depending on numbers. As a guide, plate up to 3 × 10^6 cells per mL, and place in a 37°, 5% CO_2 incubator. Feed cells every other day by replacing half of the medium, split wells only if cells become confluent. Like T cells, MAIT cells survive best when reaching around 80% confluence. MAIT cells can be used for functional assays around 5 days after sorting. Depending on numbers sorted, viable cultures of MAIT cells can be maintained over several months. When numbers decline, wells can be pooled to maintain high density. We do not recommend periodic re-stimulation of these polyclonal lines with irradiated allogeneic PBMC and PHA, but we recommend to check purity periodically (*see* **Note 6**).
5. *Alternative protocol:* MAIT cells can be expanded directly from mononuclear cells with 5-A-RU and MG. Resuspend 5-A-RU at 5-10 mg/mL in DMSO, store in small aliquots at −80°, and protect from light. Distribute PBMC at 5 × 10^6/well, 2 mL/well in complete medium and titrate 5-A-RU from 10 μM, with 100 μM MG. Feed cultures every other day for 2 weeks, check MAIT cell frequency and proceed with sorting as described above.

3.4 Purity Characterization

1. After 5 days culture, collect a sample of DC and of MAIT cells, pellet in a 96-well plate for FACS staining. All washes in this protocol can be done at 100 g, 3 min.
2. Wash cells once in PBS and stain with live/dead staining (titrated according to the manufacturer's instructions), 20 min at room temperature.
3. Wash cells once in MACS buffer and incubate pellets with the desired antibody mix (*see* Table 1 for suggested antibodies) on ice for 30 min.
4. Wash cells once in MACS buffer, acquire at the cytometer, with the appropriate controls (*see* **Note 6**).

3.5 MAIT Cell Activation

1. Collect DC and MAIT cells in 15 mL tubes, wash once with 10 mM Hepes-buffered RPMI, resuspend in 3 mL RPMI–10% FCS, and count cells using an automated cell counter or a hemocytometer.

2. Resuspend DC and MAIT at 5×10^5 cells/mL and 3×10^6 cells/mL, respectively, and distribute 100 μL of DC in the desired number of wells of a 96-well flat-bottom plate.
3. Add MAIT cell ligands at the desired concentration, 10 μL of a 20× final concentration/well. We routinely do the assay in triplicates, over a range of 5-A-RU concentrations, from 1 μg/mL to 1 ng/mL, in threefold serial dilutions. MG is kept constant at 50 μM. If needed, add anti-MR1, CD40L, or IL-12 blocking antibodies (or relevant isotype controls) at 20 μg/mL final concentrations 30 min before adding MAIT cells (*see* **Note 7**).
4. Add MAIT cells in 100 μL/well and incubate for the desired length of time in a 37°, 5% CO_2 incubator. When monitoring under the microscope, clusters will be observed between DC and MAIT cells from 6 h. By 24 h, at the highest antigen dose, the majority of DC would have been killed by MAIT cells (*see* **Note 8**).

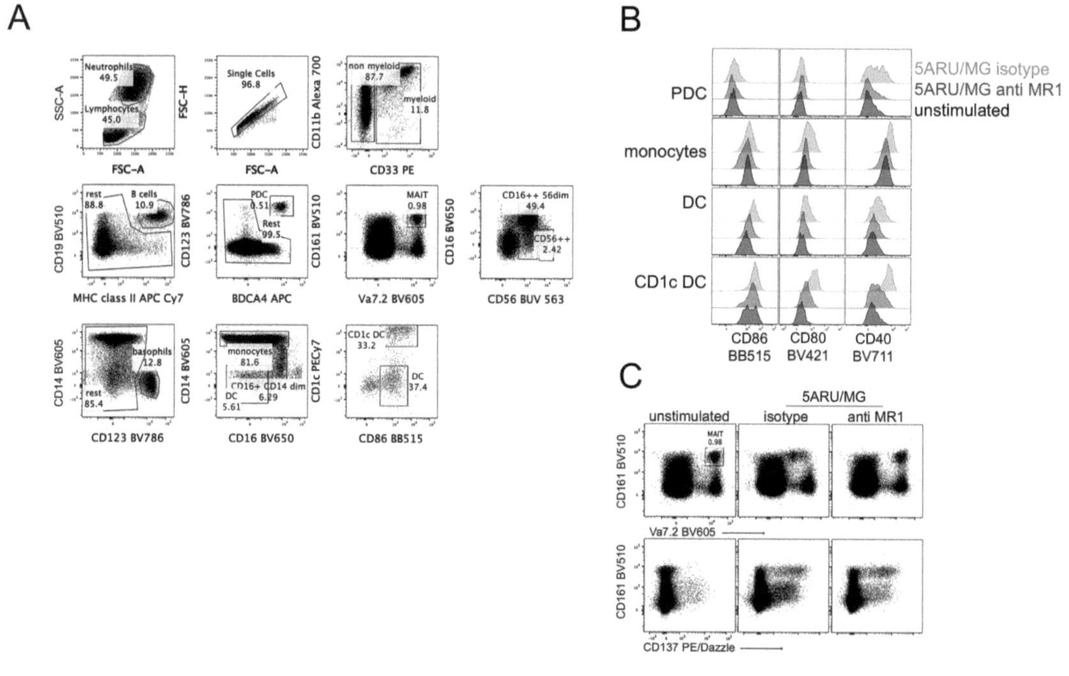

Fig. 2 Crosstalk between MAIT cells and myeloid populations in whole blood. (**a**) Gating strategy of different myeloid and lymphoid populations by multicolor flow cytometry. (**b**) Activation of monocytes, plasmacytoid DC and DC subsets in whole blood samples exposed to the MAIT cell ligand 5-A-RU. Histogram overlays of unstimulated cells (black) and cells activated in the presence of 5-A-RU and MG with isotype control (light gray) or anti-MR1 blocking antibody (dark gray). (**c**) Activation of MAIT cells in the same samples, determined by TCR Vα7.2 downregulation and CD137 upregulation. The same donor is shown in all panels, gating done according to Ref. [7]. Panels **a** and **c** are the same donor shown in **b**

5. Harvest supernatants (180 μL) and freeze at −20 °C or perform ELISA immediately.
6. *Optional:* After collecting supernatants, wash the remaining cells in PBS, stain with live/dead cell staining as described in **step 1** in Subheading 3.4 and proceed with staining to determine extent of MAIT and DC activation.

   MAIT and DC are easily separated by FSC/SSC; however, to account for size cell increase upon activation, a CD2 antibody can be included in the mix. We routinely use CD69, CD25, CD137, PDL1, CD83 antibodies to assess cell activation. Despite different autofluorescence and cell size, information on both DC and MAIT cell activation can be gained from a single stained sample (Fig. 2) (*see* **Note 9**).

3.6 MAIT Cell Activation with Bacterial Antigens

Several bacterial species, commensal and pathogenic, are known to produce intermediates of the vitamin B2 as a source of MAIT cell agonists (an initial list was provided in [2]). These antigens accumulate intracellularly and can also reach a stimulatory concentration in bacterial supernatants, and both can be used in stimulation experiments (*see* **Note 10**).

1. Grow DH5a *E. coli* (or any bacterial species in its optimal growth medium) overnight at 37° with shaking (200 rpm) in 5 mL LB broth from a single colony.
2. Next day, pellet bacteria and sterile filter supernatant. This can be titrated on DC as a source of antigen. A parallel LB culture should be used as "unstimulated" controls as antigenic vitamins are present in LB.
3. Wash bacterial pellets three times in PBS and dilute to have an OD_{600} of 0.5, corresponding to approximately 4×10^7 bacteria/mL. Live bacteria can be added to DC, at MOI 5:1, 2:1, 1:1 for 3 h in antibiotic-free medium. Remaining extracellular bacteria are then killed with 50 μg/mL gentamycin during the overnight culture. Blocking MR1 and IL12-p70 antibodies can be added at 20 μg/mL 1 h after gentamycin.

   Alternatively, bacteria can be UV inactivated at 254 nm for 10 min and added to DC in the presence of antibiotics.
4. Cells are washed three times in complete medium after 16 h; MAIT cells are added, and supernatant and cells are harvested after further 24 h for analysis, as described in **steps 5** and **6** in Subheading 3.5.

3.7 ELISA

1. Coat ELISA plates with 25 μL of capture antibody/well diluted in coating buffer at a final concentration of 4 μg/mL. Incubate overnight at 4° or 2 h at 37 °C.
2. Wash plates four times in PBS 0.1% Tween-20, block a minimum of 20 min with 200 μL PBS–10% FCS.

3. Flick buffer and add 25 μL supernatant/well, either diluted or undiluted. Have a standard curve with at least eight points, starting at the concentration recommended by the company supplying the antibody pair. Include blank wells with buffer and medium only. Incubate overnight at 4° or 2 h at 37°.
4. Wash plates four times in PBS 0.1% Tween-20, add 25 μL biotinylated antibody/well (1 μg/mL final, in PBS 10% FCS). Incubate 60–90 min at room temperature, under foil.
5. Wash plates six times in PBS 0.1% Tween-20, add 25 μL avidin peroxidase/well (2.5 μg/mL final, in PBS 10% FCS). Incubate 45 min at room temperature, under foil.
6. Wash plates eight times in PBS 0.1% Tween-20, 1× in PBS, tap dry, and develop assay: prepare 5 mL developing solution every two ELISA plates (2.45 mL 10 mM citric acid monohydrate, 2.55 mL 250 mM Na_2HPO_4—$2H_2O$, 20 μL H_2O_2, and a 5 mg tablet of o-phenylenediamine dihydrochloride).
7. Add 25 μL/well of developing solution prepared in **step 6** of Subheading 3.7 and stop development with 25 μL/well 2 N H_2SO_4 when the sixth standard well develops or when blank wells start tainting. Measure color intensity in a plate reader at 490 nm and quantify using the standard curve.

3.8 Whole Blood Assay

Given the high frequency of MAIT cells in healthy donors (up to 10% of $CD3^+$ cells), their function can be easily tested in whole blood, without the need of previous PBMC separation. This offers the advantage of investigating the crosstalk with several primary DC populations, present in the blood at low numbers, and with neutrophils, which are otherwise mostly lost in the gradient separation (Fig. 2).

We have successfully used this protocol with blood freshly drawn from colleagues and incubated with stimuli within 30 min from harvesting.

1. Dilute blood 1:2 in complete medium.
2. Distribute 1 mL/5 mL polypropylene tube. Add the stimuli (synthetic vitamin, bacterial supernatants, or bacteria) at the desired concentration and the required blocking antibodies/isotype controls at the final concentration of 20 μg/mL.
3. Incubate in a 37°, 5% CO_2 incubator overnight. If cytokine production and CD40L upregulation need to be determined, add blockers of ER/Golgi transport (Brefeldin or Golgi Plug) 1 h after stimulation, and maintain overnight.
4. Gently remove plasma layer and freeze for cytokine determination. Transfer the remaining volume to a 15 mL polypropylene tube and add 10 mL RBC lysis for 10 min.

5. Spin 100 g, 10 min. Carefully resuspend pellet in MACS buffer, block 15 min with 20 μg/mL normal immunoglobulins and surface stain with the desired mix of antibodies.
6. To detect CD40L and intracellular cytokines, proceed with fixation and permeabilization, according to the instructions provided with the Foxp3 fix/perm buffer (*see* **Note 11**).

4 Notes

1. We do not use BSA in MACS buffer, as recommended by Miltenyi, as it can contain LPS, which affects DC differentiation.
2. Prepare small (50 mL) aliquots of fresh T cell medium each time, to preserve cytokine concentrations. Likewise, have a bottle of RPMI–10% FCS for routine use, and only supplement the required amounts with GM-CSF/IL-4.
3. Before using MACS technology for the first time, read the data sheet provided by the manufacturer; our protocol is an adaptation of the original protocol and has consistently given purity over 95%.
4. For better results, use a clone of CD14 antibody different from the beads-conjugated CD14. Compare a sample pre, post, and the negative fraction of the column. Expect a purity around 95%. It is common to have a sizeable proportion of monocytes left in the negative fraction because of saturation of the column.
5. LS columns have a capacity of 1×10^8 magnetically labeled cells from up to 2×10^9 cells. We find that some labeled monocytes are not collected and will be isolated during the second separation, together with CD2+ cells. As we proceed to FACS sorting MAIT cells, this does not represent an issue; however, it should be taken into account when considering purity. For this reason, it is also essential to purify monocytes first.
6. MAIT cell frequency varies among donors, hence we routinely order 2–4 leukocyte cones and choose the one with the highest MAIT cell frequency for separation (Fig. 3). It is possible to test MAIT cell frequency on whole blood and proceed at the same time with the mononuclear cell purification.

   Method: mix 1 mL of diluted blood with 10 mL red blood cell lysis solution (we use Qiagen catalog 158904), 10 min at room temperature (or until a clear color is reached). Spin 10 min, 100 g, remove supernatant, and add antibody cocktail listed in Table 2 to cell pellet. Keep some unstained control cells and some cells for compensation controls.

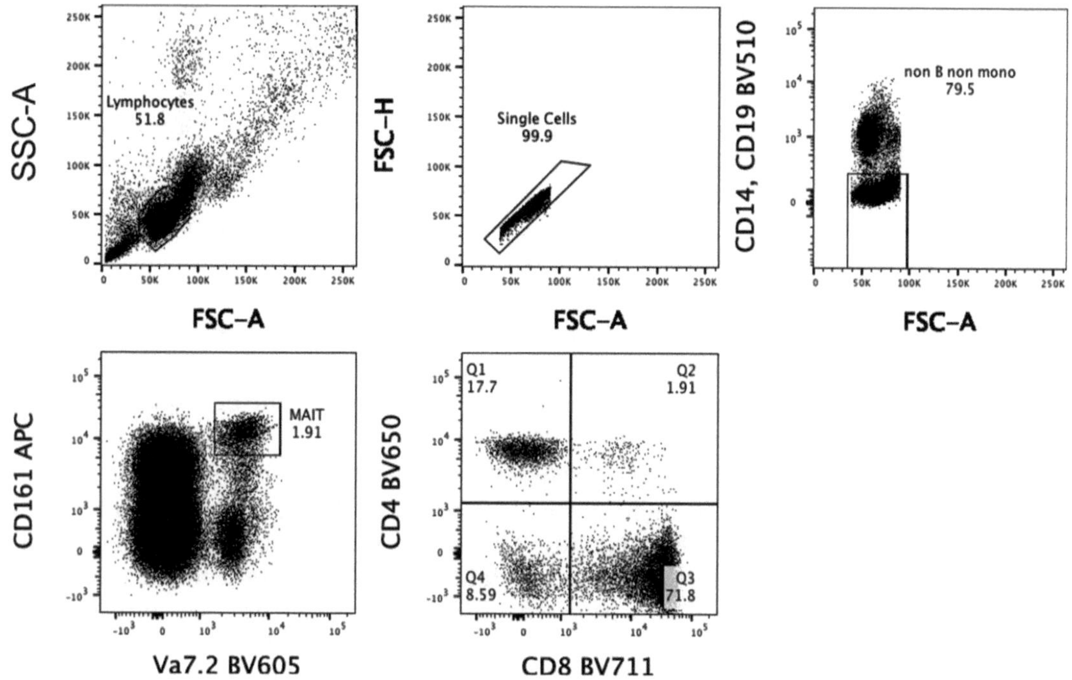

Fig. 3 Determination of MAIT cell frequency in whole blood, ahead of monocyte separation. To identify MAIT cells, lymphocytes are gated according to FSC/SSC, doublets are excluded, CD14 monocytes and CD19 B cells are removed, and Vα7.2 CD161 double-positive cells are gated. MAIT cells can be further separated on the basis of CD4 and CD8 expression, the majority of the cells being CD8, around 10% do not express any coreceptor and few cells are CD4$^+$

Table 2
Panel to check MAIT cell frequency in whole blood

	Fluorophore	Amount (in μL)
CD3	ef780	0.5
CD4	BV650	1.5
CD8	BV711	2
CD161	APC	5
Vα7.2	BV605	2
CD14, CD19	BV510	1 each
Brilliant violet stain buffer		15

Gate on lymphocytes according to FSC and SSC, on singlets, remove B cells and monocytes, gate on CD3 cells, and identify the frequency of CD161$^+$ Vα7.2$^+$ cells (Fig. 3).

The panel described in Table 2 can be used also to check MAIT purity on a weekly basis, with the omission of CD3, CD14, and CD19 antibodies. If MAIT cells are stained less than 5 days from the sorting, some of the antibodies used for the sort may still be present and affect the interpretation of the results.

7. MAIT cells express MR1 and can self-present antigen, although their efficiency is low [7]. Have control wells where MAIT cells are exposed to 5-ARU/MG in the absence of DC. Alternatively, pulse DC for a minimum of 5 and a maximum of 16 h, wash 3× in 96-well plates (200 μL medium each wash, 100 g, 3 min), and add MAIT cells.
8. MAIT cells are highly cytolytic and are also susceptible to antigen-induced cell death, particularly at high concentration of antigen. The best FACS results can be obtained at low antigen dose (below 30 ng/mL); at all other concentrations cytokine release is an excellent readout for activation.
9. The minimal panel to check DC differentiation should include CD83, CD80, CD86, CD1a, CD14, and MHC class II. Fluorophores should be adjusted to the cytometer available, and two panels of three markers can be used instead of six markers together. If testing a serum batch for optimal DC differentiation, compare immature DC versus cells treated 24–36 h with a TLR ligand (i.e., 1 μg/mL LPS, Invivogen). Immature DCs are $CD1a^{++}$, $CD14^{-}$, $CD83^{-}$, $CD80^{dim}$, $CD86^{dim}$, and $HLA\text{-}DR^{dim}$. Upon activation, CD1a is partially downregulated, while upregulation of CD83, CD80, CD86, and HLA molecules should be observed.

   Despite very efficient MAIT (and iNKT) cell activation, MR1 (and CD1d) is expressed at very low levels on human DC and often not above the background of the relevant isotype controls. We therefore do not measure their surface expression by FACS.
10. The bacterial supernatant also contains TLR agonists, hence it induces DC maturation and cytokine secretion, which can cause concomitant MR1-independent activation. However, at low TLR-ligands concentration, we have demonstrated synergy of bacterial signals and MAIT cell agonists in enhancing DC maturation in a CD40L-dependent manner [7].
11. Some antibodies do not work well after fixation. (1) The listed clones of CD80, CD86, CD40, and CD38 are among those, and it is best to stain these markers at the surface, in the absence of brefeldin. (2) The MAIT TCR, like any other TCR, is downregulated upon activation [8], and it is better stained intracellularly. (3) The example panel provided was developed for a BD Symphony equipped with five lasers.

Acknowledgments

We acknowledge the contribution of Craig Waugh and the Weatherall Institute of Molecular Medicine flow cytometry facility for cell sorting of MAIT cells.

This work was supported by the U.K. Medical Research Council (Medical Research Council Human Immunology Unit) and Cancer Research UK (Programme Grant C399/A2291 to V.C.).

References

1. Salio M, Cerundolo V (2015) Regulation of lipid specific and vitamin specific non-MHC restricted T cells by antigen presenting cells and their therapeutic potentials. Front Immunol 6:388. https://doi.org/10.3389/fimmu.2015.00388
2. Kjer-Nielsen L, Patel O, Corbett AJ, Le Nours J, Meehan B, Liu L, Bhati M, Chen Z, Kostenko L, Reantragoon R, Williamson NA, Purcell AW, Dudek NL, McConville MJ, O'Hair RA, Khairallah GN, Godfrey DI, Fairlie DP, Rossjohn J, McCluskey J (2012) MR1 presents microbial vitamin B metabolites to MAIT cells. Nature 491:717–723. https://doi.org/10.1038/nature11605
3. McWilliam HE, Eckle SB, Theodossis A, Liu L, Chen Z, Wubben JM, Fairlie DP, Strugnell RA, Mintern JD, McCluskey J, Rossjohn J, Villadangos JA (2016) The intracellular pathway for the presentation of vitamin B-related antigens by the antigen-presenting molecule MR1. Nat Immunol 17:531–537. https://doi.org/10.1038/ni.3416
4. Hartmann N, Harriff MJ, McMurtrey CP, Hildebrand WH, Lewinsohn DM, Kronenberg M (2018) Role of MAIT cells in pulmonary bacterial infection. Mol Immunol 101:155–159. https://doi.org/10.1016/j.molimm.2018.06.270
5. Ussher JE, Bilton M, Attwod E, Shadwell J, Richardson R, de Lara C, Mettke E, Kurioka A, Hansen TH, Klenerman P, Willberg CB (2014) CD161++ CD8+ T cells, including the MAIT cell subset, are specifically activated by IL-12 +IL-18 in a TCR-independent manner. Eur J Immunol 44:195–203. https://doi.org/10.1002/eji.201343509
6. Wilgenburg BV, Loh L, Chen Z, Pediongco TJ, Wang H, Shi M, Zhao Z, Koutsakos M, Nussing S, Sant S, Wang Z, D'Souza C, Jia X, Almeida CF, Kostenko L, Eckle SBG, Meehan BS, Kallies A, Godfrey DI, Reading PC, Corbett AJ, McCluskey J, Klenerman P, Kedzierska K, Hinks TSC (2018) MAIT cells contribute to protection against lethal influenza infection in vivo. Nat Commun 9:4706. https://doi.org/10.1038/s41467-018-07207-9
7. Salio M, Gasser O, Gonzalez-Lopez C, Martens A, Veerapen N, Gileadi U, Verter JG, Napolitani G, Anderson R, Painter G, Besra GS, Hermans IF, Cerundolo V (2017) Activation of human mucosal-associated invariant T cells induces CD40L-dependent maturation of monocyte-derived and primary dendritic cells. J Immunol 199:2631–2638. https://doi.org/10.4049/jimmunol.1700615
8. Salio M, Valitutti S, Lanzavecchia A (1997) Agonist-induced T cell receptor down-regulation: molecular requirements and dissociation from T cell activation. Eur J Immunol 27:1769–1773. https://doi.org/10.1002/eji.1830270726

Chapter 9

Influenza A Virus-Infected Lung Epithelial Cell Co-Culture with Human Peripheral Blood Mononuclear Cells

Liyen Loh, Marios Koutsakos, Katherine Kedzierska, and Timothy S. C. Hinks

Abstract

Sensing of influenza A virus (IAV) infection by pattern recognition receptors can occur by either direct infection of lung epithelial cells or uptake of virus-infected cells by innate cells such as dendritic cells/ monocytes. This triggers a series of downstream events including activation of the inflammasome, the production of cytokines, chemokines, and the upregulation of stress-induced ligands that can lead to the activation of innate cells. These cells include innate lymphocytes such as MAIT, NKT, NK, and γδ T cells. Here we describe a method used to allow activation of human innate lymphocytes in co-culture with an IAV-infected human lung epithelial cell line (A549) to measure ex vivo effector functions (TNF and IFNγ) in a mixed culture environment. We describe (1) infection of the human lung epithelial cell line, (2) co-culture with PBMC, and (3) measurement of activation using intracellular cytokine staining.

Key words Virus, MAIT cell, Flow cytometry, Tetramer, Infection, Human, Epithelial cell

1 Introduction

The innate immune response serves as the first line of defense during viral infections. Sensing of influenza A virus (IAV) infection by pattern recognition receptors (e.g., TLR and RIG-I) can occur by either direct infection of lung epithelial cells or uptake of virus-infected cells by innate cells such as dendritic cells/monocytes. This triggers a series of downstream events including activation of the inflammasome, the production of cytokines, chemokines, and the upregulation of stress-induced ligands that can lead to the activation of innate cells. These cells include innate lymphocytes such as MAIT, NKT, NK, and γδ T cells. These lymphocytes can be activated by non-classical MHC interactions, cytokine-mediated signals or both. This method allows for the activation of human innate lymphocytes in co-culture with IAV-infected human lung epithelial cells (A549) and is used to measure *ex vivo* effector functions (TNF and IFNγ) in a mixed culture environment [2]. The objective is to

Helen Kaipe and Isabelle Magalhaes (eds.), *MAIT Cells: Methods and Protocols*, Methods in Molecular Biology, vol. 2098, https://doi.org/10.1007/978-1-0716-0207-2_9,

measure and recapitulate the events of early IAV infection in vitro, in a co-culture system with human peripheral blood mononuclear cells (PBMC) and IAV-infected human lung epithelial cells.

The method described in this chapter comprises three main steps: (1) infection of a human epithelial cell line, (2) co-culture with PBMC to activate the virus responsive cells, and (3) intracellular cytokine staining to measure the extent of functional activation.

2 Materials

2.1 Reagents and Buffers

1. Complete RPMI (cRPMI): Roswell Park Memorial Media, 10% heat-inactivated fetal calf serum (FCS), 100 U/mL Penicillin, 100 U/mL Streptomycin, and 100 μM MEM Vitamins.
2. Human lung epithelial cell line, A549 (ATCC, VA, USA).
3. PR8 virus (influenza A strain/H1N1/Puerto Rico/1934).
4. Trypsin Versene (In-house preparation).
5. Trypan Blue and Counting Chamber.
6. Brefeldin A—Golgi PLUG (BD, CA, USA).
7. Live/dead Fixable Aqua Dead Cell Stain Kit (ThermoFisher, MA, USA).
8. Cytofix/Cytoperm Fixation/Permeabilization Kit (BD Biosciences, USA).
9. 10% Lysol or 1% Virkon.
10. Fluorescence activated flow cytometry (FACS) buffer: phosphate-buffered saline (PBS), 2 mM EDTA, 0.5% bovine serum albumin (BSA). From a 500 mL bottle of PBS, add 40 mL to a 50 mL falcon containing 2.5 g BSA powder, vortex hard, then filter-sterilize back into PBS bottle using a syringe through a 0.22 um filter.
11. 1% paraformaldehyde (PFA) solution. Dilute 16% paraformaldehyde 1:16 with FACs buffer.
12. Antibodies for flow cytometry (Table 1).
13. Phosphate-buffered saline (PBS).

2.2 Plastic

1. T75 flasks (Corning, NY, USA).
2. 50 mL Flacon tube (Fischer Scientific, MA, USA).
3. 96-well U-bottom plate polystyrene (Greiner, Germany).

2.3 Equipment

1. Flow Cytometer, BD LSR FORTESSA, or equivalent.
2. Water Bath.

Table 1
Example flow cytometric activation panel and IAV nucleoprotein expression

Marker	Fluorophore	Laser	Clone	Dilution
Surface stain activation				
Live Dead	Aqua	Violet		1/800 (stain in PBS prior to surface stain)
MR1-5-OP-RU Tet	SA-BV421	Violet		*Titrate 1/200–1/400
CD19	APC-H7	Red	HIB19	1/100
CD14	APC-H7	Red	MϕP9	1/100
CD8α	PerCP-Cy5.5	Blue	SK1	1/50
TCRγδ	FITC	Blue	2F11	1/50
CD4	BV650	Violet	OKT4	1/200
CD161	BV605	Violet	HP-3G10	1/50
CD3	PE-CF594	Yellow/Green	UCHT1	1/200
TCR Vα7.2	PE	Yellow/Green	3C10	1/200
CD56	PE-Cy7	Yellow/Green	NCAM16.2	1/100
Intracellular stain activation				
TNF	APC	Red	MAb11	1/50
IFNγ	AF700	Red	B27	1/150
Intracellular stain IAV-NP infection				
IAV nucleoprotein	FITC	Blue	1331	1/100

Typical flow cytometry panel compatible with a four-laser BD LSRII Fortessa flow cytometer, allowing identification of innate and adaptive lymphocyte subsets and assessment of activation measured by intracellular cytokine staining
*Batches of SA-conjugated and Tetramerized MR1-5-OP-RU will vary and require titration prior to usage

3 Methods

Personal protective equipment (PPE) should be worn at all times (gloves, lab coat, and eye protection) (*see* **Note 1**).

3.1 IAV Infection of Human Lung Epithelial Cell Line, A549

1. 24 h prior to infection, in two T75 flasks, seed 5 × 10^6 A549 cells in a total volume of 20 mL of media (one flask for IAV infection and the second flask for uninfected control A549s).
2. On the day of infection: leave one flask of A549 cells in the incubator (uninfected control). Wash the other flask with room

temperature PBS once, cap and gently rotate flask from side to side. Aspirate PBS with glass tissue culture pipette.

3. Thaw virus (PR8) [1] on ice and add 174 μL to 10 mL of room temperature PBS in a 50 mL falcon tube (depending on viral titer of stock) to achieve a multiplicity of infection (MOI) of ~10–30∗. Gently pipette this into the T75 containing A549 cells.

   ∗Example calculation of MO1 10:

   An MOI of 10 using 1×10^6 PBMC per well requires 1×10^7 virus particles/well. The volume required/well of a 1×10^9 plaque forming units (pfu)/mL virus titer is 1×10^7 pfu/1×10^9 pfu/mL = 0.01 mL/well or 10 μL/well.

4. Incubate flask horizontally for 1 h in the 37 °C incubator (5% CO_2).
5. Remove both T75 flasks from incubator and add 10 mL of cRPMI to the flask containing virus. Cap and gently rotate from side to side. Aspirate media from both flasks.
6. To detach A549 cells, wash flasks once with room temperature PBS, aspirate, and add 2.5 mL of Trypsin versene to each flask. Gently tilt the flask to ensure that the solution coats the entire flask.
7. Incubate for 5 min in the 37 °C incubator (5% CO_2).
8. Add 10 mL of cRPMI to T75 flasks and transfer the contents into two 50 mL falcon tubes. Centrifuge for 5 min at 500 × *g*, 25 °C. Aspirate supernatant.
9. Resuspend cells in 2 mL of cRPMI and perform cell counts using trypan blue estimation.
10. Adjust the volume of A549 cells so that the final concentration is 2×10^6 cells/mL.

3.2 Co-Culture (Start During the 1 h Incubation with Virus)

1. Thaw PBMCs in 37 °C water bath and gently pipette dropwise into 9 mL of pre-warmed cRPMI per cryovial and centrifuge at 500 × *g* for 5 min (*see* **Note 2**).
2. Aspirate media and count cells. Resuspend PBMCs at 10×10^6 cells/mL in cRPMI. For each sample aliquot 100 μL of cells (1×10^6 PBMC) into three wells of a 96-well U-bottom plate. These wells will correspond to Media Control, uninfected A549 + PBMC, and IAV-infected A549 + PBMC, respectively.

   To check IAV nucleoprotein levels, *see* **Note 3**. Add 100 μL of infected and uninfected A549 cells to separate wells in the 96-well plate.

3. Add 100 μL of uninfected A549s or IAV-infected A549s (2×10^5 cells) into wells containing PBMC. Leave one well

with PBMC only, add 100 uL of cRPMI to this well. Place this plate in the 37 °C incubator (5% CO_2).

4. After 3–4 h, add brefeldin A (BFA-GOLGI PLUG), 1:2000 to all wells and incubate for a further 6 h in the 37 °C incubator (total co-culture 10 h).
5. Remove plate and continue with intracellular cytokine (ICS) staining or place in the 4 °C covered in foil to stain the next day.

3.3 Intracellular Cytokine Staining

1. Spin down plate by centrifuging at 400 × *g* for 5 min at 4 °C. Discard supernatant in waste container containing 10% Lysol or 1% Virkon in class II biosafety cabinet.
2. Stain cells with live/dead discrimination marker Aqua (1:800) final volume of 50 μL/well. Use PBS as a diluent (*see* **Note 4**). Incubate at room temperature in the dark for 15 min.
3. Centrifuge plate at 400 × *g* for 5 min at 4 °C. Discard supernatant.
4. Add 50 μL of surface phenotype stain (Table 1) to each well. Incubate for 30 min on ice, in the dark.
5. Wash cells once with 150 μL of FACs buffer. Centrifuge for 5 min at 1500 rpm, 4 °C. Flick off supernatant in discard container in biohazard cabinet.
6. Resuspend the cells in 100 μL of cold cytofix/perm solution and incubate on ice in the dark for 20 min.
7. Wash cells with 100 μL of diluted (1:10 in dH_2O) perm/wash buffer. Centrifuge for 5 min at 450 × *g*, 4 °C.
8. Resuspend cells in 50 μL of intracellular cytokine stain, see (Table 1) below. Incubate on ice in the dark for 30 min.
9. Wash cells with 150 μL of perm/wash buffer. Centrifuge for 5 min at 450 × *g*, 4 °C.
10. Repeat with a second wash with 200 μL of FACs buffer. Centrifuge for 5 min at 450 × *g*, 4 °C.
11. Resuspend cells in 100 μL of 1% PFA and transfer to bullet tubes. Keep samples in the dark and at cold until acquisition on the flow cytometer. For suggested flow cytometric gating strategy *see* Fig. 1.

4 Notes

1. Biological Hazards—Human PBMC samples are classified as non-infectious. Influenza A virus—PR8-strain (H1N1) is a lab-adapted strain of IAV virus. Work should be risk assessed, and we recommend controls which include but are not restricted to the following: Lab coat, safety glasses, and gloves

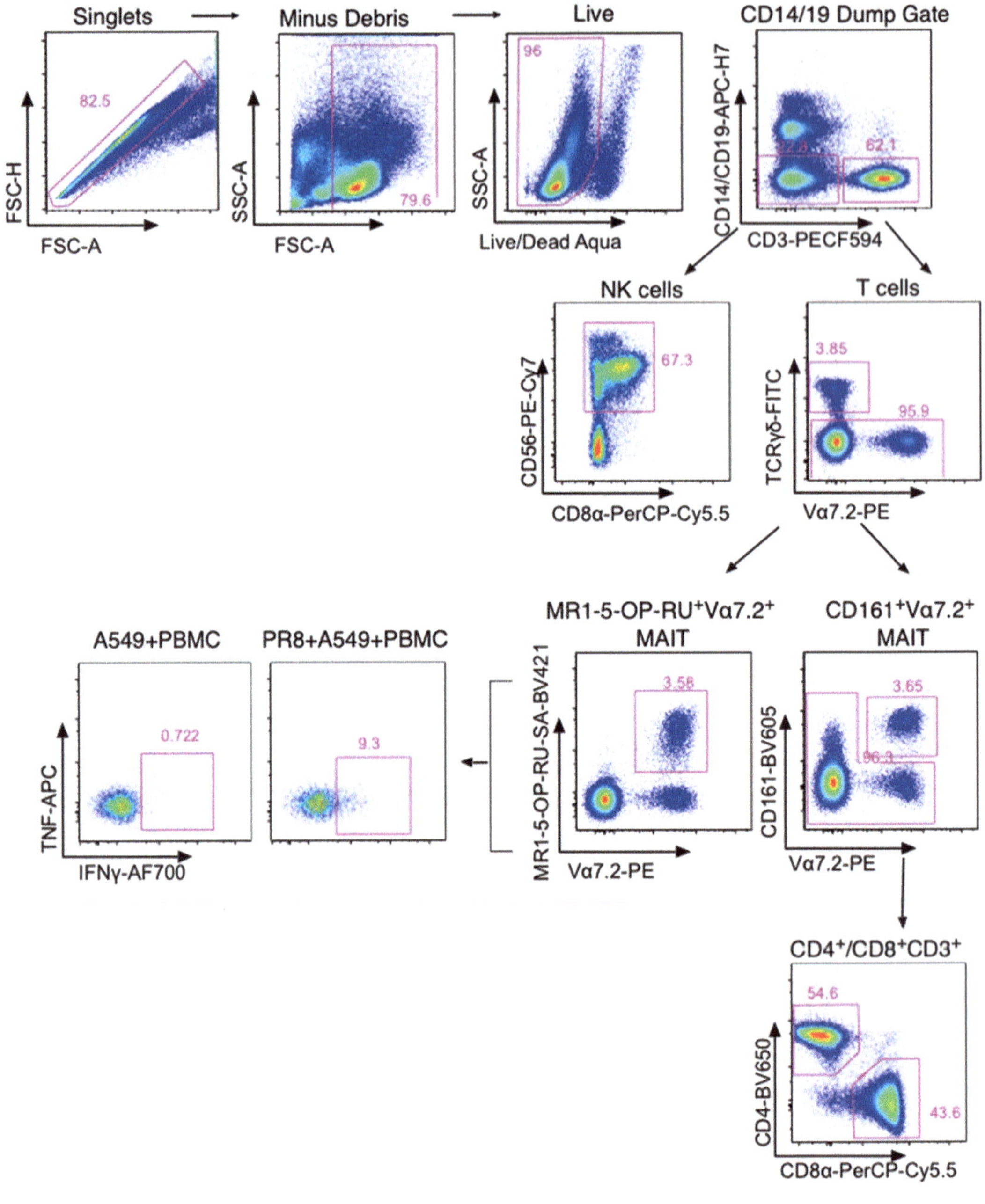

Fig. 1 Flow cytometry gating strategy for MAIT cells and other lymphocyte subsets

should be worn when performing this protocol. Work with human PBMCs and virus in a Class II biohazard cabinet. Use filter tips when working with virus. Decontaminate all pipette tips that have been used for human and virus work in 10% lysol or 1% Virkon when working in the biohazard cabinet. After use, the biohazard hood should be decontaminated by wiping down with 70% ethanol and by UV sterilization for 15 min before any further use. All waste and its container must be disposed as hazardous waste.

2. MAIT cell responses after in vitro influenza co-culture are highly variable between donors. Freshly processed PBMCs may aid in the detection of IFNγ cytokine responses after influenza co-culture.
3. To determine if influenza virus infection of lung epithelial cells is successful after 10 h of culture, intracellular cytokine staining for influenza A virus nucleoprotein is determined by flow cytometry. Follow **steps 1–3** and **6–11** of Subheading 3.3 Intracellular cytokine staining.
4. Fixable viability dyes react with exposed amine groups within permeable cells. Therefore, to prevent wasteful reaction with proteins in cytometry buffers, it is recommended to resuspend cells in protein-free media for the viability staining step.

Acknowledgments

This work was funded by grants to TSCH from the Wellcome Trust (104553/z/14/z, 211050/Z/18/z) and K.K from the National Health and Medical Research Council (NHMRC) Program Grant (1071916). K.K. is an NHMRC Senior Research Fellow (1102792). The content represents only the authors' views and not those of the European Commission.

References

1. Koutsakos M, Illing PT, Nguyen THO et al (2019) Human CD8(+) T cell cross-reactivity across influenza A, B and C viruses. Nat Immunol
2. Loh L, Wang Z, Sant S et al (2016) Human mucosal-associated invariant T cells contribute to antiviral influenza immunity via IL-18-dependent activation. Proc Natl Acad Sci U S A 113:10,133–10,138

Chapter 10

Quantification of Human MAIT Cell-Mediated Cellular Cytotoxicity and Antimicrobial Activity

Wan Rong Sia, Caroline Boulouis, Muhammad Yaaseen Gulam, Andrea Lay Hoon Kwa, Johan K. Sandberg, and Edwin Leeansyah

Abstract

The mucosa-associated invariant T (MAIT) cells represent the most abundant population of antimicrobial T cells in humans. When encountering cells infected with riboflavin-producing bacteria, this innate-like T cell population rapidly release a plethora of pro-inflammatory cytokines, mediates antimicrobial activity, and kill infected cells. Here, we describe methodological approaches and protocols to measure their cytotoxicity and antimicrobial effector function using multi-color flow cytometry-based and standard microbiological techniques. We provide specific guidance on protocols and describe potential pitfalls for each of the presented methodologies. Finally, we discuss potential applications and current limitations of our approaches to the study of human MAIT cell antimicrobial properties.

Key words Human, Mucosa-associated invariant T cells, Cytolytic molecules, Multi-color flow cytometry, Antimicrobial

1 Introduction

Mucosa-associated invariant T (MAIT) cells are an evolutionarily conserved subset of T cells that are highly abundant in mucosal tissues, the liver, and peripheral blood [1]. In humans, MAIT cells express a semi-invariant T cell receptor (TCR) Vα7.2 coupled with a restricted TCR Vβ repertoire, and recognize antigens in complex with the MHC-Ib-related protein (MR1) [2, 3]. MR1 displays an extraordinary level of evolutionary conservation among placental and marsupial mammals [2, 3], strongly supportive of the notion that MR1 and MAIT cells perform critical functions in the immune system. MAIT cells are defined by the high expression of CD161, the IL-18 receptor subunit (IL-18R), and the transcription factor PLZF [4, 5]. The vast majority of MAIT cells are $CD8^+$, with some $CD4^+$ or CD4/8 double-negative populations [6]. They also express high levels of tissue-homing chemokine receptors, such as CCR6, CCR9, and CXCR6, and are highly efficient at trafficking to

Helen Kaipe and Isabelle Magalhaes (eds.), *MAIT Cells: Methods and Protocols*, Methods in Molecular Biology, vol. 2098, https://doi.org/10.1007/978-1-0716-0207-2_10,

inflamed and infected tissues [7–10]. MAIT cells recognize microbial vitamin B_2 (riboflavin) metabolites produced by the vast majority of bacteria presented by MR1 molecules [11, 12]. In bacteria, riboflavin is a critical component in a wide variety of bacterial cellular processes [13, 14]. MAIT cells are thus able to recognize and respond to a vast number of bacteria, upon which they rapidly produce both pro-inflammatory and tissue-protective cytokines, including IFN, TNF, IL-17, and IL-22 [15–18], mediate cellular cytotoxicity [19–21], and suppress bacterial growth [22]. The potent MAIT cell antimicrobial activity contributes to the role of MAIT cells in the protection against diverse bacterial infections in both animal models and humans [4, 22–28].

In this chapter, we present flow-cytometry-based approaches to investigate human MAIT cell cytotoxic effector function and methodologies to assess MAIT cell antimicrobial activity using the *Escherichia coli* bacteria as a model microbe and the HeLa cell line as a model target cell. Throughout the chapter, we provide step-by-step guidelines on the practicalities and the need to follow standard operating procedures at certain critical steps to avoid potential pitfalls. The protocols and the methods described in detail herein allows us to extensively study functional responses of MAIT cells against various microbes that harbor the riboflavin biosynthesis pathway. The protocols can be easily adapted to answer research questions concerning their role in normal physiology, such as understanding on how MAIT cell kill bacteria-infected cells and the underlying mechanisms of MAIT cell antimicrobial activity. Furthermore, the protocols can be applied to research questions concerning MAIT cell cytolytic and antimicrobial effector function during infectious and inflammatory diseases.

The described methods are limited in several ways. First, the MAIT cell numbers needed to perform these broad assays are relatively high and may not be always be available. Secondly, resting peripheral blood MAIT cells do not kill infected cells at appreciable levels, probably reflecting the basic immunobiology of these cells, and must be primed or pre-activated via exogenous cytokines or TCR-dependent activation. Thirdly, these methods are unable to assess the state of bacterial viability following exposure to MAIT cell effector molecules, that is, we are unable to distinguish whether the bacteria are truly unviable or in a dormant state. These limitations can be potentially alleviated through the improved methodologies in MAIT cell expansion in vitro and by combining with reagents that are able to detect bacterial metabolic and viability states. Further complementary approaches may include whole-cell proteomic and transcriptomic analyses on sorted MAIT cells, infected target cells, and bacterial cells following the co-culture. These types of investigations can be guided by results obtained using the protocols described in the present paper.

2 Materials

2.1 General Lab Plastics and Equipment

1. Tubes—50 mL conical tubes, 1.5 mL Eppendorf tubes.
2. Bench-top Vortex mixer.
3. Centrifuge with cooling function and adaptors for 96-well plates, 15-mL and 50-mL Falcon tubes.
4. Bench-top microfuge with cooling function and adaptors for 1.5-mL Eppendorf tubes.
5. Bacterial incubator with continuous shaking function.
6. 37 °C incubator supplied with 5% CO_2.
7. Microscope.
8. Hemocytometer with Neubauer chamber or equivalent.

2.2 Common Reagents and Buffers

1. 0.1% Trypan blue.
2. Trypsin-EDTA, 0.25%, with Phenol Red.
3. Phosphate-buffered saline (PBS) without magnesium and calcium.
4. Complete R-10 medium: RPMI-1640 medium supplemented with 25 mM HEPES, 2 mM L-glutamine, 10% fetal bovine serum (FBS), 50 μg/mL gentamicin, and 100 μg/mL Normocin (Invivogen).
5. RFG200 medium: RPMI supplemented with 10% FBS, 2 mM L-Glutamine, 25 mM HEPES, and 200 μg/mL gentamicin.
6. Antibiotic- and serum-free (ASF)-RPMI medium: RPMI-1640 medium supplemented with 2 mM L-Glutamine and 25 mM HEPES.

2.3 Culture and Measurement of *E. coli* for Main Stock and Live Infection

1. *Escherichia coli* laboratory strains or clinical isolates.
2. Lysogeny broth (LB) and agar plates (also known as Luria-Bertani medium).
3. Bacteria freezing solution: 50% (v/v) glycerol with 50% (v/v) FBS, 0.22 μm filter-sterilized.
4. 1-μL inoculating loop.
5. 15-mL culture tube.
6. 37 °C shaking incubator.
7. Spectrophotometer capable of measuring OD_{600}: OD_{600} for bacteria measured with DeNovix DS-11 fx Series Spectrophotometer.
8. 0.22 μm filter-sterilized 0.1% (v/v) Triton-X.
9. Disposable sterile L-shaped bacterial cell spreaders.
10. Manual or computerized bacteria colony counter.

2.4 Priming and Preparing MAIT Cells for Cytotoxicity Assay

1. Human peripheral blood seronegative for HBV, HCV, HIV-1, HIV-2, HTLV-1, HTLV-2, and syphilis.
2. Magnetic beads and columns for isolating MAIT cells as previously described [16].
3. Recombinant human cytokines—rhIL-2, IL-7, IL-12, IL-15, and IL-18 for 21 days as previously described [29].
4. CellTrace™ Violet (CTV) Cell Proliferation Kit supplemented with dimethyl sulfoxide.

2.5 Culturing and Preparing Target Cells for Live E. coli Infection

1. Target cell lines: HeLa, 293T-hMR1, A549, and other appropriate cell lines.
2. 96-well flat-bottom plates.

2.6 Flow Cytometry Measurements of MAIT Cells and Cytotoxicity

1. 96-well V-bottom plates.
2. Polypropylene FACS tubes.
3. FACS buffer: PBS supplemented with 2 mM EDTA and 2% FBS.
4. BD Cytofix/Cytoperm™ Fixation/Permeabilization Kit.
5. Fluorescently conjugated monoclonal antibodies (Table 1) (*see* **Note 1**).
6. Flow Cytometry Hardware and Software: samples stained with the described antibody panels are suitable for acquisition in a LSRFortessa flow cytometer (BD Biosciences) equipped with 355 nm, 405 nm, 488 nm, 561 nm, and 640 nm lasers, and by using FACSDiva™ (BD Biosciences) as a software platform and graphical user interface.
7. Anti-mouse IgG κ beads (BD™ CompBeads Set Anti-Mouse IgG κ; BD Biosciences).
8. FlowJo analysis software (TreeStar).

3 Methods

3.1 Culture of E. coli for Main Stock and Live Infection

1. Culture *E. coli* at 1×10^5 colony-forming units (CFU)/mL in 1 L of LB for 16 h at 37 °C with shaking at 225 rpm.
2. Collect into 50-mL conical tubes and centrifuge bacterial culture at 1500 × *g* for 20 min at room temperature (RT) with no brake.
3. Resuspend pellet in 10–20 mL of bacteria freezing solution, then aliquot, freeze, and store in −80 °C until needed.
4. One day before infection, prepare a 3-mL starter culture of bacteria. Thaw a frozen aliquot of *E. coli*. Using a sterile

Table 1
Monoclonal antibodies used in protocol for cytotoxicity assay including the fluorochrome, clone, and laser and filters suitable for data acquisition

Marker	Fluorochrome	Clone	Laser and filter (nm)
MR1-tetramer for pre-staining before surface and intracellular staining			
hMR1 5-OP-RU[a]	APC [PE]	Not applicable	640 (R670/14) 561 [(YG586/15)]
Cocktail for surface staining before intracellular staining			
CD3	FITC	UCHT1	488 (B530/30)
CD4 (*see* **Note 21**)	APC-H7	RPA-T4	640 (R780/60)
CD8α	BV570	RPA-T8	405 (V610/20)
Vα7.2	PE-Cy7 [PE]	3C10 3C10	561 (YG780/60) 561 [(YG586/15)]
CD161	PE-Cy5	DX12	561 (YG670/30)
Dead cell marker (DCM)	Near-infrared [AQUA]	Not applicable	640 (R780/60) 405 [(V525/50)]
CD107a[b]	BUV395	H4A3	355 (UV379/28)
Cocktail for intracellular staining			
Granzyme B	AF700	GB11	640 (R730/45)
Granulysin	PE [AF647]	DH2 DH2	561 (YG586/15) 640 [(R670/14)]
Active caspase 3	BV650	C92-605	405 (V670/30)

[], Alternate fluorochrome and instrument settings; *V* Violet laser, *B* Blue laser, *YG* Yellow-green laser, *R* Red laser, *BV* Brilliant Violet™

[a]The MR1 tetramer technology was developed jointly by Dr. James McCluskey, Dr. Jamie Rossjohn, and Dr. David Fairlie, and the material was produced by the NIH Tetramer Core Facility as permitted to be distributed by the University of Melbourne.

[b]CD107a added at the beginning of the assay and included in the surface staining cocktail (*see* Subheading 3.6).

inoculation loop, inoculate 1 μL of thawed bacteria into 3 mL of LB in a 15-mL culture tube.

5. Incubate at 37 °C for 14 h with continuous shaking at 225 rpm.
6. On the day of the assay, subculture 200 μL of bacteria into 2 mL of LB in a 15-mL culture tube.
7. Further incubate at 37 °C with shaking at 225 rpm for 30 min.
8. Determine optical density reading at 600 nm (OD_{600}) using a spectrophotometer.
9. Continue incubation until bacterial turbidity reaches $OD_{600} = 0.5$.

10. Estimate *E. coli* cell numbers by converting the measured OD_{600} value with the pre-determined cell number conversion factor (*see* **Note 2**).
11. Centrifuge 1 mL of bacteria subculture in a 1.5-mL Eppendorf tube at 5000 × *g* for 3 min at 4 °C.
12. Wash bacteria pellet twice with 1 mL PBS at 5000 × *g*, 3 min, 4 °C.
13. Resuspend bacteria pellet in 1 mL ASF-RPMI (*see* **Note 3** and Subheading 3.3).

3.2 Priming of MAIT Cells

In general, resting and/or freshly isolated MAIT cells have very low cytotoxic capacity and presumably antimicrobial activity as well. Thus, priming and/or activation of MAIT cells through MR1-TCR-dependent interaction or through exogenous cytokines is frequently required to detect MAIT cell cytotoxic and antimicrobial capacity. The methods to prime or activate MAIT cells have been previously reported and will not be described herein. Nonetheless, we include below a general description on how to prime or activate MAIT cells and several references to the protocols.

1. Isolate MAIT cells using magnetic beads and prime MAIT cells with 10-25 ng/mL rhIL-7 for 48–72 h as previously described [16].
2. Alternatively, activate and expand MAIT cells in vitro using 5-OP-RU and rhIL-2, IL-7, IL-12, IL-15, and IL-18 for 21 days as previously described [29]. Determine MAIT cells purity by flow cytometry on day before assay for E:T ratio adjustment (*see* **Notes 4** and **5**).

3.3 Preparing HeLa Cells for Live E. coli Infection

1. On the day before the assay, seed HeLa cells into two 96-well flat-bottom plates at 5 × 10^4 cells in 200 μL Complete R-10 medium per well (*see* Subheading 3.6). Label plates as Plate A and Plate B. Plate A will be dedicated for enumeration of live bacterial counts as a measurement of MAIT cell antimicrobial activity and Plate B will be dedicated for measurement of MAIT cell cytolytic activity and target (HeLa) cell killing by flow cytometry.
2. Incubate plated HeLa cells overnight at 37 °C/5%CO_2, for at least 14 h (*see* **Note 6**).
3. On day of assay, determine cell number in each well to normalize E:T ratio. Remove overnight medium from 3 to 5 representative wells and wash once with 200 μL PBS.
4. Trypsinize HeLa cells from these wells to determine cell numbers by adding 30 μL trypsin-EDTA, incubate for 5 min at 37 °C, and add 170 μL R-10 medium to inactivate trypsin.

5. Detach HeLa cells with vigorous pipetting and perform cell count in the presence of 0.1% Trypan blue using a hemocytometer.
6. Using the average HeLa cell counts, determine the amount of *E. coli* prepared under Subheading 3.1 for the live infection at the intended numbers of bacteria per cell.
7. In addition, using the average HeLa cell counts, determine the numbers of effector MAIT cells needed for an effector to target cells (E:T) ratio of 5:1 (*see* **Note 7**).

3.4 Preparation of MAIT Cells for the Cytotoxicity Assay

1. Collect in vitro primed, activated, or expanded MAIT cells into a 50 mL conical tube (*see* **Note 8**).
2. Wash MAIT cells once with up to 50 mL of PBS and centrifuge cells at 390 × *g* at RT for 5 min.
3. Resuspend cell pellet with 1 mL of PBS, count with hemocytometer, and adjust concentration of MAIT cells to 5×10^6 cells/mL.
4. Add 20 μL of dimethyl sulfoxide (DMSO) to a lyophilized tube of CTV, then prepare a 1:50 CTV working stock in PBS.
5. Add the CTV working stock to the resuspended MAIT cells at a 1:100 dilution, mix well with a pipette, and vortex (*see* **Note 9**).
6. Incubate in the dark at 37 °C, 5%CO_2 for 30 min.
7. Quench reaction with 20 mL pre-warmed Complete R-10 medium at 37 °C for 15 min, then centrifuge labeled MAIT cells at 390 × *g* at RT for 5 min.
8. Wash labeled MAIT cells twice with 15 mL PBS, centrifuge cells at 390 × *g* at RT for 5 min (*see* **Note 10**).
9. Determine the numbers of MAIT cells required for an E:T ratio of 5:1 (*see* Subheading 3.3 and **Note 7**). Resuspend MAIT cells in 90 μL of ASF-RPMI medium per reaction such that each reaction contains MAIT cell numbers that yield 5:1 E:T ratios.

3.5 E. coli Infection of HeLa Cells

1. Remove overnight medium from all wells and gently wash HeLa cells twice with 200 μL of pre-warmed ASF-RPMI.
2. Remove medium completely and add 90 μL of ASF-RPMI medium per well (*see* **Note 11**).
3. Dilute the *E. coli* subculture prepared under Subheading 3.1 with ASF-RPMI to the required numbers of bacteria per 10 μL per reaction.
4. Add 10 μL of diluted *E. coli* subculture to selected washed HeLa cells. Mix well.
5. Add 10 μL of ASF-RPMI medium to non-*E. coli*-infected control wells.

6. Gently centrifuge the 96-well plate at 58 × *g* with minimal brake for 2 min at RT to increase contact between *E. coli* and HeLa cells and enhance *E. coli* infection of HeLa cells (*see* **Note 12**).
7. Incubate at 37 °C, 5%CO_2 for 3 h.
8. Remove media and wash cells vigorously twice with 200 μL pre-warmed RFG200 medium to remove extracellular *E. coli* as much as possible.
9. Add 200 μL of RFG200 medium and incubate at 37 °C, 5% CO_2 for 1 h to kill off any residual extracellular *E. coli* (*see* **Note 13**).

3.6 Co-Culture of Effector (MAIT) Cells with Target (HeLa) Cells

After encountering riboflavin-producing bacteria-infected cells, activated MAIT cells kill infected target cells and reduce live bacterial loads. The following protocols are designed to assess multiple functional readouts of MAIT cell effector activity from the same experiment, viz. MAIT cell cytolytic capacity, killing of infected target cells, and enumeration of the live bacterial loads. To perform this complex experimental setup, we will divide into two parallel experiments: MAIT cell cytotoxicity and killing of infected cells in one experiment, and enumeration of live bacterial counts as a measurement of MAIT cell antimicrobial activity on another experiment. Careful attention to the details in the protocols is essential to successfully execute these experiments simultaneously. The entire workflow is summarized in Fig. 1 (*see* **Note 14**).

1. Following gentamicin treatment of infected HeLa cells with *E. coli* in Subheading 3.5 **step 9**, remove media from the wells, wash cells twice with 200 μL pre-warmed ASF-RPMI medium, and add 100 μL ASF-RPMI medium into each well. *Optional:* check efficiency of gentamicin treatment in killing residual extracellular bacteria by sampling on a well. Collect the supernatant after the final wash as described for Subheading 3.7 (*see* **Note 15**).
2. For the plate dedicated to enumerating live bacterial counts and measurements of MAIT cell antimicrobial activity (Plate A): add 10 μL of ASF-RPMI medium into 90 μL labeled MAIT cells. Mix well and add the mixture into the appropriate HeLa cell wells.
3. For the plate dedicated to measuring MAIT cell cytotoxicity and HeLa cell killing (Plate B): prepare a 1:10 dilution of anti-CD107a-BUV395 in 10 μL ASF-RPMI medium, then add 10 μL of 1:10 diluted anti-CD107a-BUV395 into 90 μL labeled MAIT cells. Mix well and add the mixture into the appropriate HeLa cell wells (*see* **Note 16**).

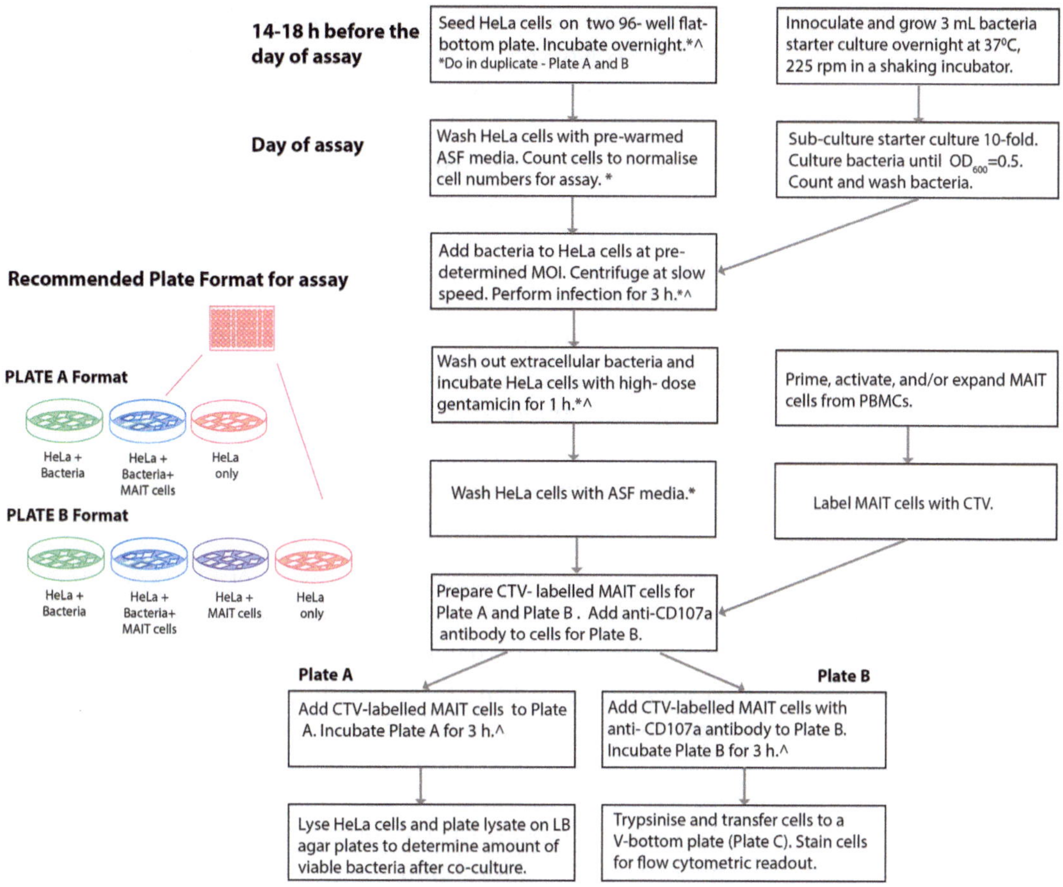

Fig. 1 Flowchart of the methods described in this chapter

4. For both Plates A and B: mix the co-culture gently without dislodging HeLa cells, then gently centrifuge plates A and B at 58 × *g* with minimal brake for 2 min at RT to increase contact between HeLa cells and MAIT cells.
5. Incubate Plates A and B for 3 h at 37 °C/5%CO_2. Continue to Subheading 3.7 for Plate A and Subheading 3.8 for Plate B (*see* **Note 17**).

3.7 Enumeration of Live E. coli Cells

1. After the end of 3-h incubation, transfer supernatants from each well into sterile 1.5-mL Eppendorf tubes.
2. Wash the wells with 200 μL sterile PBS, then transfer the wash and combine with the corresponding supernatants in the 1.5-mL Eppendorf tubes from the previous step. Keep on ice.
3. Add 100 μL of 0.22 μm filter-sterilized 0.1% Triton-X into each well of Plate A. Mix vigorously to allow complete lysis of the cells (*see* **Note 18**).

4. Incubate for 10 min at RT, and then add 100 μL LB medium to stop the reaction.
5. Mix and transfer the lysates into new 1.5-mL Eppendorf tubes. Keep on ice.
6. When possible, immediately plate 20 μL of the lysates onto the LB agar plates in triplicates (*see* **Note 19**). Thoroughly spread over the agar surfaces using sterile L-shaped bacterial cell spreaders or equivalent. Depending on the numbers of original bacterial input, it may be necessary to prepare 1:10 serial dilutions in LB medium prior to plating (*see* **Note 20**).
7. *Optional:* plate 20 μL of the supernatants from **step 2** onto the LB agar in triplicates.
8. Incubate plates overnight at 35–37 °C and count the bacterial colonies using a manual or computerized bacteria colony counter. *See* Subheading 3.11 for data analysis.

3.8 Measurements of MAIT Cell Cytotoxic Capacity and Target Cell Killing by Flow Cytometry

1. Mix cells and transfer supernatants from Plate B to a V-bottom 96-well plate (Plate C), then spin Plate C at 604 × *g* for 2 min at RT. Discard supernatants.
2. Wash the cells in Plate B once with 200 μL PBS, then transfer and combine the supernatants into the corresponding wells of Plate C. Spin Plate C at 604 × *g* for 2 min at RT. Place Plate C on ice.
3. Add 30 μL trypsin-EDTA into each well of Plate B, then incubate at 37 °C for 5 min.
4. Terminate reaction by adding 100 μL pre-warmed Complete R-10 medium to each well. Mix vigorously.
5. Discard the supernatants from Plate C, then transfer the trypsinized cells from plate B into the corresponding wells of plate C.
6. Spin Plate C at 604 × *g* for 2 min at 4 °C and discard supernatant.
7. Incubate the cells with 50 μL of the MR1 tetramer cocktail for 40 min at RT in the dark.
8. Wash cells once with 150 μL FACS buffer and spin at 604 × *g* for 2 min at 4 °C. Discard supernatant and incubate the cells with 50 μL of the antibodies against the surface markers of interest (*see* Table 1, *see* **Note 21**) for 20 min at 4 °C in the dark.
9. Wash cells once with 150 μL FACS buffer and spin at 604 × *g* for 2 min at 4 °C. Discard supernatant and fix the cells with 100 μL of 1× BDCytofix/Cytoperm for 30 min at 4 °C in the dark.

10. Wash away the fixative solution with 100 μL 1× BD Perm/Wash, then spin at 755 × *g* for 2 min at 4 °C and discard supernatant. Wash once more with 200 μL of 1× BD Cytofix/Cytoperm, repeat the centrifugation step, and discard the supernatant.
11. Incubate the cells with mAbs against the cytolytic molecules of interest (*see* Table 1) for 30 min at 4 °C in the dark.
12. Wash once with 100 μL 1× BD Perm/Wash, then spin at 755 × *g* for 2 min at 4 °C.
13. Discard supernatant and resuspend in 100–200 μL FACS Buffer. Run the samples on the flow cytometer or store at 4 °C in the dark until needed (*see* **Note 22**).

3.9 Multi-Color Flow Cytometry Compensation

We have previously provided detailed instruction on how to perform spectral overlap compensation for multi-color flow cytometry in a previous issue of this series [30].

3.10 Multi-Color Flow Cytometry Analysis

1. Use the FACSDiva software to export the acquired sample files in Flow Cytometry Standard (FCS) 3.0 files, required for further analysis in the FlowJo analysis software (TreeStar).
2. In FlowJo, use the *Compensation Wizard* tool to compensate the sample files with the bead files. A compensation matrix will be generated by FlowJo indicating the amount of spillover of one fluorochrome into the other, for all fluorochrome combinations in the panel used. This matrix might need to be adjusted, as previously described [30].
3. Start by plotting the time vs. side scatter area (SSC-A) in order to check the stability of the acquisition.
4. Proceed to the single cells gating by plotting FSC-H vs FSC-A and/or SSC-H vs SSC-A.
5. Separate effector from target cells by plotting CTV vs. SSC-A. Gate on effector cells on the CTV-positive population, and target cells on the CTV-negative population.
6. On the target cell population, plot FSC-A vs. SSC-A and create a relatively tight gate of the effector cells. Follow by plotting active caspase 3 vs. dead cell marker (DCM) to measure the levels of target cell killing by MAIT cells.
7. On the effector cell population, plot SSC-A vs. DCM and create a gate on viable (DCM^-) cells. Follow by plotting FSC-A vs. SSC-A to create a lymphocyte gate.
8. From within the lymphocyte gate, plot CD3 vs. Vα7.2 and create a T cell gate on $CD3^+V\alpha7.2^{-/+}$ cells. Continue by creating MAIT cell gate based on dual positive expression of Vα7.2 and MR1 5-OP-RU, regardless of CD161 expression (*see* **Note 23**).

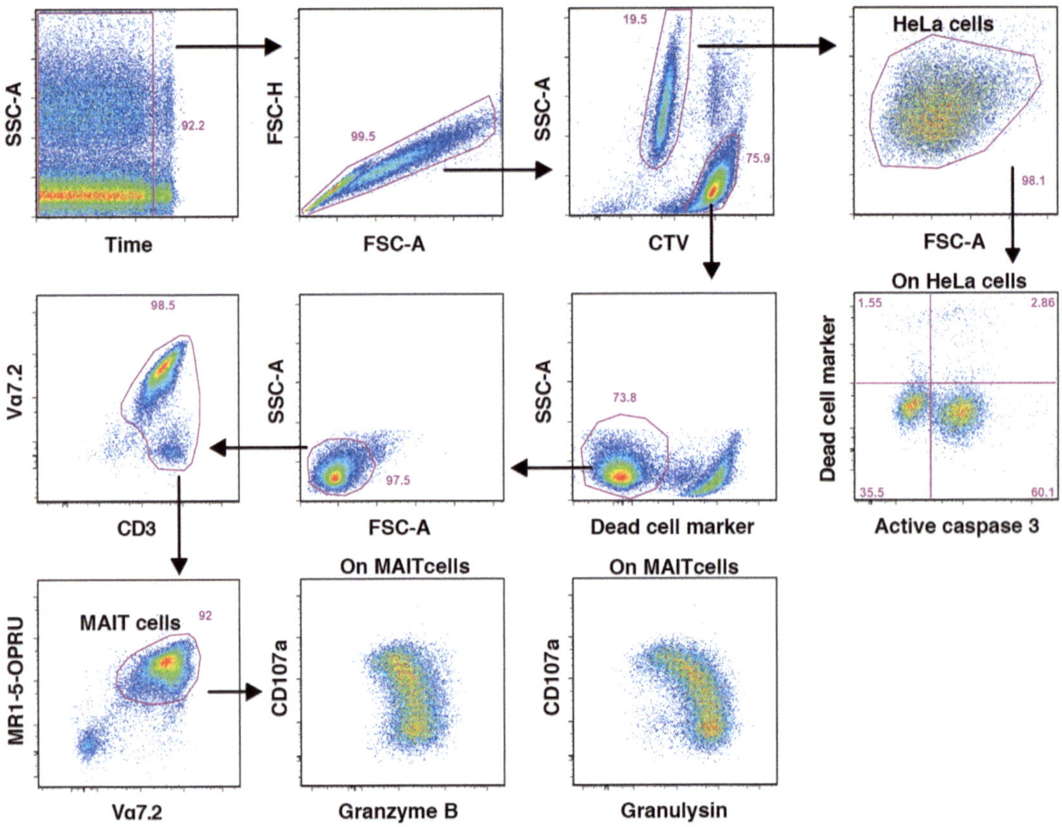

Fig. 2 Gating strategy to identify MAIT cells and HeLa cells in the co-culture system. Apoptosis of target (HeLa) cells was determined by active caspase-3 expression and dead cell marker. MAIT cell cytolytic capacity was determined by the degranulation (CD107a), granulysin, and granzyme B levels

9. From within the MAIT cell gate, measure the levels of degranulation (CD107a$^+$), granzyme B, and granulysin to determine MAIT cell cytotoxic activity.
10. Figure 2 illustrate and summarize our gating strategy for the MAIT cells and HeLa cell co-culture functional readouts.

3.11 Enumeration and Analysis of Live Bacteria Counts

1. Average the colony counts and express the results as CFU/mL by multiplying the counts by 50. Note that each user should determine the lower and upper limits of detection, although in general a range of 20–250 CFU per plate is acceptable, depending on the colony size. In the event that the CFU >250/plate at neat, then tenfold serial dilutions are recommended. Note that the lower limit of detection is numerically increased (i.e., reduced sensitivity) when serially diluted samples are used.
2. When comparing the numbers of live bacteria among experimental groups, they can be directly expressed as CFU/mL, or expressed as a percent CFU inhibition over the nil MAIT cell control groups as followed:

$$\%\text{CFU inhibition} = 100\% - \left[\left(\frac{\text{CFU/mL in the presence of MAIT cells}}{\text{CFU/mL in nil MAIT cell controls}}\right) \times 100\%\right]$$

4 Notes

1. Antibodies listed in the staining panel on Table 1 is a non-exhaustive list and can be changed according to different needs.
2. Different OD_{600} values will be reported on different spectrophotometer setup and for different bacteria. It is recommended to prepare own cell number conversion factor against measured OD_{600} values. Cell number conversion factor can be determined by correlating measured OD_{600} values to colony-forming units (CFU) grown on agar plates. If the measured OD_{600} value exceeds 0.5, dilute the culture and repeat incubation with shaking until $OD_{600} = 0.5$.
3. If target cells are not ready for infection before the end of the bacteria final wash, the prepared bacteria culture can be placed on ice. This however, could compromise the efficiency of the live bacterial infection.
4. 5-OP-RU is the prototypical of MAIT cell antigen and not commercially available but may be requested from Dr. David. P. Fairlie, University of Queensland, Brisbane, Australia [31].
5. If using MAIT cells expanded from PBMC, a purity of at least 75% MAIT cells in culture is recommended. Purity of expanded culture can be determined on day before assay for optimal E:T ratio adjustment.
6. Target cell line used for assay should be in passaged for at least once after thawing, and 1-week should elapse prior to the experiments.
7. Effector to Target (E:T) ratio for the assay should be optimized for each user depending on the MAIT cell culture condition.
8. To enhance MAIT cell cytotoxic capacity, MAIT cells can be re-primed with cytokines 24 h before the assay.
9. When labeling MAIT cells with CTV, mixing really well is crucial to ensure universal staining. This procedure can be done during the 3 h live infection of bacteria on the target cells of Subheading 3.5, **step** 7.
10. After CTV labeling, always keep labeled MAIT cells in dark.
11. Before live *E. coli* infection, it is critical to completely remove any residual medium and washing the target cells twice to remove any antibiotics that can affect efficiency of the live infection. We recommend to have the target cells prepared

before the final wash of bacteria culture. This can be performed during the 30 min incubation of the diluted bacteria subculture under Subheading 3.5, **step** 7.

12. Slow spin to increase the contact time of the bacteria and target cells for enhanced infection efficiency: recommended speed for acceleration is 6 and deceleration is 2 (based on Sorvall ST40R centrifuge).
13. Treat the target cells to kill any residual extracellular *E. coli*, usually with an aminoglycoside-class antibiotic, such as gentamicin. Timing and concentration for the gentamicin incubation may need to be optimized for different bacteria.
14. This complex killing assay protocol is designed for a co-culture of three different cells types (effector cells—MAIT cells, target cells—cell line of interest, and the bacteria of interest) with two different endpoint readouts (CFU enumeration and flow cytometry analysis). We recommend users familiarize with the protocol and workflow illustrated on Fig. 1 first before starting on the assay. This will ensure a coordinated experimental execution.
15. After the final wash with ASF media in Subheading 3.6, **step 1**, collect the supernatant and plate the collected supernatant on LB agars. The plated supernatant should yield no colonies after an effective gentamicin wash and incubation.
16. After adding MAIT cells to co-culture, minimize Plate B from light exposure as CTV and fluorochrome-conjugated anti-CD107a-BUV395 are light sensitive.
17. The duration of target and effector cell co-culture can be optimized. We have established that incubation of co-culture at 3 h gives the best early apoptotic events-readout in our assay. Significant late apoptotic events will eventually occur as denoted by positive expression of both active caspase 3 and amine-reactive dead cell marker if the co-culture is performed much longer than 3 h.
18. 0.1% Triton will lyse mammalian cells but will have minimal impact on *E. coli* cells.
19. Triton-lysates and the supernatants should be stored on ice immediately after collection to prevent growth of the bacteria. Though recommended to be plated as soon as possible for the best results, these can be stored on ice for up to 24 h prior to plating. If not, it is recommended to freeze any remaining lysates and supernatants. To do this, add equal volume of bacterial freezing solution to the remaining lysates and supernatants, then store at −80 °C.
20. It is recommended that users initially determine the optimal number of bacteria required to infect the target cells for each

bacterial strain, so that it will yield at least 1000 CFU/mL (i.e., min 20 CFU per plate at neat) in the absence of MAIT cells.

21. Since the majority of MAIT cells are CD4 negative, CD4 antibody can be omitted or included in a "dump-channel" with Dead Cell Marker.
22. When possible, the samples should be run on the flow cytometer no later than 24 h after staining.
23. In the absence of MR1-5-OP-RU tetramers, MAIT cells can be identified through the combination of high levels of CD161 and expression of the TCR Vα7.2 (Vα7.2^{+} CD161^{hi}). However, in some circumstances, CD161 will be markedly downregulated, rendering identification of MAIT cells through this strategy difficult.

References

1. Le Bourhis L, Mburu YK, Lantz O (2013) MAIT cells, surveyors of a new class of antigen: development and functions. Curr Opin Immunol 25(2):174–180. https://doi.org/10.1016/j.coi.2013.01.005
2. Huang S, Martin E, Kim S, Yu L, Soudais C, Fremont DH, Lantz O, Hansen TH (2009) MR1 antigen presentation to mucosal-associated invariant T cells was highly conserved in evolution. Proc Natl Acad Sci U S A 106(20):8290–8295. https://doi.org/10.1073/pnas.0903196106
3. Treiner E, Duban L, Bahram S, Radosavljevic M, Wanner V, Tilloy F, Affaticati P, Gilfillan S, Lantz O (2003) Selection of evolutionarily conserved mucosal-associated invariant T cells by MR1. Nature 422(6928):164–169. https://doi.org/10.1038/nature01433
4. Le Bourhis L, Martin E, Peguillet I, Guihot A, Froux N, Core M, Levy E, Dusseaux M, Meyssonnier V, Premel V, Ngo C, Riteau B, Duban L, Robert D, Huang S, Rottman M, Soudais C, Lantz O (2010) Antimicrobial activity of mucosal-associated invariant T cells. Nat Immunol 11(8):701–708. https://doi.org/10.1038/ni.1890
5. Tsukamoto K, Deakin JE, Graves JA, Hashimoto K (2013) Exceptionally high conservation of the MHC class I-related gene, MR1, among mammals. Immunogenetics 65 (2):115–124. https://doi.org/10.1007/s00251-012-0666-5
6. Dias J, Boulouis C, Gorin JB, van den Biggelaar R, Lal KG, Gibbs A, Loh L, Gulam MY, Sia WR, Bari S, Hwang WYK, Nixon DF, Nguyen S, Betts MR, Buggert M, Eller MA, Broliden K, Tjernlund A, Sandberg JK, Leeansyah E (2018) The CD4(−)CD8(−) MAIT cell subpopulation is a functionally distinct subset developmentally related to the main CD8(+) MAIT cell pool. Proc Natl Acad Sci U S A 115(49):E11513–E11522. https://doi.org/10.1073/pnas.1812273115
7. Dusseaux M, Martin E, Serriari N, Peguillet I, Premel V, Louis D, Milder M, Le Bourhis L, Soudais C, Treiner E, Lantz O (2011) Human MAIT cells are xenobiotic-resistant, tissue-targeted, CD161hi IL-17-secreting T cells. Blood 117(4):1250–1259. https://doi.org/10.1182/blood-2010-08-303339
8. Leeansyah E, Ganesh A, Quigley MF, Sonnerborg A, Andersson J, Hunt PW, Somsouk M, Deeks SG, Martin JN, Moll M, Shacklett BL, Sandberg JK (2013) Activation, exhaustion, and persistent decline of the antimicrobial MR1-restricted MAIT-cell population in chronic HIV-1 infection. Blood 121 (7):1124–1135. https://doi.org/10.1182/blood-2012-07-445429
9. Martin E, Treiner E, Duban L, Guerri L, Laude H, Toly C, Premel V, Devys A, Moura IC, Tilloy F, Cherif S, Vera G, Latour S, Soudais C, Lantz O (2009) Stepwise development of MAIT cells in mouse and human. PLoS Biol 7(3):e54. https://doi.org/10.1371/journal.pbio.1000054
10. Walker LJ, Kang YH, Smith MO, Tharmalingham H, Ramamurthy N, Fleming VM, Sahgal N, Leslie A, Oo Y, Geremia A, Scriba TJ, Hanekom WA, Lauer GM, Lantz O, Adams DH, Powrie F, Barnes E, Klenerman P (2012) Human MAIT and CD8alphaalpha cells develop from a pool of type-17

precommitted CD8+ T cells. Blood 119 (2):422–433. https://doi.org/10.1182/blood-2011-05-353789

11. Corbett AJ, Eckle SB, Birkinshaw RW, Liu L, Patel O, Mahony J, Chen Z, Reantragoon R, Meehan B, Cao H, Williamson NA, Strugnell RA, Van Sinderen D, Mak JY, Fairlie DP, Kjer-Nielsen L, Rossjohn J, McCluskey J (2014) T-cell activation by transitory neo-antigens derived from distinct microbial pathways. Nature 509(7500):361–365. https://doi.org/10.1038/nature13160
12. Kjer-Nielsen L, Patel O, Corbett AJ, Le Nours J, Meehan B, Liu L, Bhati M, Chen Z, Kostenko L, Reantragoon R, Williamson NA, Purcell AW, Dudek NL, McConville MJ, O'Hair RA, Khairallah GN, Godfrey DI, Fairlie DP, Rossjohn J, McCluskey J (2012) MR1 presents microbial vitamin B metabolites to MAIT cells. Nature 491(7426):717–723. https://doi.org/10.1038/nature11605
13. Bacher A, Eberhardt S, Fischer M, Kis K, Richter G (2000) Biosynthesis of vitamin b2 (riboflavin). Annu Rev Nutr 20:153–167. https://doi.org/10.1146/annurev.nutr.20.1.153
14. Fischer M, Bacher A (2008) Biosynthesis of vitamin B2: structure and mechanism of riboflavin synthase. Arch Biochem Biophys 474 (2):252–265. https://doi.org/10.1016/j.abb.2008.02.008
15. Dias J, Leeansyah E, Sandberg JK (2017) Multiple layers of heterogeneity and subset diversity in human MAIT cell responses to distinct microorganisms and to innate cytokines. Proc Natl Acad Sci 114(27):E5434–E5443. https://doi.org/10.1073/pnas.1705759114
16. Dias J, Sobkowiak MJ, Sandberg JK, Leeansyah E (2016) Human MAIT-cell responses to Escherichia coli: activation, cytokine production, proliferation, and cytotoxicity. J Leukoc Biol 100(1):233–240. https://doi.org/10.1189/jlb.4TA0815-391RR
17. Gibbs A, Leeansyah E, Introini A, Paquin-Proulx D, Hasselrot K, Andersson E, Broliden K, Sandberg JK, Tjernlund A (2017) MAIT cells reside in the female genital mucosa and are biased towards IL-17 and IL-22 production in response to bacterial stimulation. Mucosal Immunol 10(1):35–45. https://doi.org/10.1038/mi.2016.30
18. Leeansyah E, Loh L, Nixon DF, Sandberg JK (2014) Acquisition of innate-like microbial reactivity in mucosal tissues during human fetal MAIT-cell development. Nat Commun 5:3143. https://doi.org/10.1038/ncomms4143
19. Kurioka A, Ussher JE, Cosgrove C, Clough C, Fergusson JR, Smith K, Kang YH, Walker LJ, Hansen TH, Willberg CB, Klenerman P (2015) MAIT cells are licensed through granzyme exchange to kill bacterially sensitized targets. Mucosal Immunol 8(2):429–440. https://doi.org/10.1038/mi.2014.81
20. Le Bourhis L, Dusseaux M, Bohineust A, Bessoles S, Martin E, Premel V, Core M, Sleurs D, Serriari NE, Treiner E, Hivroz C, Sansonetti P, Gougeon ML, Soudais C, Lantz O (2013) MAIT cells detect and efficiently lyse bacterially-infected epithelial cells. PLoS Pathog 9(10):e1003681. https://doi.org/10.1371/journal.ppat.1003681
21. Leeansyah E, Svard J, Dias J, Buggert M, Nystrom J, Quigley MF, Moll M, Sonnerborg A, Nowak P, Sandberg JK (2015) Arming of MAIT cell cytolytic antimicrobial activity is induced by IL-7 and defective in HIV-1 infection. PLoS Pathog 11(8): e1005072. https://doi.org/10.1371/journal.ppat.1005072
22. Meierovics A, Yankelevich WJ, Cowley SC (2013) MAIT cells are critical for optimal mucosal immune responses during in vivo pulmonary bacterial infection. Proc Natl Acad Sci U S A 110(33):E3119–E3128. https://doi.org/10.1073/pnas.1302799110
23. Chua WJ, Truscott SM, Eickhoff CS, Blazevic A, Hoft DF, Hansen TH (2012) Polyclonal mucosa-associated invariant T cells have unique innate functions in bacterial infection. Infect Immun 80(9):3256–3267. https://doi.org/10.1128/iai.00279-12
24. Georgel P, Radosavljevic M, Macquin C, Bahram S (2011) The non-conventional MHC class I MR1 molecule controls infection by Klebsiella pneumoniae in mice. Mol Immunol 48(5):769–775. https://doi.org/10.1016/j.molimm.2010.12.002
25. Gold MC, Cerri S, Smyk-Pearson S, Cansler ME, Vogt TM, Delepine J, Winata E, Swarbrick GM, Chua WJ, Yu YY, Lantz O, Cook MS, Null MD, Jacoby DB, Harriff MJ, Lewinsohn DA, Hansen TH, Lewinsohn DM (2010) Human mucosal associated invariant T cells detect bacterially infected cells. PLoS Biol 8 (6):e1000407. https://doi.org/10.1371/journal.pbio.1000407
26. Grimaldi D, Le Bourhis L, Sauneuf B, Dechartres A, Rousseau C, Ouaaz F, Milder M, Louis D, Chiche JD, Mira JP, Lantz O, Pene F (2014) Specific MAIT cell behaviour among innate-like T lymphocytes in critically ill patients with severe infections. Intensive Care Med 40(2):192–201. https://doi.org/10.1007/s00134-013-3163-x
27. Leung DT, Bhuiyan TR, Nishat NS, Hoq MR, Aktar A, Rahman MA, Uddin T, Khan AI,

Chowdhury F, Charles RC, Harris JB, Calderwood SB, Qadri F, Ryan ET (2014) Circulating mucosal associated invariant T cells are activated in Vibrio cholerae O1 infection and associated with lipopolysaccharide antibody responses. PLoS Negl Trop Dis 8(8):e3076. https://doi.org/10.1371/journal.pntd.0003076

28. Smith DJ, Hill GR, Bell SC, Reid DW (2014) Reduced mucosal associated invariant T-cells are associated with increased disease severity and Pseudomonas aeruginosa infection in cystic fibrosis. PLoS One 9(10):e109891. https://doi.org/10.1371/journal.pone.0109891

29. Gherardin NA, Loh L, Admojo L, Davenport AJ, Richardson K, Rogers A, Darcy PK, Jenkins MR, Prince HM, Harrison SJ, Quach H, Fairlie DP, Kedzierska K, McCluskey J, Uldrich AP, Neeson PJ, Ritchie DS, Godfrey DI (2018) Enumeration, functional responses and cytotoxic capacity of MAIT cells in newly diagnosed and relapsed multiple myeloma. Sci Rep 8(1):4159. https://doi.org/10.1038/s41598-018-22130-1

30. Dias J, Sandberg JK, Leeansyah E (2017) Extensive phenotypic analysis, transcription factor profiling, and effector cytokine production of human MAIT cells by flow cytometry. Methods Mol Biol 1514:241–256. https://doi.org/10.1007/978-1-4939-6548-9_17

31. Mak JY, Xu W, Reid RC, Corbett AJ, Meehan BS, Wang H, Chen Z, Rossjohn J, McCluskey J, Liu L, Fairlie DP (2017) Stabilizing short-lived Schiff base derivatives of 5-aminouracils that activate mucosal-associated invariant T cells. Nat Commun 8:14599. https://doi.org/10.1038/ncomms14599

Chapter 11

MAIT Cell Activation by Fungal Pathogens

Susanne Jahreis, Sarah Boettcher, and Marie von Lilienfeld-Toal

Abstract

MAIT cells can recognize and respond to several pathogens in a distinct manner. However, many studies are focused on activation by bacteria and by viruses, while reports about molds and MAIT cells are rare. Here, we describe a method to investigate MAIT cell interaction with filamentous fungi from *Aspergillus* and *Mucorales* species. This includes preparation of fungal conidia, cell isolation, and analysis by flow cytometry.

Key words MAIT cell, Mold, *Aspergillus*, *Mucorales*, Flow cytometry, Antigen-presenting cell, Activation

1 Introduction

MAIT cells are unconventional T cells with an invariant T cell receptor (TCRVα7.2) recognizing riboflavin metabolites [1]. Until now, two major activation pathways have been described: a TCR-dependent one involving antigen presentation by antigen-presenting cells and a TCR-independent way induced by cytokines. The latter is described for stimulation by viruses [2, 3], while bacteria and molds, which are able to synthesize riboflavin, can induce TCR-dependent activation [4, 5]. The biology of MAIT cells is mainly studied by using either synthetic ligands [6] or stimulation by bacterial pathogens [7]. However, activation by fungal microbes has been described until now only for yeasts like *Candida albicans*, *Candida glabrata*, and *Saccharomyces cerevisiae* [4]. We recently described a rapid MAIT activation by filamentous fungi from the species *Aspergillus* [8] and *Mucorales* (Böttcher et al., submitted) in an antigen-presenting cell-dependent manner. A fast method to investigate MAIT stimulation by fungal pathogens is to use human peripheral blood mononuclear cells (PBMC) containing antigen-presenting cells like monocytes as well as MAIT cells. For deeper analysis, separation of both cell types might be useful. Furthermore, fungal conidia (or germlings) need to be prepared and previously tested for swelling and germination

Helen Kaipe and Isabelle Magalhaes (eds.), *MAIT Cells: Methods and Protocols*, Methods in Molecular Biology, vol. 2098,
https://doi.org/10.1007/978-1-0716-0207-2_11,

behavior. As read-out, flow cytometry is quite common as it enables to check several descriptive and functional MAIT cell parameters simultaneously (as well as viability and antigen-presenting cells). These three parts are described in the following chapter.

2 Materials

2.1 Preparation of Fungal Conidia

1. Consumables: Petri dishes, 15 mL tubes, 30 μm cell strainer.
2. Malt agar: Dissolve 40 g malt extract, 4 g yeast extract, and 15 g agar in 1 L *Aqua dest*. Adjust pH to 5.7–6.0 and sterilize at 121 °C for 35 min.
3. PBS containing 0.01% Tween20: Slowly add 50 μL Tween20 to 500 mL PBS and swirl. Wait until detergent and formed foam is dissolved.
4. 4% paraformaldehyde (PFA): Add 4 g PFA in 100 mL PBS and dissolve by heating at 80 °C for 3–5 h. Store aliquots at −20 °C.
5. Sabouraud 2% Glucose Bouillon: Dissolve 30 g Sabouraud 2% Glucose Bouillon in 1 L *Aqua dest* and sterilize at 121 °C for 35 min.
6. Thoma chamber.

2.2 Isolation of PBMC, MAIT Cells, and Monocytes

1. Consumables: 15 mL tubes, 50 mL tubes, 96-well U-bottom plate.
2. Separating solution (density of 1.077 g/mL).
3. 1× erythrocyte lysis buffer: Dissolve 4.15 g NH_4Cl, 0.5 g $KHCO_3$, and 0.0179 g EDTA in 500 mL *Aqua dest*. 10× erythrocyte lysis buffer can be prepared and stored at 4 °C. To obtain 1× erythrocyte lysis buffer, dilute 1:10 with *Aqua dest*.
4. Neubauer chamber.
5. Medium: Add 50 mL heat-inactivated fetal calf serum (FCS) to 500 mL RPMI1640.
6. PBS containing 2 mM EDTA (PBS–EDTA): Dissolve 0.358 g EDTA in 500 mL PBS for 1–2 h at room temperature.
7. MACS buffer: PBS with 0.5% BSA and 2 mM EDTA. Dissolve 2.5 g BSA and 0.358 g EDTA in 500 mL sterile PBS for 1–2 h at room temperature. Filter sterile and store at 4 °C.
8. CD14 MicroBeads, CD8 MicroBeads, LS columns, magnet for isolation from Miltenyi Biotec.
9. CD3 FITC (clone SK7), CD8 APC-H7 (clone SK1), CD161 PE-Vio770 (clone REA631).
10. Fluorescent Activated Cell Sorter.

11. Anti-MR1 antibody (clone 26.5) and the respective isotype control (mouse IgG2a).
12. Brefeldin A (monensin).

2.3 Flow Cytometry

1. Saponin buffer: PBS with 0.1% BSA, 0.1% NaN_3, 0.5% saponin. Prepare 10% BSA solution by adding 1 g BSA to 10 mL PBS, dissolve by vortexing, filter sterile, and store 500 µL aliquots at −20 °C. Dissolve 0.05 g NaN_3 and 0.25 g saponin in 49.5 mL PBS, and add 500 µL of 10% BSA solution. Store at 4 °C.
2. PBS–EDTA: see Subheading 2.2, **item 6**.
3. Mouse serum.
4. FACS-Fix: PBS with 0.1% BSA, 0.1% NaN_3, 1% PFA. Dissolve 0.05 g NaN_3 in 37 mL PBS, add 500 µL of 10% BSA solution, and 12.5 mL of 4% PFA. Store at 4 °C.
5. Antibodies for flow cytometry (Table 1).

3 Methods

3.1 Preparation of Fungal Conidia

Prepare fungal conidia under sterile conditions preferentially under a separate laminar hood (*see* **Note 1**).

1. Grow wild-type conidia of *Aspergillus* (*A. fumigatus*, *A. flavus*, *A. terreus*, *A. nidulans*) or *Mucorales* (*Lichtheimia corymbifera*, *Rhizopus arrhizus*, *Rhizopus microsporus*) on malt agar plates at 37 °C or 30 °C (*Mucor circinelloides*) without CO_2 in the dark until sporulation occurs after 4–7 days.
2. For harvesting add 15 mL sterile PBS with 0.01% Tween20 and detach conidia from the plate using a Drigalski spatula. Filter conidia suspension through a 30-µm cell strainer, centrifuge at 3000 × *g* for 5 min at room temperature, and resuspend in sterile PBS (*see* **Note 2**).
3. Swell conidia in RPMI (*Aspergillus* species, *L. corymbifera*) or Sabouraud 2% Glucose Bouillon (*M. circinelloides*, *R. arrhizus*, and *R. microsporus*) at 37 °C without CO_2 under continuous shaking (*see* **Notes 3** and **16**).
4. Wash two times with 10 mL PBS with 0.01% Tween20, centrifuge at 3000 × *g* for 5 min at room temperature. Each time resuspend the pellet by vortexing vigorously.
5. Resuspend in 10 mL PBS with 0.01% Tween20, filter conidia suspension through a 30-µm cell strainer, centrifuge at 3000 × *g* for 5 min, resuspend in sterile PBS, and store at 4 °C until used (*see* **Note 2**).
6. Count conidial cell number in Thoma chamber prior to use.

Table 1
Flow cytometry antibodies

Antibody	Isotype	Clone
CD3	Mouse IgG1	SK7
CD4	Mouse IgG1	SK3
CD8	Mouse IgG1	SK1
CD14	Mouse IgG2a	M5E2
CD25	Mouse IgG1	M-A251
CD69	Mouse IgG1	FN50
CD107a	Mouse IgG1	H4A3
CD161	Mouse IgG1	HP-3G10
TCR Vα7.2	Mouse IgG1	3C10
Granzyme A	Mouse IgG1	CB9
Granzyme B	Mouse IgG1	GB11
IFN-γ	Mouse IgG1	4S.B3
Perforin	Mouse IgG2b	dG9
TNF	Mouse IgG1	MAb11

7. Optional: Fix conidia with 4% PFA for 1 h under continuous shaking. Wash three times with 10 mL PBS with 0.01% Tween20, resuspend in sterile PBS, and store at 4 °C until use (*see* **Note 4**).

3.2 Isolation of PBMC, MAIT Cells, and Monocytes

Conduct cell isolation under sterile conditions at room temperature.

3.2.1 Isolation of PBMC

1. Mix 5 mL blood from a buffy coat with 30 mL PBS at room temperature. Carefully layer 35 mL blood–PBS mixture over 15 mL separating solution in a 50-mL tube and centrifuge at 500 × *g* for 25 min at room temperature, *without brake* (*see* **Notes 5–7**).
2. After centrifugation, carefully transfer the PBMC ring into a fresh 50 mL tube prefilled with 10 mL PBS. Add up to 50 mL with PBS, centrifuge at 300 × *g* for 10 min at room temperature with brake (*see* **Note 8**).
3. Discard supernatant, resuspend pellet in 1 mL 1× erythrocyte lysis buffer, incubate 90 s, and add up to 50 mL with PBS. Centrifuge at 210 × *g* for 10 min at room temperature.

4. Discard supernatant, resuspend pellet in 25 mL PBS. Centrifuge at 210 × *g* for 8 min at room temperature.
5. Discard supernatant, resuspend in 1 mL medium (or MACS buffer), and count cell number in Neubauer chamber.

3.2.2 Isolation of Monocytes

Use whole blood or PBMC to isolate CD14+ monocytes with microbeads according to manufacturer's instructions. Here, the isolation from PBMC using CD14 MicroBeads and LS columns from Miltenyi Biotec is described (*see* **Note 9**).

1. Resuspend PBMC in 80 μL MACS buffer per 1 × 10^7 cells in a 15-mL tube (*see* **Note 10**). Add 20 μL CD14 MicroBeads, mix, and incubate for 15 min at 4 °C.
2. Wash by adding 10 mL MACS buffer and centrifuge at 300 × *g* for 8 min at 4 °C.
3. Remove supernatant completely and resuspend up to 10^8 cells in 500 μL MACS buffer.
4. Place LS column in magnet and prepare by rinsing with 3 mL MACS buffer. Add cell suspension and collect flow-through (contains unlabeled non-CD14 cells). Wash three times with 3 mL MACS buffer. Each time wait until column reservoir is empty before adding new buffer.
5. Remove column from magnet and place into new 15 mL tube, apply 5 mL MACS buffer, and immediately flush out CD14+ cells by firmly pushing the plunger into the column.
6. Use a small aliquot to determine purity by flow cytometry by staining for CD14. Centrifuge at 300 × *g* for 5 min at room temperature and discard supernatant. Resuspend in medium at desired concentration, plate in 96-well U-bottom plate, and keep at 37 °C and 5% CO_2 in the incubator until needed.

3.2.3 Isolation of MAIT Cells

Use whole blood or PBMC to isolate CD8+ cells with microbeads according to manufacturer's instructions. Here, the isolation from PBMC with CD8 MicroBeads and LS columns from Miltenyi Biotec is described (*see* **Note 11**).

1. Resuspend PBMC in 80 μL MACS buffer per 1 × 10^7 cells in a 15-mL tube (*see***Note 12**). Add 20 μL CD8 MicroBeads, mix, and incubate for 15 min at 4 °C.
2. Wash by adding 10 mL MACS buffer and centrifuge at 300 × *g* for 8 min at 4 °C.
3. Remove supernatant completely and resuspend up to 10^8 cells in 500 μL MACS buffer.
4. Place LS column in magnet and prepare by rinsing with 3 mL MACS buffer. Add cell suspension and collect flow-through (contains unlabeled non-CD8 cells). Wash three times with

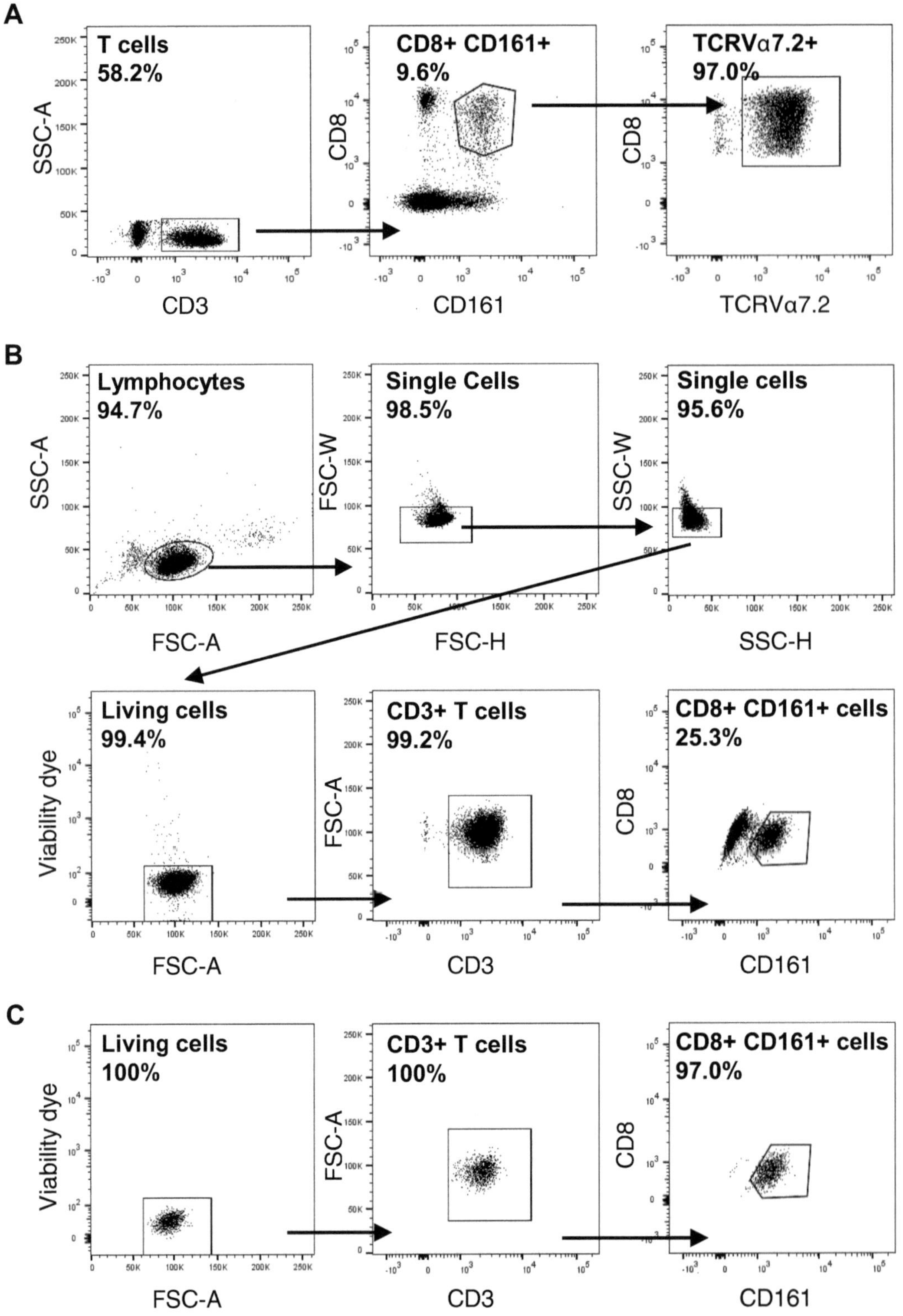

Fig. 1 Sorting of MAIT cells. (**a**) Staining strategy for sorting of MAIT cells. (**b**) Gating strategy for sorting MAIT cells. (**c**) Analysis of cell purity after sorting

3 mL MACS buffer. Each time wait until column reservoir is empty before adding new buffer.

5. Remove column from magnet and place into new 15 mL tube, apply 5 mL MACS buffer, and immediately flush out CD8+ cells by firmly pushing the plunger into the column.
6. Centrifuge at 300 × *g* for 5 min at room temperature, discard supernatant, and resuspend cells in 1 mL sterile PBS–EDTA.
7. Stain isolated CD8+ cells for sorting with CD3, CD161, and CD8 antibodies for 15 min in the dark at room temperature. Wash with 10 mL PBS–EDTA and resuspend in 2 mL PBS–EDTA with 1.5% FCS (*see* **Notes 13** and **14**).
8. Shortly before sorting add a viability dye (e.g., Sytox Blue). Purify MAIT cells with a cell sorter by gating as follows: lymphocyte population by FSC-A/SSC-A properties—single cells by FSC-H/FSC-W and SSC-W/SSC-H—viable cells by viability dye/FSC-A—T cells by CD3/FSC-A and finally MAIT cells by CD161/CD8. Analyze cell purity after sorting (*see* Fig. 1b, c).

3.3 Coincubation

3.3.1 Total PBMC

1. Resuspend cells and fungalconidia in medium (*see* **Note 15**).
2. Prepare PBMC suspension of 6.6×10^6 cells/mL and add 150 μL (1×10^6 cells) per well in a 96-well U-bottom plate.
3. Add 50 μL medium for unstimulated control or 50 μL of conidia suspension (40×10^6/mL = 2×10^6 cells, multiplicity of infection (MOI) = 2) and coincubate at 37 °C and 5% CO_2 for 4 or 24 h (*see* **Note 16**).

3.3.2 Isolated MAIT Cells and Monocytes

1. Resuspend monocytes at 2×10^6 cells/mL and add 50 μL (1×10^5 cells) per well in a 96-well U-bottom plate.
2. Resuspend MAIT cells at 0.4×10^6 cells/mL and add 50 μL (0.2×10^5 cells) per well in a 96-well U-bottom plate.
3. Add 10 μL conidia suspension (20×10^6/mL = 2×10^5 cell, MOI = 2) and coincubate at 37 °C and 5% CO_2 for 4 or 24 h.

Optional: For blocking TCR-dependent activation, add anti-MR1 antibody (clone 26.5) or the respective isotype control (mouse IgG2a) at 10 μg/mL directly before adding the conidia suspension.

For checking MAIT cell degranulation add CD107a antibody directly into the coculture for 4 h. Add Brefeldin A for the last 3 h of the coculture.

For intracellular staining of cytokines add brefeldin A (and monensin) during the last 4 h of coculture.

3.4 Flow Cytometry

Consider general flow cytometry recommendations like choice of antibody clones, combination of fluorochromes, antibody titration, and perform instrument settings (compensation) in advance.

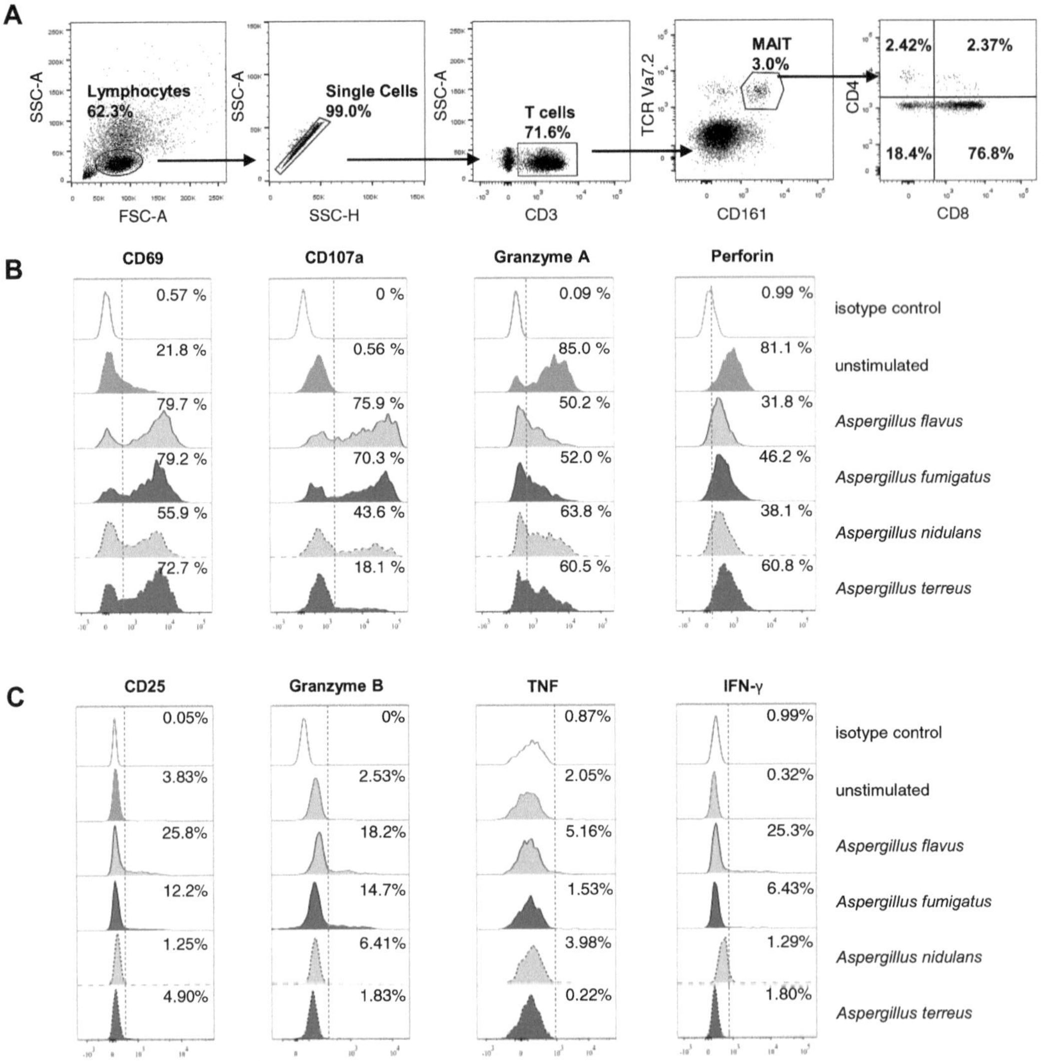

Fig. 2 Flow cytometric analysis of MAIT cell activation by *Aspergillus* species. (**a**) Gating strategy for MAIT cells. (**b**) Stimulation of human PBMC with 2 h swollen conidia from different *Aspergillus* species for 4 h. Expression of extracellular CD69, CD107a and intracellular perforin and granzyme A is shown. (**c**) Stimulation of human PBMC with 4 h swollen fixed conidia from different *Aspergillus* species for 24 h. Expression of extracellular CD25 and intracellular granzyme B, TNF, and IFN-γ is shown* (*Jahreis *et al.*, Human MAIT cells are rapidly activated by *Aspergillus* spp. in an APC-dependent manner. Eur J Immunol. 2018;48 (10):1698–1706, Copyright Wiley-VCH Verlag GmbH & Co. KGaA. Reproduced with permission)

1. After coincubation, centrifuge 96-well U-bottom plate at 300 × *g* for 5 min at room temperature and remove supernatant. Add 150 μL PBS–EDTA, transfer into 96-well V-bottom plate for staining and centrifuge at 300 × *g* for 5 min.
2. Perform extracellular staining for MAIT cells (CD3, CD161, TCRVα7.2, and if needed CD4 and CD8) and activation marker (e.g., CD25 and CD69). Prepare antibody master mix

by adding needed antibody amount in 50 μL PBS–EDTA per well. Add 50 μL antibody mix, incubate for 15 min at room temperature in the dark (*see* **Notes 17** and **18**).

3. Add 150 μL PBS–EDTA and centrifuge at 300 × *g* for 5 min. Discard supernatant and either resuspend in PBS–EDTA for analysis or proceed with intracellular staining.
4. Fix cells in 200 μL 4% PFA by incubating for 5 min at room temperature in the dark. Centrifuge at 300 × *g* for 5 min and discard supernatant.
5. Wash by adding 200 μL PBS–EDTA, centrifuge at 300 × *g* for 5 min, and discard supernatant.
6. Block with 25 μL saponin buffer and 5% mouse serum for 10 min at room temperature in the dark (*see* **Note 19**).
7. Prepare antibody mix by adding needed antibody amount in 25 μL saponin buffer per well (e.g., perforin, granzymes, and cytokines). Without washing add 25 μL antibody mix, incubate for 30 min at room temperature in the dark (*see* **Note 20**).
8. Wash by adding 150 μL saponin buffer, centrifuge at 300 × *g* for 5 min, and discard supernatant.
9. Wash by adding 150 μL PBS–EDTA, centrifuge at 300 × *g* for 5 min, and discard supernatant.
10. Resuspend in PBS–EDTA for direct or in FACS-Fix for later analysis.

Representative data of MAIT cells after stimulation with different *Aspergillus* species analyzed by flow cytometry is shown in Fig. 2b, c.

4 Notes

1. Before starting your experiments, it is useful to check conidia size and germination behavior. Pre-swelling of *Aspergillus* conidia facilitates recognition and uptake by phagocytes [9]. However, germlings and hyphae are not phagocytosed anymore and can also lead to clogging of the flow cytometer [8].
2. The fungal stock solution of resting conidia can be stored at 4 ° C for 1 month; swollen conidia can be stored at 4 ° C for 1 week.
3. Take at least five times more conidia than needed for coincubation. There is a high loss of fungal cells due to the many washing steps.
4. Shorter fixation time is possible, test efficiency by streaking fixed conidia on a malt agar plate and check growth after

overnight incubation at 37 ° C without CO_2. Fungal conidia can also be fixed by heat inactivation for 1 h at 90 ° C.

5. It is recommended to use commercially available sterile PBS with low endotoxin levels.
6. Use isotone separating solution with a density of 1.077 g/mL, commercially available formats are recommended.
7. If more cells are needed, up to 10 mL buffy coat can be resuspended with 25 mL PBS, for higher blood volumes prepare more 50 mL tubes with separating solution. If fresh blood is used, dilute 1:2 in PBS.
8. Removing 10–15 mL plasma first may make removal of the PBMC ring easier.
9. According to the manufacturer LS columns can be used for a maximum of 10^8 labeled cells and 2×10^9 total cells.
10. The number of monocytes is donor dependent; however, approximately 15–20% of the PBMC are CD14+ cells.
11. Consider that the percentage of MAIT cells is highly donor dependent and can range from 1% to 10% of peripheral T cells [10]. If several donors are available, checking the MAIT frequency in advance by staining leucocytes with CD3, CD161, and TCRVα7.2 is helpful.
12. The number of CD8+ cells is donor dependent; however, approximately 15% of the PBMC are CD8+ cells.
13. Using this method CD4−CD8− MAIT cells and CD4+ MAIT cells are excluded.
14. For sorting, staining of TCRVα7.2+ is also possible. However, 97% of the CD3+CD161+CD8+ lymphocyte population is TCRVα7.2+ (*see* Fig. 1a) and possible pre-activation of the TCR by antibody staining is prevented.
15. For coincubation times longer than 24 h addition of antibiotics into the culture medium (e.g., 1% penicillin/streptomycin) is recommended.
16. For short coincubation of 4 h alive 2 h pre-swollen *Aspergillus* or resting *Mucorales*conidia can be used (for *A. fumigatus* maximal time for swelling and coincubation in total is 10 h). For longer coincubation use fixed conidia to prevent overgrow of cells with fungal mycelium, e.g., perform 24 h stimulation with 4 h pre-swollen fixed *Aspergillus* and 2 h pre-swollen fixed *Mucorales*conidia.
17. Gating of MAIT cells is usually distinct (*see* Fig. 2a). For setting gates of all investigated markers it is recommended to prepare a control staining containing antibodies for MAIT gating plus all respective isotype controls which are added during extracellular and intracellular staining steps.

18. For gating CD107a either use the respective isotype control (mouse IgG1) which is also directly added to the coculture or set gates according to the unstimulated control. Staining of extracellular MAIT cell marker is performed after coincubation as described. As Brefeldin A might interfere with expression of activation and other markers, it is recommended to perform CD107a staining separately.
19. Use serum from the same species as antibodies used for intracellular staining are raised in (e.g., if antibodies are of murine origin use mouse serum).
20. Cytokine detection after short fungal stimulation is not expected, thus it is recommended to check after 24 h fungal stimulation.

Acknowledgment

This work was supported by the German Federal Ministry of Education and Health (BMBF) Germany, (Center for Sepsis Control and Care (01E01002); InfectoGnostics Research Campus (13GW0096D)) and Gilead (Förderprogramm Infektiologie).

References

1. Patel O, Kjer-Nielsen L, Le Nours J et al. (2013) Recognition of vitamin B metabolites by mucosal-associated invariant T cells. Nat Commun 4:2142. https://doi.org/10.1038/ncomms3142
2. Loh L, Wang Z, Sant S et al. (2016) Human mucosal-associated invariant T cells contribute to antiviral influenza immunity via IL-18-dependent activation. dependent activation. Proceedings of the National Academy of Sciences of the United States of America 113 (36):10133-10138. https://doi.org/10.1073/pnas.1610750113
3. Dias J, Leeansyah E, Sandberg JK (2017) Multiple layers of heterogeneity and subset diversity in human MAIT cell responses to distinct microorganisms and to innate cytokines. Proceedings of the National Academy of Sciences of the United States of America 114 (27): E5434–E5443. https://doi.org/10.1073/pnas.1705759114
4. Le Bourhis L, Martin E, Peguillet I et al. (2010) Antimicrobial activity of mucosal-associated invariant T cells. Nat Immunol 11(8):701–708. https://doi.org/10.1038/ni.1890
5. Kjer-Nielsen L, Patel O, Corbett AJ et al. (2012) MR1 presents microbial vitamin B metabolites to MAIT cells. Nature 491 (7426):717–723. https://doi.org/10.1038/nature11605
6. Gherardin NA, Keller AN, Woolley RE et al. (2016) Diversity of T Cells Restricted by the MHC Class I-Related Molecule MR1 Facilitates Differential Antigen Recognition. Immunity 44 (1):32–45. https://doi.org/10.1016/j.immuni.2015.12.005
7. Kurioka A, Ussher JE, Cosgrove C et al. (2015) MAIT cells are licensed through granzyme exchange to kill bacterially sensitized targets. Mucosal Immunol 8 (2):429–440. https://doi.org/10.1038/mi.2014.81
8. Jahreis S, Bottcher S, Hartung S et al. (2018) Human MAIT cells are rapidly activated by Aspergillus spp. in an APC-dependent manner. Eur J Immunol 48(10):1698–1706. https://doi.org/10.1002/eji.201747312
9. Hartung S, Rauh C, Hoang TNM et al. (2018) Fast and Quantitative Evaluation of Human Leukocyte Interaction with Aspergillus fumigatus Conidia by Flow Cytometry. Cytometry Part A : the journal of the International Society for Analytical Cytology. https://doi.org/10.1002/cyto.a.23653
10. Dusseaux M, Martin E, Serriari N et al. (2011) Human MAIT cells are xenobiotic-resistant, tissue-targeted, CD161hi IL-17-secreting T cells. Blood 117 (4):1250–1259. https://doi.org/10.1182/blood-2010-08-303339

Chapter 12

A Flow Chamber Assay for Studying MAIT Cell Trafficking

Farhat Parween, Hongwei H. Zhang, and Joshua M. Farber

Abstract

Human MAIT cells show little expression of the selectin CD62L and the chemokine receptor CCR7, which are important for entering lymph nodes, and high expression of selectin ligands and chemokine receptors that mediate trafficking into inflamed tissue. Extravasation of leukocytes into tissue requires sequential steps including rolling, firm arrest, crawling, and transendothelial migration, and can be modeled using endothelial cell monolayers in flow chambers that approximate the sheer stress found in post-capillary venules. Using MAIT cells purified from elutriated lymphocytes by fluorescence-activated cell sorting, we have used flow chambers to demonstrate roles for individual chemokine receptors in specific steps required for extravasation. These methods provide a general way to study the molecular mechanisms underlying MAIT cell trafficking from blood into tissue.

Key words MAIT cells, Transendothelial migration, Flow chamber, Lymphocytes, Chemokines

1 Introduction

In humans, MAIT cells are found in large numbers in liver and intestine but not in lymph nodes [1], and migrate into sites of infection [2]. Consistent with their preferential localization in peripheral tissue, MAIT cells express little CD62L and CCR7, but high levels of selectin ligands and the chemokine receptors CCR6, CCR5, and CXCR6 [1, 3]. Nonetheless, there have been few studies characterizing the mechanisms of MAIT cell trafficking.

Extravasation of leukocytes at sites of inflammation occurs due to cytokine-induced expression of selectins and integrin ligands, and display of chemoattractants/chemokines on endothelial cells [4, 5]. The multistep paradigm of leukocyte extravasation on the activated endothelium involves the sequential steps of rolling, mediated by selectin ligands/selectins, followed by firm arrest, crawling, and transendothelial migration, mediated by chemoattractant receptors, integrins, and their ligands, as well a variety of adhesion proteins [5–7]. These steps can be studied using flow chambers in which are grown monolayers of endothelial cells,

Helen Kaipe and Isabelle Magalhaes (eds.), *MAIT Cells: Methods and Protocols*, Methods in Molecular Biology, vol. 2098,
https://doi.org/10.1007/978-1-0716-0207-2_12,

such as those obtained from human umbilical veins [8]. Unlike trans-well-based assays for chemotaxis, flow chambers using video-microscopy allow for the quantitative, real-time analysis of each of the steps the leukocytes' extravasation under conditions of physiological shear stress, which is an important promoter of leukocyte migration across endothelium [8–10]. As a reductionist, in vitro assay flow chambers provide a great deal of flexibility. Sources of endothelial cells can be varied, as can the populations of leukocytes. Cells can be manipulated genetically and/or subjected to stimuli and/or inhibitors for the investigation of molecular mechanisms. A limitation of the assay is that it only monitors the initial interactions between leukocytes and endothelial cells, and cannot fully recapitulate the process of extravasation through a bona fide post-capillary venule.

Using a stringent assay for extravasation in vivo, we showed that human MAIT cells were significantly better at migrating into the inflamed skin of mice as compared with other subsets of $CD8\alpha^+$ T cells [3], suggesting that MAIT cells serve as excellent model cells for understanding mechanisms of efficient extravasation of effector-capable T cells into inflammatory sites. The protocol below describes how the flow chamber assay can be used to investigate the MAIT cell:endothelial cell interactions that initiate extravasation in inflammation.

2 Materials

Materials and procedures include human cells. Although we typically use cells from healthy donors, it is always possible that samples contain blood-borne pathogens, so that proper precautions must be taken in working with any human material, both in their handling and in their disposal. Obtaining samples from humans must be done with informed consent and under protocols approved after appropriate institutional review consistent with the Declaration of Helsinki principles. All manipulations with cells prior to the flow chamber assays should be performed under sterile conditions, typically in a Class II biosafety cabinet. We have listed suppliers and catalog numbers for many reagents that we use, but alternative reagents can usually be substituted.

2.1 Cells

1. Primary human umbilical vein endothelial cells (HUVEC) from ATCC, Manassas, VA cat#: PCS-100-013 (*see* **Notes 1** and **2**).
2. Human whole blood and elutriated lymphocytes from healthy donors (*see* **Note 3**).

2.2 Cell Culture Media and Reagents

1. HUVEC culture medium: Vascular cell basal medium (ATCC cat# PCS-100-030) containing endothelial vascular cell growth kit-VEGF (ATCC cat# PCS-100-041) and 100 U/mL penicillin–streptomycin.
2. RPMI complete medium: RPMI-1640 with L-glutamine (Lonza cat# 12-702F) containing 10% fetal bovine serum (FBS), 100 U/mL penicillin–streptomycin, and 10 mM HEPES.
3. FACS Buffer: HBSS containing 4% FBS.
4. Cell freezing medium: FBS containing 8% DMSO.
5. Ficoll-Paque™ PLUS.
6. Red blood cell lysis buffer, Ammonium-Chloride-Potassium (ACK) Lysing Buffer or equivalent.

2.3 Cell Culture Plastics

1. 75 cm^2 U-shaped canted neck cell culture flask with vent cap.
2. 24-well flat bottom surface modified cell culture plate with lid.
3. Flow chambers: μ-Slide I 0.4 mm Luer ibiTreat microscopy chambers # 1.5 polymer coverslip, tissue culture treated, sterilized (Ibidi, LLC, Germany).
4. 70 μm cell strainer.
5. 50 mL Falcon tubes.
6. 15 mL Falcon tubes.
7. 5 mL polystyrene round-bottom tubes.

2.4 Recombinant Proteins

1. Human TNFα.
2. Human fibronectin.

2.5 Antibodies

1. RosetteSep for human $CD8^+$ T cell enrichment (Stem Cell Technologies cat# 15063).
2. Brilliant Violet 421 anti-human CD8α (clone RPA-T8).
3. Brilliant Violet 605 anti-human CD45RO (clone UCHL1).
4. PE-Cy5 anti-human CD62L (clone DREG-56).
5. FITC anti-human TCRVα7.2 (clone 3C10).
6. Allophycocyanin anti-human CD161 (clone NKR-P1A).
7. PE-Cy7 anti-human CCR6 (clone G034E3).
8. Fluorescence-activated cell sorter, such as Aria cytometer (BD Biosciences).

2.6 Flow Chamber Assay

1. Tube adapter set (Ibidi, LLC, Germany cat# 10831).
2. 60 mL syringe with Luer lock tip (Monoject™ cat# 414635).
3. 12 mL syringe with Luer lock tip (Monoject™ cat# 875713).
4. Stopcock with Luer connections, 1-way, male lock (Cole-Parmer cat# EW-30600-00) or equivalent.

5. 3-way stopcock with two female Luer lock ports and one SPIN-LOCK connector.
6. Monoject™ blunt cannula 16 gauge with Luer lock (Cardinal cat# 8881202322).
7. 2 mL microcentrifuge tubes.
8. 44 Programmable Syringe Pump (Harvard Apparatus Model 55-114).
9. Integrated Live Cell Workstation Leica AF 6000LX (Leica Microsystems, Inc.) or equivalent.
10. Imaris software (Bitplane).

3 Methods

Performing flow chamber experiments includes the following major steps:

1. Preparation of the flow chambers with HUVEC monolayers.
2. FACS sorting of MAIT cells from elutriated lymphocytes.
3. Running the flow chamber assay with videomicroscopy.
4. Analysis and quantification of MAIT cell:endothelial cell interactions.

3.1 Preparation of HUVEC Monolayers

1. Culture cells from a frozen vial of HUVEC in a 75-cm^2 tissue culture flask in a 37 °C, 5% CO_2 incubator to approximately 85% confluency. Growth to confluency typically occurs after 3 days, so that cells should be plated at Day-2, 2 days before obtaining the elutriated lymphocytes from the apheresis donor, which we set as Day 0.
2. HUVEC are harvested on Day-1, one day before using the flow chamber in an experiment. One hour before harvesting the HUVEC, carefully coat the flow chamber with fibronectin. The flow chambers used here have a channel of length 50 mm, width 5.0 mm, and height 0.4 mm, with a channel volume of 100 μL. Apply 100 μL of 50 μg/mL fibronectin in PBS by hand with a constant and rapid flow to avoid air bubbles and incubate for 1 h at 37 °C (*see* **Note 4**).
3. Harvest HUVEC as described in **Note 2** and resuspend cells in 10 mL of HUVEC culture medium and centrifuge in a table-top centrifuge at 1500 rpm (500 × *g*) for 5 min. Resuspend the cell pellet in 1 mL HUVEC culture medium. Count cells in a hemocytometer, adjust volume to achieve a concentration of 1.5–2 × 10^6 cells/mL, and add TNFα to a concentration of 40 ng/mL.
4. Wash the chambers by flushing with 150 μL of HUVEC culture medium containing 40 ng/mL TNFα into one end and

simultaneously aspirate it from the opposite side. Add 100 μL HUVEC suspension from **step 3** at one end by tilting the chamber slightly and aspirating 100 μL of buffer from the opposite side. Incubate the chamber flat in a 37 °C, 5% CO_2 incubator for 30 min for HUVEC to settle into a monolayer.

5. After HUVEC settle, fill the chamber with an additional 60 μL of HUVEC culture medium containing 40 ng/mL TNFα and continue incubating for 18–20 h.

3.2 FACS Sorting of MAIT Cells from Elutriated Lymphocytes

MAIT cells typically constitute 1–4% of peripheral T cells in human blood. In order to obtain sufficient numbers of cells for the flow chamber assays, we typically start with approximately $1.5–3 \times 10^8$ elutriated lymphocytes, collected from the donor on Day 0. On the same day, we obtain 20 mL of freshly donated whole blood, which need not be from the lymphocyte donor. The blood products are left at room temperature overnight on a rocker set at 55 rpm.

1. Collect the apheresis product in 50 mL Falcon tubes and centrifuge in a table-top centrifuge at 2000 rpm (900 × *g*) for 10 min at room temperature. Remove the supernatant, resuspend the pellet in 20 mL whole blood, and filter with the 70 μ cell strainer into a fresh 50 mL Falcon tube. Add 1 mL of RosetteSep human $CD8^+$ T cell enrichment cocktail and incubate at room temperature for 20–30 min (*see* **Note 5**).
2. For each 10 mL of the above mixture dilute with 20 mL FACS buffer in separate Falcon tubes. Pipette 10 mL of Ficoll-Hypaque (density = 1.078) into each of two 50 mL Falcon tubes and carefully add 30 mL of diluted blood sample on top. Centrifuge in a table-top centrifuge at 2000 rpm (900 × *g*) for 20 min with brake off at room temperature (*see* **Note 6**).
3. After centrifugation, granulocytes and aggregated erythrocytes are found at the bottom of the tube, $CD8^+$ T cells form at the interface above the Ficoll layer, and plasma is on top. Carefully collect $CD8^+$ T cells from the interface of both the tubes and pool. Fill the tube with the pooled cells with FACS buffer and centrifuge without delay at 1500 rpm (500 × *g*) for 5 min (*see* **Note 7**).
4. Add 5 mL of ACK lysis buffer to the cell pellet. After 2 min, wash cells by adding 45 mL of FACS buffer and centrifuge in a table-top centrifuge at 1500 rpm (500 × *g*) for 5 min at room temperature.
5. Resuspend the cell pellet in 500 μL FACS buffer and add anti-human CD8α (5 tests, 25 μL), anti-TCRVα7.2 (5 tests, 25 μL), anti-CD62L (5 tests, 25 μL), anti-CD45RO (5 tests, 25 μL), and anti-CD161 (10 tests, 50 μL) antibodies, and incubate in the dark for 30 min (*see* **Note 8**). Wash cells by adding FACS buffer to fill the tube (50 mL) and centrifuge at 1500 rpm

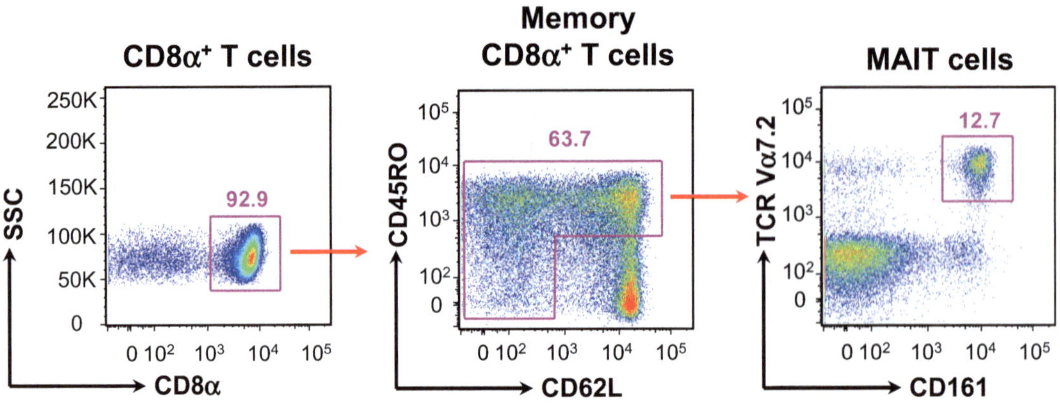

Fig. 1 Sequential gating of RosetteSep-purified CD8$^+$ T cells for purification of MAIT cells and (if desired) non-MAIT memory-phenotype T cells. Headings indicate the types and numbers indicate the percentages of cells within the demarcated areas

($500 \times g$) for 5 min at room temperature. Resuspend the cell pellet in 1 mL FACS buffer. Count cells in a hemocytometer, adjust volume to achieve a concentration of 0.5×10^6 cells/mL. Transfer cells to a 5 mL polystyrene round-bottom tube and leave on ice and covered with aluminum foil while the cell sorter is readied. For sorting cells, use standard procedures to gate for lymphocytes based on forward and side scatter and singlets. Cells are then gated sequentially for CD8α and CD45RO and CD62L for cells with the memory phenotype (CD45RO$^+$ CD62L$^{+/-}$ or CD45RO$^-$ CD62L$^-$). Memory-phenotype cells are then gated based on TCRVα7.2 and CD161 in order to purify MAIT cells (TCRVα7.2$^+$CD161$^+$) and non-MAIT cells that can be used in the flow chamber assays as comparators if desired (Fig. 1). Sorting should be continued in order to obtain at least 1×10^6 MAIT cells (or other subsets) for each flow chamber assay. Subsets of sorted cells are collected by centrifugation in the FACS collection tube or 15 mL Falcon tube at 1500 rpm ($500 \times g$) for 5 min at room temperature. Resuspend the cell pellet in 1 mL RPMI complete medium. Count cells in a hemocytometer, adjust volume to achieve a concentration of 1×10^6 cells/mL in RPMI complete medium, transfer to a 24-well plate (1 mL per well), and leave overnight in a 37 °C, 5% CO_2 incubator.

3.3 Running the Flow Chamber Assay

Use the Integrated Live Cell Workstation Leica AF 6000LX (Leica Microsystems Inc.) or equivalent, which has features to control and maintain temperature, pH, and humidity during an experiment. The complete set-up used for the flow chamber experiment is shown in Fig. 2, which can be referred to for clarifying the steps described below.

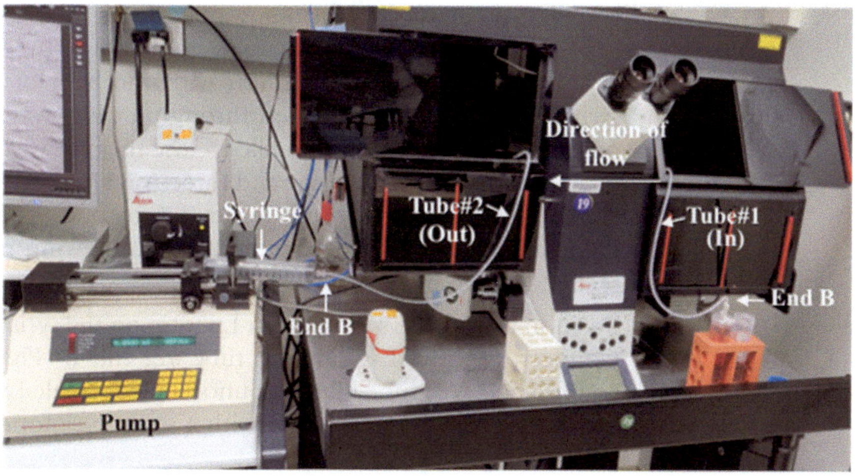

Fig. 2 Integrated Live Cell Workstation Leica AF 6000LX set-up for flow chamber experiment. Labels of tubing correspond to those in Fig. 3 and as described in the text

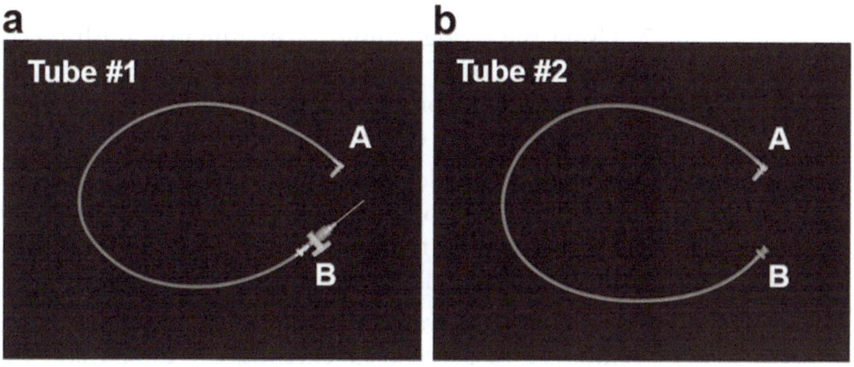

Fig. 3 Tube adapter set for flow chamber assays. (**a**) Tube #1 connects flow chamber to the 50 mL Falcon tube containing 37 °C RPMI complete medium with or without cells and (**b**) tube #2 connects the flow chamber to the 60 mL syringe on the pump. Ends of each tube are labeled for reference to text and Fig. 2

1. Remove the HUVEC-coated flow chambers from the incubator and check under a phase contrast microscope to be sure that the endothelial cells have formed a confluent monolayer.
2. Transport the HUVEC-coated flow chambers and the plate of lymphocytes to the microscope room in an insulated container.
3. Bring the internal temperature of the microscope workstation to 37 °C. With the HUVEC-coated flow chamber outside the microscope workstation, remove the caps, check for any air bubbles inside the chamber, and fill the reservoir with RPMI complete medium until there is a small dome of liquid.
4. Connect end B of tube #1 to the 16 gauge blunt cannula (Fig. 3a) using the 1-way stopcock and place the tip of the cannula into a 50-mL Falcon tube containing 37 °C RPMI

complete medium with the one-way stopcock resting on the lip of the tube in order to remain accessible. Fill tube #1 with 37 °C RPMI complete medium with the help of a 12-mL syringe with Luer tip attached through a three-way stopcock on end A. Once the tubing is filled, close the one-way stopcock on end B, remove the syringe and three-way stopcock from end A and connect tube #1 through end A to one of the female Luer fittings of the flow chamber with a screwing motion to seat end A completely.

5. Position the 60 mL syringe with Luer tip on the syringe pump. Connect the 60 mL syringe to tube #2, end B (Fig. 3b) and connect end A of tube #2 to the unoccupied female Luer fitting of the flow chamber, as above with a screwing motion to seat end A completely.
6. Place the flow chamber under the microscope equipped with a 20 × DIC objective (Fig. 2). Focus and position the chamber to record from a field with a fully confluent monolayer. Adjust the microscope to record data for 20 min per sample.
7. Using the syringe pump as an aspirator, wash the monolayer with 2 mL of 37 °C RPMI complete medium at a flow rate of 3.2 mL/min, corresponding to a shear stress of 5 dyn/cm^2 to rid the chamber of any floating debris.
8. For each sample/assay transfer 0.8 mL, corresponding to 8×10^5 cells from one well of the 24-well plate, into a 2-mL microcentrifuge tube and add 1.2 mL of 37 °C RPMI complete medium, mix cells well, and move expeditiously to **step 9**.
9. Following the wash in **step** 7, close the stopcock at end B of tube #1 to prevent air entering the tubing and transfer the cannula to the tube containing cells.
10. Open the stopcock and using the syringe pump begin a flow of 0.53 mL/min, corresponding to a shear stress of 0.75 dyn/cm^2. Monitor the chamber under the microscope and begin video recording at a rate of four frames/s when cells enter the chamber (Time 0). Just before the cell suspension is exhausted, transfer the end B of tube #1 to the 50-mL Falcon tube containing 37 °C RPMI complete medium, which is replenished as necessary during the experiment. At the 4-min time point, increase the flow rate to 3.2 mL/min, corresponding to a shear stress of 5 dyn/cm^2 for an additional 16 min (*see* **Note** 9).

3.4 Analysis and Quantification of Videomicroscopy Data

For the analysis of MAIT cell (or other T cell) trafficking, use the Imaris software. Imaris software helps to categorize cells and create a time-coded track for each cell. Identifications of cells made by the software need to be checked for accuracy. Count rolling cells as all cells that rolled, whether or not they subsequently arrested.

Arrested cells are those that remained stopped on the endothelial monolayer for 10 s or longer. Transmigrating cells are those that migrate under the endothelial monolayer, which results in the cells' stepwise darkening (*see* **Note 10**).

4 Notes

1. As a model of inflamed vascular endothelium, a monolayer of TNFα-stimulated HUVEC is used because these cells express the major adhesive ligands found on many inflamed venules [11]. Although this protocol focuses on MAIT cell trafficking, it could be adapted to study the transendothelial migration of other leukocytes.
2. For expanding HUVEC for future use, thaw the vial containing 5×10^5 viable HUVEC from ATCC into 10 mL of HUVEC culture medium, pellet by centrifugation in a table-top centrifuge at 1500 rpm ($500 \times g$) for 5 min, and culture in 20 mL of fresh HUVEC culture medium in a 75-cm^2 flask in a 37 °C in 5% CO_2 incubator. Grow cells to confluence and split 1:3 using trypsinization with 5 mL of 0.05% trypsin-EDTA per flask and collect cells by gentle pipetting. Cells are grown for two passages before freezing, one confluent flask into three vials using freezing medium. Cells are stored at −80 °C for months or longer term in liquid nitrogen.
3. For isolating MAIT cells, we usually start with approximately 2×10^8 lymphocytes, which are typically ¼ of the cells obtained from one donor.
4. Use a separate flow chamber to assay each MAIT cell sample, for example cells treated with a specific inhibitor vs. untreated control cells.
5. RosetteSep isolates cells by negative selection, by crosslinking unwanted cells to red blood cells. The pre-purification of cells significantly shortens the time needed for cell sorting.
6. Ficoll and FACS buffer should be at room temperature. The cells in blood should be pipetted slowly and close to the Ficoll layer, down the side of the tube that is tilted slightly in order to prevent disrupting the interface.
7. Remove cells from the interface without disturbing the gradient and avoid collecting Ficoll along with cells.
8. The vast majority of human MAIT cells express CD8α, and have generally been identified by their co-expression of TCRVα7.2 and CD161. We found that CCR6 and CD161 marked identical cells within the $CD8\alpha^+$ $TCRV\alpha7.2^+$ subset, so that CCR6 and CD161 could be used interchangeably for identifying MAIT cells [3]. If staining for CCR6, use 10 tests,

50 μL of the PE-Cy7 anti-human CCR6 antibody in place of anti-human CD161.

9. It is critical to be vigilant when setting up the flow chamber, transferring the cannula, and running the assay in order to prevent air from entering the tubing or the chamber.

10. Virtually all cells that roll and arrest do so during the first 4 min of the recording at shear stress of 0.75 dyn/cm^2. Some transmigrating cells may return to the upper side of the endothelial monolayer, usually to transmigrate again. These cells are scored as undergoing transendothelial migration. Under the experimental conditions described here using TNFα-activated HUVEC, a typical video recording should show approximately 18 MAIT cells rolling, 13 cells arresting, and 7 cells undergoing transendothelial migration [3].

Acknowledgment

This research was supported by the Intramural Research Program of the NIH, National Institute of Allergy and Infectious Diseases.

References

1. Dusseaux M, Martin E, Serriari N, Peguillet I, Premel V, Louis D, Milder M, Le Bourhis L, Soudais C, Treiner E, Lantz O (2011) Human MAIT cells are xenobiotic-resistant, tissue-targeted, CD161hi IL-17-secreting T cells. Blood 117:1250–1259
2. Le Bourhis L, Martin E, Peguillet I, Guihot A, Froux N, Core M, Levy E, Dusseaux M, Meyssonnier V, Premel V, Ngo C, Riteau B, Duban L, Robert D, Huang S, Rottman M, Soudais C, Lantz O (2010) Antimicrobial activity of mucosal-associated invariant T cells. Nat Immunol 11:701–708
3. Lee CH, Zhang HH, Singh SP, Koo L, Kabat J, Tsang H, Singh TP, Farber JM (2018) C/EBPdelta drives interactions between human MAIT cells and endothelial cells that are important for extravasation. Elife 7:e32532
4. Pober JS, Sessa WC (2007) Evolving functions of endothelial cells in inflammation. Nat Rev Immunol 7:803–815
5. Nourshargh S, Alon R (2014) Leukocyte migration into inflamed tissues. Immunity 41:694–707
6. Springer TA (1994) Traffic signals for lymphocyte recirculation and leukocyte emigration: the multistep paradigm. Cell 76:301–314
7. Vestweber D (2015) How leukocytes cross the vascular endothelium. Nat Rev Immunol 15:692–704
8. Cinamon G, Alon R (2003) A real time in vitro assay for studying leukocyte transendothelial migration under physiological flow conditions. J Immunol Methods 273:53–62
9. Kitayama J, Hidemura A, Saito H, Nagawa H (2000) Shear stress affects migration behavior of polymorphonuclear cells arrested on endothelium. Cell Immunol 203:39–46
10. Cinamon G, Shinder V, Alon R (2001) Shear forces promote lymphocyte migration across vascular endothelium bearing apical chemokines. Nat Immunol 2:515–522
11. Luscinskas FW, Ding H, Tan P, Cumming D, Tedder TF, Gerritsen ME (1996) L- and P-selectins, but not CD49d (VLA-4) integrins, mediate monocyte initial attachment to TNF-alpha-activated vascular endothelium under flow in vitro. J Immunol 157:326–335

Part III

Monitoring and In Vitro Amplification of MAIT Cells

Chapter 13

Production of MR1 Tetramers Loaded with Microbial Ligands

Nicole A. Ladd, Cara A. Froyd, and Erin J. Adams

Abstract

In lieu of peptides, the monomorphic MHC-I-like molecule MR1 presents small molecule antigens to stimulate a subset of αβ T cells known as mucosal-associated (semi-) invariant T (MAIT) cells or, more broadly, MR1-restricted (MR1T) cells. The MR1 ligands identified to date are limited to derivatives and intermediates of the riboflavin and folate biosynthesis pathways and their presentation is therefore thought to be an indicator of infection by microbial species that can synthesize riboflavin. MAIT cells have, in recent years, been studied and isolated using a tetrameric reagent of recombinant MR1 loaded with the canonical ligand 5-OP-RU due to its potency toward MAIT clones. However, new evidence has shown that the repertoire of MR1 ligands is much more diverse than previously appreciated and, consistent with this, that the 5-OP-RU tetramer does not bind all MR1T cells. To study MR1-restricted T cell clones in the context of unique bacterial infection, we have generated a tetramer of MR1 loaded with diverse microbial antigens. The production of this reagent is detailed in this chapter.

Key words MR1, MAIT, Recombinant protein, Microbes, Mucosal immunity, T Cell Receptor, MR1T

1 Introduction

One cornerstone of the canonical adaptive immune system is defined by an interaction axis with three components: a molecule belonging to the major histocompatibility complex (MHC), a peptidic antigen bound to the MHC, and the αβ T cell receptor (TCR) that is reactive toward this MHC–ligand complex. In contrast to this dogma, several MHC-like molecules have evolved to present antigens of non-peptide origin. Among these is MHC-I related protein 1 (MR1), which is capable of presenting small molecule antigens to αβ T cells during the course of microbial infection [1, 2]. The cognate T cells, known as mucosal-associated (semi-) invariant T (MAIT) cells, are mucosal resident lymphocytes that use a semi-conserved rearrangement of the TRAV1-2 gene with a small number of TCRα junctional genes (TRAJ20/33) paired with a

Helen Kaipe and Isabelle Magalhaes (eds.), *MAIT Cells: Methods and Protocols*, Methods in Molecular Biology, vol. 2098, https://doi.org/10.1007/978-1-0716-0207-2_13,

limited number of TCRβ variable genes (TRBV6/20) [3, 4]. The semi-invariant nature of the TCRα chain rearrangement ensures encoding of a tyrosine residue in the CDR3α loop; it is this tyrosine that contacts the canonical MR1 ligands in a conserved fashion [5, 6].

The activating MR1 ligands identified to date are limited to derivatives of the riboflavin biosynthetic pathway, which is expressed by some bacterial, plant, and fungal species but has been lost in animals. 5-Amino-6-D-ribitylaminouracil (5-A-RU), an intermediate in this pathway, undergoes spontaneous condensation with small molecules in the cellular environment to produce ribityllumazine and ribityluracil compounds with variable potential to activate cognate T cells [1]. One ribityluracil derivative, 5-(2-oxopropylideneamino)-6-D-ribityluracil (5-OP-RU), is the most potent MR1 antigen identified to date and has therefore been used as the standard for potency and defining MR1-restricted T cell populations. Since 5-OP-RU has only been described through synthetic chemistry, it is unclear whether it is physiologically relevant.

When MAIT cells were first described, it was believed that they were a homogenous population with reactivity toward a small number of ligands in a fashion similar to invariant Natural Killer T (iNKT) cells, which respond to lipid antigens presented by the MHC-like molecule CD1d. Isolating MR1-restricted T cells using a tetrameric MR1 ligand loaded with the potent canonical ligand 5-OP-RU ($MR1^{+5\text{-}OP\text{-}RU}$) reinforces this understanding of the MR1-MAIT axis, as the majority of tetramer^{+} cells are TRAV1-2^{+} [7]. In order to assess both the ligand repertoire of MR1 and characteristics of its cognate T cells, we have developed a recombinant MR1 tetramer reagent that can be loaded either with a heterogeneous mixture of microbial ligands ($MR1^{+Bact}$) or a single, synthetic, or isolated antigen ($MR1^{+Ag}$). Strikingly, in several donors, $MR1^{+Bact}$ tetramers stain a larger pool of the total CD3^{+} CD4^{-} cells than $MR1^{+5\text{-}OP\text{-}RU}$, and, in general, a greater percentage of the $MR1^{+Bact}$ positive cells are TRAV1-2^{-} than of the $MR1^{+5\text{-}OP\text{-}RU}$ positive cells [2]. We therefore refer to these cells as MR1-restricted T (MR1T) cells, a population within which MAIT cells exist as a subset. These data challenge the model that the MR1-MAIT axis is tightly restricted in both ligand repertoire and TRAV1-2 usage.

In this chapter, we detail the production of this tetrameric recombinant MR1 ligand. First, we describe a protocol for the production and purification of recombinant, chimeric (bovine-human) MR1 in insect and mammalian cells and how to load this protein with a heterogeneous population of microbial ligands through introduction of microbes into the expression medium. We then present a purification strategy for recovery of the recombinant protein. Additionally, we describe a protocol for producing and refolding chimeric MR1 from bacterial inclusion bodies around selected MR1 ligands. Finally, we detail the tetramerization of these protein reagents for use in downstream applications.

2 Materials

General reagents and cell strains. All chemical reagents are filtered using two glass fiber prefilters before use (1.6 μm pore size). Any reagents used for column chromatography in a fast protein liquid chromatography (FPLC) system are filtered through 0.22 μm nitrocellulose membrane filters. Store all reagents at room temperature unless otherwise noted.

2.1 General Reagents

1. *Mycobacteriumsmegmatis* strain: mc^2155 (ATCC).
2. Sterile 7H9 media (*M. smegmatis* medium): suspend 2.35 g 7H9 broth base in 450 mL distilled water and add 1 mL of glycerol. Heat gently with stirring to dissolve the media. Cool to room temperature and add 50 mL of Middlebrook ADC Enrichment. Use a vacuum filtration system with a 0.22-μm pore size to sterilize the medium. Store at 4 °C.
3. *Escherichia coli* (*E. coli*) strain: BL21(DE3)pLysS.
4. Sterile Luria-Bertani (LB) broth (*E. coli* medium): standard LB is composed of 10 g tryptone, 10 g NaCl, and 5 g yeast extract per 1 L of media. Adjust pH to 7.0 before bringing up to volume. Autoclave to sterilize.
5. Sterile, vented test tubes and 50 mL Erlenmeyer flasks.
6. Refrigerated centrifuges and stir plates.
7. S200 Increase 10/300 GL (GE Healthcare) size exclusion chromatography column, AKTA FPLC system.
8. Glass fiber prefilters (1.6 μm pore size).
9. 0.45 μm centrifugal filters.
10. Ni-NTA Agarose.
11. 10× HEPES Buffered Saline (HBS): 100 mM HEPES pH 7.4, 1.5 M NaCl in ddH_2O. Prepare 1× by diluting in ddH_2O.
12. 10× Tris-Buffered Saline (TBS): 100 mM Tris–HCl, 1.5 M NaCl in ddH_2O, pH 8.0. Prepare 1× by diluting in ddH_2O.
13. Additives for direct Ni-NTA binding: 200 mM $NiCl_2$, 2 M imidazole, 1 M $MgCl_2$, and 5 M NaCl.
14. SDS-PAGE gels: 4% stacking gel/12% resolving gel (polyacrylamide). (We handcast our gels, but 4–15% Criterion™ Tris–HCl Protein Gel, 18 well, 30 μL (Bio-Rad) is a similar product; use the running buffer recommended by the manufacturer.)
15. 5× SDS-PAGE running buffer: Fill a 2 L bucket with 1600 mL ddH_2O. Add 121.14 g Tris–HCl Base and 187.68 g glycine

and stir until dissolved. Bring the volume to 2 L with ddH_2O.- Filter the solution with two glass fiber filters. Add 10 g sodium dodecyl sulfate (SDS), using caution as it is a respiratory and skin irritant. Stir until dissolved. Prepare 1× by diluting in ddH_2O.

2.2 Expression of Recombinant MR1 in Eukaryotic Cell Lines

2.2.1 Expression of Bovine Human MR1 (bhMR1) in Insect (Hi-Five) Cells

1. pAcGP67-A plasmid (BD Biosciences).
2. *Spodoptera frugiperda* (Sf9) cells.
3. BestBac 2.0 Δ v-cath/chiA Linearized Baculovirus DNA (Expression Systems).
4. *Trichoplusia ni* (High Five) cells.
5. Orbital shakers kept at 27 and 37 °C.
6. Lonza serum-free media supplemented with 2 mM L-glutamine (*see* **Note 1**). Do not add antibiotics such as gentamicin for the production of recombinant MR1 with co-infection with selected microbes (or for the co-infection control culture). Store at 4 °C and warm to 27 °C before adding to growing cultures.
7. Baculovirus with appropriate construct in frame with the gp67 secretion signal sequence (Fig. 1 and Table 1). See Subheading 3.1.1 for references regarding baculovirus production for insect cell expression.
8. Sterile 2.8 L non-baffled flasks appropriate for insect cell culture.

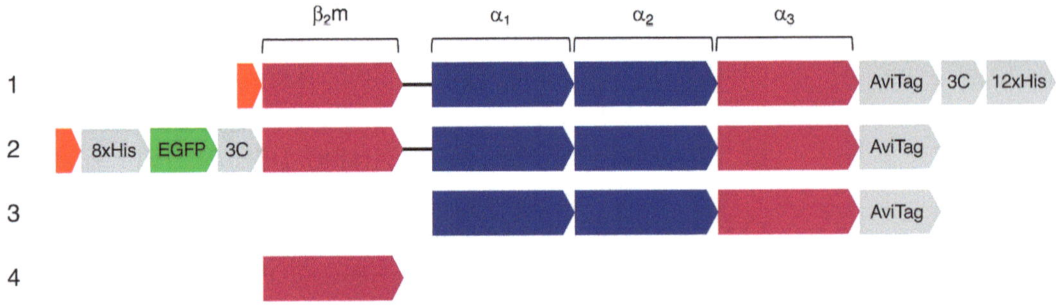

Fig. 1 Constructs used for the production of MR1 in three expression systems. In all figures, magenta blocks indicate that the domain is derived from bovine MR1, navy blocks indicate that the domain is derived from human MR1, red blocks indicate the signal peptide (gp67 for construct 1 and human IL-2 for construct 2), and the black line represents a glycine/serine linker. "3C" is short for a 3C cleavage site. Construct 1 is used for baculoviral expression in Hi-Five cells and is cloned into the pAcGP67a vector. Construct 2 is for baculoviral expression in HEK293T cells and is cloned into the pVLAD6 vector. Constructs 3 and 4 are used for preparation of MR1 inclusion bodies; each is cloned into the pET28a(+) vector. We typically include a short (3–4) glycine/serine linker between each tag on the protein (not between the MR1 domains)

Table 1
Amino acid sequences of each component used to design the constructs in Fig. 1

gp67 Signal Peptide	MVSAIVLYVLLAAAAHSAFA
Human IL-2 Signal Peptide	MYRMQLLSCIALSLALVTNS
GS Linker	GGGGSGGSGSGGGGS
AviTag	GLNDIFEAQKIEWHE
3C Cleavage Site	LEVLFQGP
Bovine β_2m (bhMR1 construct, contains signal peptide)	MARFVALVLLGLLSLSGLDAIQRPPKIQVYSRHPPEDGKPN YLNCYVYGFHPPQIEIDLLKNGEKIKSEQSDLSFSKDWSF YLLSHAEFTPNSKDQYSCRVKHVTLEQPRIVKWDRDL
Bovine β_2m (hpMR1 construct)	IQRPPKIQVYSRHPPEDGKPNYLNCYVYGFHPP QIEIDLLKNGEKIKSEQSDLSFSKDWSFYLLSHAEFTPN SKDQYSCRVKHVTLEQPRIVKWDRDL
Human α_1 Domain	RTHSLRYFRLGVSDPIHGVPEFISVGYVDSHPITTYD SVTRQKEPRAPWMAENLAPDHWERYTQLLRG WQQMFKVELKRLQRHYNHS
Human α_2 Domain	GSHTYQRMIGCELLEDGSTTGFLQYAYDGQDFLIFNKDTL SWLAVDNVAHTIKQAWEANQHELLYQKNWLEEECIA WLKRFLEYGKDTLQRT
Human α_3 Domain	EPPLVRVNRKETFPGVTALFCKAHGFYPPEIYM TWMKNGEEIVQEIDYGDILPSGDGTYQAWASIELDP QSSNLYSCHVEHCGVHMVLQVP
Bovine α_3 Domain	EPPKVRVNHKETFPGITTLYCRAYGFYPPEISIN WMKNGEEIFQDTDYGGILPSGDGTYQTWVSVELDP QNGDIYSCHVEHGGVHMVLQGFQESETIL
EGFP	MDSTESLFTGVVPILVELDGDVNGHKFSVRGEGEGDA TNGKLTLKFICTTGKLPVPWPTLVTTLTYGVQCF SRYPDHMKQHDFFKSAMPEGYVQERTITFKDDGTYK TRAEVKFEGDTLVNRIELKGIDFKEDGNILGHKLEYNFN SHNVYITADKQKNGIKANFKIRHNVEDGSVQLADH YQQNTPIGDGPVLLPDNHYLSTQSKLSKDPNEKRDHM VLLEFVTAAGITH

We typically include a short (3–4 residue) glycine/serine linker between each tag on the protein (not between the MR1 domains)

2.2.2 Expression of Human Platform MR1 (hpMR1) in HEK293T Cells

1. pVLAD6 plasmid (Addgene).
2. *Spodoptera frugiperda* (Sf9) cells.
3. BestBac 2.0 Δ v-cath/chiA Linearized Baculovirus DNA (Expression Systems).
4. HEK293T cells (ATCC).
5. Complete HEK293T media: Dulbecco's Modification of Eagle's Medium supplemented with either 10% or 2% heat

inactivated fetal bovine serum and 2 mM L-glutamine; sterile filtered using a vacuum filtration system.

6. Trypsinization media: 0.25% Trypsin, 0.53 mM EDTA in Ca^{2+}/Mg^{2+} free Hanks Balanced Salt Solution with phenol red.
7. Sterile T-25 and T-75 flasks.
8. Complete Freestyle293 Media: Freestyle293 Expression Medium supplemented with 2% heat-inactivated fetal bovine serum, 4 mM L-glutamine, and 10 mM HEPES; sterile filtered using a vacuum filtration system.
9. Sterile 125 mL and 1 L PET Bottles.
10. Baculovirus with appropriate construct in frame with a mammalian signal sequence. See Subheading 3.1.2 for references regarding baculovirus production for mammalian expression.
11. 1 M sodium butyrate made fresh for each inoculation. Sterile (syringe) filter in a biosafety cabinet before addition to cell culture.

2.3 Purification of MR1 from Baculoviral Expression

1. Gravity-flow chromatographycolumns, reservoir volume 10 mL.
2. 30 kDa centrifugal membrane filters.
3. Recombinant HRV 3C protease with an N-terminal 6×His tag.
4. 200 mM phenylmethylsulfonyl fluoride (PMSF) in 100% EtOH. Store at −20 °C.

2.4 Bacterial Production and Purification of Human Platform MR1 and Bovine β_2m Inclusion Bodies

1. pET28a(+) plasmid (Millipore).
2. 50 mg/mL stock of kanamycin, syringe filtered and stored at −20 °C until use.
3. 1 M IPTG stock, syringe filtered and stored at −20 °C until use.
4. DNase I.
5. Lysozyme.
6. Homogenizer and microfluidizer.
7. Lysis Buffer: 50 mM Tris–HCl pH 8.0, 1% (v/v) Triton X-100, 1% (w/v) sodium deoxycholate, 100 mM NaCl, 5 mM $MgCl_2$, 1 mM DTT. If premade, add fresh DTT to 1 mM before use.
8. Inclusion Body Wash Buffer #1: 50 mM Tris–HCl, pH 8.0, 0.5% (v/v) Triton X-100, 100 mM NaCl, 1 mM EDTA pH 8.0, 1 mM DTT. If premade, add fresh DTT to 1 mM before use.
9. Inclusion Body Wash Buffer #2: 50 mM Tris–HCl, pH 8.0, 1 mM EDTA pH 8.0, 1 mM DTT. If premade, use fresh DTT to 1 mM before use.

2.5 Refolding Recombinant Protein from Inclusion Bodies

1. Resuspension Buffer: 20 mM Tris–HCl pH 8.0, 8 M Urea, 0.5 mM EDTA pH 8.0, 1 mM DTT. Make aliquots and store at −80 °C. Add fresh DTT to 1 mM to the resuspension buffer before use.
2. Refolding Buffer: 0.1 M Tris–HCl pH 8.5, 2 mM EDTA pH 8.0, 0.4 M L-arginine, 5 mM reduced glutathione (GSH), 0.5 mM oxidized glutathione (GSSG). Make fresh for each refold. Add GSH/GSSG after bringing the mixture to volume and immediately before adding the inclusion bodies.
3. Dialysis Buffer: 10 mM Tris–HCl pH 8.0.
4. Fisherbrand Regenerated Cellulose Dialysis Tubing.
5. DEAE Sepharose Fast Flow Resin.
6. 5 M NaCl in ddH_2O.
7. DEAE Wash Buffer: 10 mM Tris–HCl pH 8.0, 50 mM NaCl.

2.6 Protein Biotinylation and Tetramerization

1. 10× BioMix A: 0.5 M bicine pH 8.3.
2. 5× BioMix B: 50 mM ATP, 50 mM MgOAc, 250 μM D-biotin.
3. Biotinylation Buffer: 10 mM Tris–HCl pH 8.0, 50 mM NaCl.
4. Recombinant BirA Ligase.
5. Streptavidin or a derivative thereof (*see* **Note 3**).
6. MR1 Ligand(s) (*see* **Note 2**).
7. 1 M HEPES pH 7.2 in ddH_2O.
8. 1.5 M NaCl in ddH_2O.

3 Methods

3.1 Expression of Recombinant MR1 in Eukaryotic Cell Lines

3.1.1 Expression of Bovine Human MR1 (bhMR1) in Insect (Hi-Five) Cells

1. A full description of the production of baculovirus is not within the scope of this publication. In brief, clone the construct outlined in Fig. 1 and Table 1 into the pAcGP67-A MCS, in frame with the gp67 secretion signal sequence. Transfect Sf9 cells with a 10:1 ratio of plasmid DNA to BestBac 2.0 Δ v-cath/chiA Linearized Baculovirus DNA and allow P0 baculovirus to amplify for 1 week at 27 °C. Clear the virus of cells by centrifugation. Perform a pilot expression test by adding the P0 baculovirus at a 1:1000 ratio to 2 mL High Five cells at a density of 2×10^6 cells/mL and shaking at 27 °C for 72 h. Harvest the supernatants by centrifugation and use a small amount of Ni-NTA to pull down the His-tagged protein, then visualize by SDS-PAGE. At the same time, amplify the P1 baculovirus in Sf9 cells, allowing the virus to propagate for 1 week at 27 °C. Again, clear the virus of cells by centrifugation. Titer the P1 baculovirus in 2 mL Hi-Five cell cultures at a density of 2×10^6 cells/mL to determine the appropriate

multiplicity of infection (MOI). For more detail on baculovirus production and amplification, see ref. 8.

2. If performing a co-infection, in which live microbes are added to protein expression medium to capture ligands in recombinant MR1, start an overnight culture of the desired bacterium such that the bacteria is at or reaching its stationary phase at the time of co-infection.
3. Add bhMR1 baculovirus at the optimized MOI to High Five cells at a density of 2×10^6 cells/mL. Shake at 27 °C and 130 rpm on an orbital shaker. If performing a co-infection, proceed immediately to **step 4**. Else, perform expression for 68–72 h.
4. For bacterial co-infection during protein production, add appropriate number of live bacteria from overnight culture to insect cell culture at optimized time after baculoviral induction (*see* **Note 4**). Shake at 27 °C for 68–72 h from the time of baculoviral induction. At the end of the expression, the co-infected culture should look distinct from the control expression (Fig. 2a).
5. Pellet the cells by centrifugation for 15 min at 800 × *g*; keep the supernatant.
6. Filter supernatant using two glass fiber prefilters to remove any cells or large particulates.
7. Concentrate the supernatant through 30 kDa membranes to about 200 mL (we use tangential flow filtration for large cultures). Buffer exchange by diluting sixfold with 1× HBS and concentrating back to about 200 mL. Repeat buffer exchange once more. Wash the concentrated protein out of the pump using 200 mL 1× HBS (*see* **Note 5**).
8. Centrifuge concentrated supernatant for 30 min at 9000 × *g* and 4 °C to pellet any remaining debris.
9. To the supernatant, add imidazole to a final concentration of 40 mM and 1 mL Ni-NTA resin (bed volume) per liter of expression culture to bind the available protein. The supernatant of the co-infected culture should look distinctly tan/yellow as compared to the control culture (Fig. 2b).
10. Stir 4 h to overnight at 4 °C.

3.1.2 Expression of Human Platform MR1 (hpMR1) in HEK293T Cells

1. Adapt a stock of 293T cells stepwise from Complete 10% FBS HEK293T media to Complete 2% FBS HEK293T media. When the cells are doubling every 24 h, they are ready to adapt to suspension culture.
2. Grow 293T cells to 70–80% confluency in two T-75 flasks in Complete 2% FBS HEK293T media. Do not use cells that are overgrown; they will grow very slowly in suspension.

Fig. 2 Visualizing co-infected and control Hi-Five expressions. Here, the culture/supernatant is labeled as schpMR1 (single chain human platform MR1), but it is referenced as bhMR1 in the text. (**a**) 70 h after baculoviral induction, the *M. smegmatis* co-infected culture (right) has hallmarks of contamination, including a cloudy, yellow appearance, a distinct odor, less foami ng, and a ring of dead cells near the top of the flask as compared to the control culture (left). (**b**) After removing all cells, filtering and buffer exchanging the protein, the supernatant of the co-infected culture (right) appears distinct in color as compared with the control supernatant (left)

3. Aspirate the media and bang the plates ten times on the surface.
4. Add 10 mL of Complete Freestyle293 Media to each T-75 flask and dislodge the cells from the plate surface by pipetting against the corner of the flask.
5. Pool all cells together and bring up final volume to 25 mL with Complete Freestyle293 Media. Transfer the 25 mL to a sterile 125 mL PET bottle.

6. To ventilate the cells, loosen the cap by half a turn and tape securely.
7. Tape the base of the bottle to an orbital shaker in a CO_2 incubator set at 37 °C, 5–8% CO_2, 130 rpm (1″ orbital).
8. Passage the cells at least three times, not allowing them to surpass 2.5×10^6 cells/mL and splitting to 0.25×10^6 cells/mL each time. After these passages, the cells should have a doubling time of approximately 24 h.
9. Transfer the cells to a 37 °C warm room environment and grow while shaking at 110 rpm (1.4″ orbital). Since the Complete Freestyle293 Media is supplemented with 10 mM HEPES, they can be passaged outside of a 5% CO_2 environment, though not indefinitely. We suggest passaging at least three times in the warm room environment before baculoviral inoculation, monitoring growth and morphology of the cells each day.
10. As described in Subheading 3.1.1, **step 1**, clone the MR1 construct shown in Fig. 1 into the pVLAD6 MCS, in frame with a mammalian signal sequence for secretion, and produce baculovirus for mammalian transduction.
11. If performing a co-infection, in which live microbes are added to protein expression medium to capture ligands in recombinant MR1, start an overnight culture of the desired bacterium such that the bacteria is at or reaching its stationary phase at the time of co-infection.
12. Add hpMR1 baculovirus at the optimized MOI to HEK293T cells at a density of 2×10^6 cells/mL. Shake at 37 °C and 130 rpm on an orbital shaker. If performing a co-infection, proceed immediately to **step 4**. Else, perform expression for 68–72 h.
13. 0–24 h past induction, add sodium butyrate (a histone deacetylase inhibitor) to 10 mM. Without this step, protein expression is low or undetectable. This time point will need to be determined empirically for each construct to optimize for the highest expression.
14. For bacterial co-infection during protein production, add appropriate number of live bacteria from overnight culture to insect cell culture at optimized time after baculoviral induction (*see***Note 4**). Shake at 37 °C for 68–72 h from the time of baculoviral induction.
15. Pellet the cells by centrifugation for 10 min at 800 × *g* and 4 °C; keep the supernatant.
16. Filter supernatant using two glass fiber prefilters to remove any cells or large particulates.

17. Dilute the supernatant 3× in 1× HBS and add $MgCl_2$ to 10 mM, $NiCl_2$ to 0.5 mM, imidazole to 25 mM, and an appropriate volume of Ni-NTA (*see***Note 6**).
18. Stir at 4 °C, then add PMSF to 0.5 mM. PMSF is a serine protease inhibitor that will help prevent degradation of the recombinant protein.

3.2 Purification of MR1 from Baculoviral Expression

1. Collect Ni-NTA resin using a glass filter attached to a side-arm Erlenmeyer flask and transfer to a gravity flow chromatography column. Typically, the protein harvested from a co-infected culture will be bright yellow (many MR1 ligands fluoresce) and therefore the resin will appear green rather than blue. Protein harvested from the control culture will not be significantly colored (Fig. 3).
2. Wash the resin thoroughly with 25 mM imidazole in 1× HBS.
3. Elute with resin with 4–5 column volumes (CV) 200 mM imidazole in 1× HBS, then 3–4 CV 400 mM imidazole in 1× HBS.
4. Run the protein on SDS-PAGE to determine which fractions contain recombinant protein.
5. Combine MR1-containing Ni fractions and add optimized amount of HRV 3C protease to protein. Incubate overnight at 4 °C.

F1 F2 F3 F4 F5 F6 F7

-Bacteria

+*M. smeg*

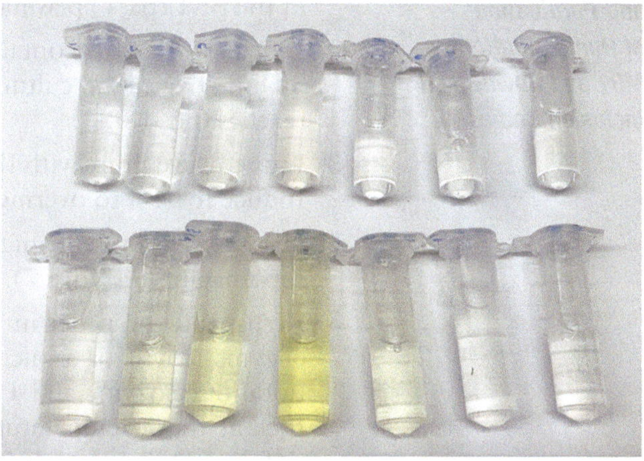

Fig. 3 Fractions of bhMR1 (produced with and without co-infection) eluted from Ni-NTA. MR1 purified from the supernatant of co-infected cells typically fluoresces yellow, consistent with the nature of the MR1 ligands known to date (i.e., riboflavin). Protein purified from the supernatant of control cells do not display this strong fluorescence. This does not indicate that the ligands captured are stimulatory

6. Use a 30-kDa centrifugal membrane filter to buffer exchange the protein into 25 mM imidazole in 1× HBS.
7. Prepare a column of Ni-NTA resin using a similar CV as for the initial Ni-NTA binding. Wash the column with 25 mM imidazole in 1× HBS.
8. Set aside a small amount of protein (10 μL) before applying it to the Ni-NTA column.
9. Add remaining protein directly to the column, and collect the flowthrough in a 15-mL conical tube. This should contain MR1 protein that has successfully had its His tag cleaved off.
10. Wash Ni-NTA resin with 3–4 CV 1× HBS and collect flowthrough.
11. Transfer the column into a second 15-mL conical tube and elute the bound protein from the column with 3–4 CV 400 mM imidazole in 1× HBS.
12. Visualize the unpurified protein, "Ni-clean" protein, and eluted protein by SDS-PAGE to make sure that the protein was cleaved thoroughly. MR1 should be present exclusively in the flowthrough. This clean-up step removes some proteins that nonspecifically bind the Ni-NTA resin and protein that is uncut by the 3C protease and may therefore be prone to aggregation or multimerization per its histidine tag.

3.3 Bacterial Production and Purification of Human Platform MR1 and Bovine β_2m Inclusion Bodies

1. Inoculate 50 mL LB containing 50 μg/mL kanamycin with BL21 *E. coli* freshly transformed with MR1 heavy chain or β_2m in the pET28a(+) plasmid. Shake overnight at 37 °C.
2. Inoculate 1 L LB containing 50 μg/mL kanamycin with the 50 mL overnight culture. Shake at 37 °C until the OD_{600} reaches 0.6–1.0.
3. Induce the culture with IPTG to a final concentration of 1 mM. Induce for 4 h to overnight at 30 °C.
4. Harvest cells by centrifugation for 15 min at 4000 × *g*; discard supernatant (*see* **Note 7**).
5. Resuspend the pellet in 25 mL Lysis Buffer. Lyse cells using enzymatic lysis or a microfluidizer. For enzymatic lysis, add a tipful of lysozyme and 1 mM PMSF to resuspended cells and rock at room temperature for a minimum of 30 min. Add a tipful of DNase I to lysis mixture and homogenize. For lysis with the microfluidizer, add DNase I to resuspended cells and homogenize; pass cells through microfluidizer at least three times under sufficient pressure to lyse *E. coli*.
6. Before centrifugation, weigh empty tubes that will be used for centrifuging inclusion bodies, noting the mass of each one.

7. Pellet inclusion bodies by centrifugation for 30 min at 8000 × *g* and 4 °C.
8. Wash inclusion body pellet with 25 mL Inclusion Body Wash Buffer #1, then pellet inclusion bodies by centrifugation for 20 min at 8000 × *g* and 4 °C.
9. Repeat **step 8** until inclusion body pellets are white.
10. Wash pellet with 25 mL Inclusion Body Wash Buffer #2, then pellet inclusion bodies by centrifugation for 20 min at 8000 × *g* and 4 °C.
11. Remove as much liquid from inclusion body pellet as possible.
12. Weigh the tube containing the inclusion body pellet; subtract the original mass of the tube (**step 6**) to obtain the mass of the inclusion body.
13. Resuspend inclusion body pellets in Inclusion Body Wash Buffer #2 such that MR1 is concentrated to 60 mg/mL and β_2m to 30 mg/mL. Make 1 mL aliquots in 1.7 mL microcentrifuge tubes. Pellet the inclusion bodies on a tabletop centrifuge for 10 min at 16,000 × *g* and 4 °C.
14. Remove supernatant and store pellets at −70 °C for long-term storage.

3.4 Refolding Recombinant Protein from Inclusion Bodies

1. Thaw inclusion bodies on ice, then resuspend each in 1 mL Resuspension Buffer with freshly added DTT.
2. To solubilize, either incubate at 50 °C for 10 min or overnight at 4 °C.
3. Pellet any insoluble material by centrifugation for 10 min at 16,000 × *g* and 4 °C.
4. During the spin, prepare fresh Refolding Buffer (typically 400 mL).
5. With the Refolding Buffer stirring at room temperature, add the supernatant of the resuspended inclusion bodies.
6. Cover with saran wrap and stir overnight at 4 °C.
7. Dialyze the refolding reaction into 10 mM Tris–HCl pH 8.0 at least four times over the next 24–48 h using a regenerated cellulose dialysis tubing. Volume of the dialysis buffer should be ten times greater than the volume of the Refolding Buffer.
8. Wash 10 mL DEAE resin with 15–20 mL 10 mM Tris–HCl pH 8.0, 50 mM NaCl.
9. Filter dialyzed protein into a clean bottle (of appropriate volume) using two glass fiber prefilters.
10. Add NaCl to the filtered protein to a final concentration of 50 mM.

11. Cap the chromatography column containing DEAE resin at the bottom and add 2–4 mL filtered protein to the top of the resin.
12. Attach a siphon to the top of the column, uncap, and allow the protein to flow over the DEAE column, collecting the flow-through in case a second bind needs to be performed. This step should be performed at 4 °C.
13. After all of the protein has been siphoned over the column, wash with 3 CV 10 mM Tris–HCl pH 8.0, 50 mM NaCl.
14. Elute the protein from the column using:
    (a) 4 × 1.5 mL fractions 10 mM Tris–HCl pH 8.0, 200 mM NaCl.
    (b) 5 × 1.5 mL fractions 10 mM Tris–HCl pH 8.0, 400 mM NaCl.
    (c) 5 × 1.5 mL fractions 10 mM Tris–HCl pH 8.0, 600 mM NaCl.
15. Run fractions on SDS-PAGE in non-reducing dye to determine which fractions contain recombinant, refolded protein.

3.5 Protein Biotinylation and Tetramerization

1. Concentrate protein and buffer exchange into 10 mM Tris–HCl, pH 8.0, 50 mM NaCl (*see* **Note 8**) if it is in a different buffer.
2. Prepare a biotinylation reaction by mixing 1 part 10× BioMix A, 2 parts 5× BioMix B, and 7 parts substrate in 10 mM Tris–HCl pH 8.0, 50 mM NaCl. The final concentration of substrate in this reaction should be as close to 40 μM as possible.
3. Add 2.5 μg BirA Ligase for every 10 nmol of substrate used (*see* **Note 9**). Incubate overnight at 4 °C.
4. For insect cell or mammalian cell-derived MR1, buffer exchange the biotinylated protein back into 1× HBS, then purify by size exclusion chromatography using an S200 Increase 10/300 GL column in 1× HBS. For refolded MR1, do not buffer exchange, but purify by size exclusion chromatography using an S200 Increase 10/300 GL column in 1× TBS.
5. Combine fractions containing MR1 and concentrate to about 10 mg/mL using a centrifugal membrane filter.
6. Perform a small-scale tetramerization to test whether the protein was thoroughly biotinylated. Prepare three samples: substrate, streptavidin, and substrate + streptavidin. Use 2 μg of the substrate and a two-fold molar excess of streptavidin where appropriate. Incubate for 15 min at room temperature, then

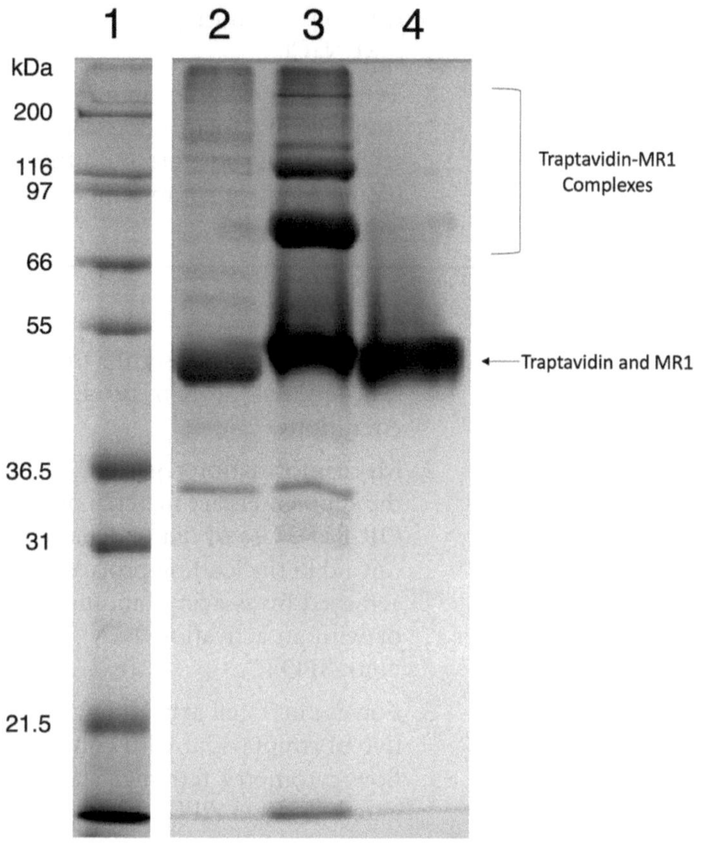

Fig. 4 Performing a Traptavidin binding test allows confirmation of complete biotinylation of the target protein. Lane 1: Ladder (Mark12 Unstained Standard, ThermoFisher Scientific). Lane 2: mammalian-derived hpMR1. Lane 3: hpMR1 and Traptavidin. Lane 4: Traptavidin. The presence of higher molecular weight bands corresponds to biotinylated MR1 binding to Traptavidin in varying ratios (1:1–4:1). Purified MR1 and Traptavidin co-migrate making it difficult to tell whether MR1 is truly fully biotinylated. Native streptavidin may be more useful for this analysis

run each sample on SDS-PAGE in non-reducing dye. Both the streptavidin tetramer and the streptavidin/biotin interaction are SDS-resistant and therefore complete biotinylation should be indicated by the dissolution of the substrate band and appearance of higher-order molecular weight bands (Fig. 4).

7. If biotinylation was sufficient and working with insect cell or mammalian cell-derived MR1, proceed to **step 8**. If working with refolded hpMR1 bβ2m, load the protein with the ligand of choice. The appropriate conditions will vary by ligand (*see* **Note 2**). Typically, we incubate refolded protein with a 100-fold molar excess of ligand in 100 mM HEPES, pH 7.2, 150 mM NaCl, for 18 or more hours at 4 °C in the dark. Stop the

reaction by diluting 5× with 100 mM HEPES pH 7.2, 150 mM NaCl.

8. Tetramerize the remaining protein. Determine the volume of streptavidin variant necessary to obtain a 1.1:4 molar ratio of streptavidin: substrate. Add this volume in 8 equal parts, incubating 5 min at room temperature between each addition.

4 Notes

1. We find that other serum-free insect cell media varieties yield less optimal protein production and bacterial co-infection conditions.
2. More information regarding loading of ligands can be found in the supplementary material of ref. 2 (including for loading of 5-OP-RU). Use of dilute denaturants (0.5 M GdnHCl or Urea) can aid in the loading process. Testing loading efficiency can be achieved by assaying binding to a recombinant T cell receptor or with an activation assay. The latter is described in ref. 2 as "tetraSPOT".
3. For use in T cell activation assays, we use traptavidin, a derivative of streptavidin with enhanced binding kinetics. For use in flow cytometry tetramer staining, we use streptavidin conjugated to either APC or PE.
4. Time of addition and number of bacteria should be optimized for each bacterial species. For BL21 *E. coli*, we find that 4×10^1 *E. coli* cells/mL of insect cell culture should be added at 18–20 h post-baculoviral induction. For *M. smegmatis*, we find that 4×10^3 *M. smegmatis* cells/mL of insect cell culture should be added within the first 30 min post-baculoviral induction. We always do a negative control expression of recombinant MR1 in antibiotic-free media without bacterial co-infection to ensure that the stock of cells was not contaminated. We have not yet determined the optimal conditions for co-infection of mammalian cells, but our data thus far indicate that significantly more bacteria need to be added to the culture (and at later time points) to obtain mammalian-derived MR1 loaded with stimulatory ligands. The appropriate loading conditions can be determined by assessing either binding of recombinant MR1 to recombinant MAIT TCRs, activation potential toward MAIT cell clones, or tetramer staining of MAIT cell clones (flow cytometry).
5. With bacterial co-infection of insect cells, direct binding of recombinant protein to Ni-NTA in Lonza media results in severe salt precipitation, making recovery of the protein difficult and decreasing the yield.

6. Unlike insect cell media, salt will not crash out of the Free-Style293 Media at high pH, so the binding can be done at pH 8–8.5, at which MR1 is stable.
7. At this point, the cell pellet can be frozen at −20 °C for up to a week or at −70 °C indefinitely.
8. Recombinant MR1 precipitates in no-salt conditions, so biotinylation should always be performed in the presence of a low concentration of NaCl. Starting with 50 mM NaCl before addition of Biomix A and Biomix B should be sufficient although starting with a concentration of up to 100 mM NaCl has yielded successful biotinylation. Incubation time may need to be adjusted for higher salt concentrations.
9. This is a useful starting guide although we have found that a higher amount of BirA enzyme is often required to yield 90–100% biotinylation.

Acknowledgments

We would like to acknowledge the Bill and Melinda Gates Foundation for funding this protocol development, grant # OPP1131709, and the T32 Training Grant to the University of Chicago: GM007183. We thank members of the Lewinsohn Lab at Oregon Health and Science University, Dr. Melanie Harriff, Dr. Adrienne Luoma, Dr. Sobhan Roy, and Dr. Caitlin Castro for helpful discussions during protocol development.

References

1. Corbett AJ, Eckle SBD, Birkinshaw RW et al (2014) T-cell activation by transitory neo-antigens derived from distinct microbial pathways. Nature 509:361–365
2. Harriff MJ, McMurtrey C, Froyd CA et al (2018) MR1 displays the microbial metabolome driving selective MR1-restricted T cell receptor usage. Sci Immunol 3:eaao2556. https://doi.org/10.1126/sciimmunol.aao2556
3. Treiner E, Duban L, Bahram S et al (2003) Selection of evolutionarily conserved mucosal-associated invariant T cells by MR1. Nature 422:164–169
4. Porcelli S, Yockey CE, Brenner MB et al (1993) Analysis of T cell antigen receptor (TCR) expression by human peripheral blood $CD4^{-}8^{-}$ alpha/beta T cells demonstrates preferential use of several V beta genes and an invariant TCR alpha chain. J Exp Med 178:1–16
5. Lopez-Sagaseta J, Dulberger CL, McFedries A et al (2013) MAIT recognition of a stimulatory bacterial antigen bound to MR1. J Immunol 191:5268–5277
6. Lopez-Sagaseta J, Dulberger CL, Crooks JE et al (2013) The molecular basis for Mucosal-Associated Invariant T cell recognition of MR1 proteins. Proc Natl Acad Sci 110:E1771–E1778
7. Gherardin NA, Keller AN, Woolley RE et al (2016) Diversity of T cells restricted by the MHC class I-related molecule MR1 facilitates differential antigen recognition. Immunity 44:32–45
8. Murhammer DW (ed) (2007) Baculovirus and insect cell expression protocols, Methods in Mol Biol, vol 388. Springer, Heidelberg

Chapter 14

Immune Monitoring of Human Mucosal-Associated Invariant T Cells by Quantitative Proteomics

Isabel Bernal, Marco van Ham, and Lothar Jänsch

Abstract

Molecular phenotypes of mucosal-associated invariant T (MAIT) cells are correlating with individual susceptibilities and outcomes in human diseases. Quantitative proteome strategies can examine such variations in the functional and druggable inventory of MAIT cells comprehensively, but protocols for the support of translational and clinical studies are still rare. Here, we describe a protocol in which MR1-restricted MAIT cells were isolated from blood donations by FACS and are then characterized by quantitative proteomics (iTRAQ-LC-MS/MS) to complement information about their unique effector phenotype and to investigate donor-/patient- or disease-specific variations in protein networks with high precision.

Key words MR1 tetramer, Mucosal-associated invariant T cells, Clinical proteomics, 8-plex iTRAQ™

1 Introduction

Proteomics constitutes a prime method to characterize the molecular basis of immune cell phenotypes. Numerous protocols were introduced to examine conventional CD3$^+$ T cells before, that discovered novel markers [1] or characterized protein networks involved in various T cell processes [2]. However, proteome studies at the level of primary and unconventional T cell subsets are still rare. We have recently characterized the proteomes of primary human MAIT cells in comparison to those of NK and CD8$^+$ T cells in healthy individuals. This proteome strategy could complement information on MAIT-specific but donor-independent immune functions that are linked with their unique innate-like effector phenotype [3]. In this chapter, we are now describing a variant proteome strategy that exclusively is focusing on the detection of donor-/patient-specific variations of MAIT cell immune functions that for instance are correlating with disease states or individual susceptibilities. The strategy allows total proteome analyses with high analytical depth (>5000 proteins) and the reliable

Helen Kaipe and Isabelle Magalhaes (eds.), *MAIT Cells: Methods and Protocols*, Methods in Molecular Biology, vol. 2098, https://doi.org/10.1007/978-1-0716-0207-2_14, © Springer Science+Business Media, LLC, part of Springer Nature 2020

quantitative comparison of MAIT cell proteomes from up to eight different samples (8-plex iTRAQ™). In summary, the described strategy is instrumental for immune monitoring studies aiming on the identification of potential MAIT cell biomarkers or drug targets in representative patient samples.

To provide unambiguous information on the proteome of primary human MAIT cells, these cells have to be isolated with high purity from blood donations by using fluorescence-activated cell sorting (FACS) based on suitable molecular markers. To reveal even moderate donor-/patient-specific variations (1.5-fold change of protein amounts), the purity of the MAIT cell subset has to be consistent and of at least 90%. Human MAIT cells are basically defined by the expression of a semi-invariant TCRα chain, Vα7.2, associated with a restricted TCRβ repertoire [4, 5]. Furthermore, they elicit high expression of C-type lectin CD161, the multidrug resistant protein MDR1, and the dipeptidyl peptidase 4 (CD26) [6, 7]. Hence, FACS of $CD3^+$ $CD161^+$ $V\alpha7.2^+$ MAIT cells was used in previous studies, although expression of CD161 and Vα7.2 is downregulated in activated MAIT cells [8, 9]. The more recent sorting strategy and the one described in this chapter enriches human MAIT cells based on their ability to bind antigen-loaded MR1 tetramers and their specific expression of the TCR Vα7.2 chain [5]. Additionally, fluorescence-coupled antibodies against CD14 and CD19 are used as negative markers to exclude monocytes and B cells, respectively, resulting in a sorting strategy to isolate $CD19^-CD14^-CD3^+V\alpha7.2^+$MR1-5-OP-RU $tetramer^+$ MAIT cells [10]. Still, additional markers might be considered on demand. For instance, MAIT cells can be further dissected in $CD8^+$ and $CD8^-CD4^-$ MAIT cells, whereby $CD8^-CD4^-$ MAIT cell are developmentally more related to $CD8^+$ MAIT cells than $CD4^+$ MAIT cells [11]. Furthermore, there is a small proportion of MAIT cells that express CD4 instead of CD8 [12].

We recommend to sort MAIT cells from fresh human peripheral blood mononuclear cell (PBMC) fractions obtained from commercially available blood collection tubes that include a gel matrix with Ficoll-gradient functionality. MAIT cell numbers can vary significantly in different donors and disease conditions [13, 14] and robust proteome analyses in this protocol can be achieved by about 50.000 MAIT cells per donor/patient. The comparative inspection of MAIT cell proteomes is accomplished by quantitative peptide sequencing using the combination of liquid chromatography and accurate mass spectrometry (LC-MS).

For that, proteomes are extracted by urea cell lysis and proteins are subsequently digested into peptides using trypsin/Lys-C endoproteases. MAIT cell peptides from eight individuals can then be labeled differentially with isobaric iTRAQ molecules [15]. 8-plex iTRAQ kits are commercially available and allow to label and combine MAIT cell peptides from up to eight individuals [16]. The big advantages of this approach are that (1) basically all sequenced and identified peptides will provide quantitative information for all conditions and (2) the combined peptide sample can be fractionated prior to LC-MS-runs that significantly increase the analytical depth. In this protocol we describe off-line fractionation by basic reverse phase chromatography, which is suitable for relatively small protein amounts in primary cells.

2 Materials

2.1 MAIT Cell Isolation from Patient Blood

1. BD Vacutainer® CPT™ mononuclear cell preparation tubes with sodium heparin.
2. Phosphate-buffered saline (PBS).
3. FACS buffer: PBS containing 2% fetal calf serum (FCS) and 2 mM ethylenediaminetetraacetic acid (EDTA).
4. Neubauer counting chamber.
5. Trypan blue.
6. Mouse antibodies against human CD14 (clone: M5E2, BV510-coupled), CD19 (clone: HIBI19, BV421-coupled), CD3 (clone: UCHT1, FITC-coupled), and Vα7.2 (clone: 3C10, PE-Cy7-coupled) are commercially available. Human MR1 tetramer coupled to the MAIT cell activation ligand 5-(2-oxopropylideneamino)-6-D-ribitylaminouracil (MR1-5-OP-RU tetramer; APC-coupled) can be ordered from National Institutes of Health (USA) (*see* **Note 1**). Fluorescent live/dead discrimination dye (LIVE/DEAD Fixable near-IR Dead Cell Stain Kit, APC-Cy7-coupled) is used to exclude dead cells.

Prepare following solutions in ultrapure water and store at −20 °C unless indicated otherwise.

2.2 Protein Isolation, Digestion, and Peptide Purification

1. LoBind Eppendorf tubes.
2. Lysis buffer: 100 mM triethylammonium bicarbonate (TEAB) in 8 M urea. For 1 L, take 100 mL 1 M TEAB, weigh 8 M urea, and add water to 1 L, stir. Prepare freshly before use.
3. 250.000 U/mL benzonase stock solution.
4. 50 mM Tris-(2-carboxyethyl)-phosphine (TCEP).
5. 200 mM *S*-methyl methanethiosulfonate (MMTS).
6. Trypsin/Lys-C reagent.
7. 100 mM TEAB.
8. Oasis HLB 1 cc (10 mg) Extraction Cartridge column (Waters).
9. 10% Trifluoroacetic Acid (TFA).

10. RP buffer A: 0.2% TFA, 60% Acetonitrile (ACN).
11. RP buffer B: 0.2% TFA, 3% ACN.

2.3 iTRAQ™ Labeling and Basic Reversed Phase (b-RP) Chromatography

1. iTRAQ™ label reagents.
2. 100% ethanol.
3. Vacuum dry centrifugation, CHRIST, RVC 2-18 CO*plus*.
4. Dissolution buffer: 0.5 M TEAB.
5. 100% methanol.
6. 200 mM ammonium formate.
7. b-RP buffer: 5 mM ammonium formate.
8. Oasis HLB 1 cc (10 mg) Extraction Cartridge column (Waters).
9. 80% ACN in b-RP buffer.
10. 15% ACN in b-RP buffer.
11. 12% ACN in b-RP buffer.
12. 9% ACN in b-RP buffer.
13. 6% ACN in b-RP buffer.
14. 3% ACN in b-RP buffer.
15. 0.2% TFA.
16. Ultracentrifuge tubes.
17. Ultracentrifuge. Sorvall Discovery MS120SE ultra centrifuge, Hitachi.

2.4 Mass Spectrometric Analysis

1. C18 pre-column (3 μm RP18 beads, Acclaim™, 75 mm × 20 mm, Dionex, Sunnyvale, USA).
2. MS buffer A: 0.1% Formic acid (FA).
3. MS buffer B: 0.1% FA, 80% ACN.
4. Stainless emitter (Thermo Scientific).

3 Methods

Use LoBind Eppendorf tubes to ensure higher protein and peptide recovery. Carry out all procedures at room temperature unless otherwise specified. Store each waste and flow-through for trouble shooting.

3.1 MAIT Cell Isolation from Donor/Patient Blood

1. Collect 2 × 8 mL donor/patient blood into BD Vacutainer® CPT™ tube (*see* **Note 2**).
2. Mix CPT tube ten times by flipping upside down.
3. Centrifuge at 1650 × *g* for 25 min at 22 °C with brake.

4. Collect all PBMCs from the interphase and transfer into a 15-mL falcon tube (*see* **Note 3**). Combine PBMCs from the same donor/patient in one tube.
5. Add PBS to 15 mL and centrifuge at 350 × *g* for 10 min at 4 °C, and omit supernatant.
6. Resuspend cell pellet in 1 mL FACS buffer and count cells using a Neubauer counting chamber.
7. Add FACS buffer to 15 mL, centrifuge at 350 × *g* for 10 min at 4 °C, and omit supernatant.
8. Prepare antibody master mix by diluting antibodies according to manufacturer's recommendations in FACS buffer (*see* **Note 4**).
9. Resuspend cell pellet with antibody mix to final concentration of 2×10^8 cells/mL and incubate for 15 min at 4 °C on a laboratory shaker.
10. Add FACS buffer to 15 mL, centrifuge at 350 × *g* for 10 min at 4 °C, and omit supernatant.
11. Resuspend cells in 1 mL PBS buffer and sort cells using an appropriate equipped cell sorter (*see* **Note 5**). Sort at least 50.000 MAIT cells by gating on lymphocytes (SSC-A and FSC-A plot)/single cells (FSC-A and FSC-H)/$CD14^-$ and $CD19^-$/live $CD3^+$ T cells/$Va7.2^+$ and MR1-5-OP-RU tetramer$^+$ cells (*see* **Note 6**).
12. Add PBS to 15 mL, centrifuge at 350 × *g* for 10 min at 4 °C, and omit supernatant.
13. Repeat **step 12**.
14. Resuspend cells in 1 mL PBS, transfer into LoBind Eppendorf tube, and centrifuge at 350 × *g* for 10 min. Omit supernatant.
15. MAIT cell pellets can be stored at −80 °C.

3.2 Cell Lysis

1. Add 50 μL lysis buffer to a cell pellet from 50.000 sorted MAIT cells and resuspend thoroughly (*see* **Note** 7).
2. Add 1 μL of 1:10 diluted benzonase stock solution and incubate for 30 min.
3. For protein reduction, add 5 μL 50 mM TCEP and incubate for 60 min at 35 °C (*see* **Note 8**).
4. For alkylation, add 2.5 μL 200 mM MMTS and incubate for 15 min (*see* **Note 9**).
5. If necessary, samples can be stored at −20 °C.

3.3 Protein Digestion

1. Add Trypsin/Lys-C reagent at a ratio of 1:20 weight per weight (e.g., 0.5 μg Trypsin/Lys-C for 10 μg protein).
2. Incubate samples for 5 h at 37 °C.

3. Add 180 μL 100 mM TEAB to dilute the samples from 8 M urea to 1 M urea and proceed with incubation at 37 °C overnight (*see* **Note 10**).

3.4 Peptide Purification and Desalting

1. Acidify samples to a pH = 5 by adding 10% TFA (*see* **Note 11**).
2. Centrifuge at 16.000 × *g* for 10 min using a tabletop centrifuge (*see* **Note 12**).
3. Transfer peptide-containing supernatant into a new 1.5 mL LoBind Eppendorf tube.
4. Pre-activate Oasis HLB 1 cc (10 mg) Extraction Cartridge columns by applying three bed volumes of 100% acetonitrile and let flow-through by gravity.
5. Activate columns by applying three bed volumes of RP buffer A and let flow-through by gravity.
6. Equilibrate columns by applying three bed volumes of RP buffer B and let flow-through by gravity.
7. Apply samples onto columns and let flow-through by gravity, and collect flow-through.
8. For optimal peptide loading, apply the flow-through from **step** 7 onto the same columns and let flow-through by gravity. Collect flow-through (*see* **Note 13**).
9. Wash the columns twice with three bed volumes of RP buffer B.
10. Elute peptides from the columns by applying three bed volumes of RP buffer A and collect samples in LoBind Eppendorf tubes.
11. Vacuum dry samples (*see* **Note 14**).
12. If necessary, samples can be stored at −20 °C.

3.5 iTRAQ™ Labeling

1. Solve peptides from Subheading 3.4, **step 11** in 30 μL dissolution buffer (*see* **Note 15**).
2. Label Trypsin/Lys-C digested peptides with isobaric iTRAQ™ reagents according to manufacturer's guidelines. Let iTRAQ™ reagents reach room temperature. Dilute iTRAQ™ labels in 70 μL ethanol (*see* **Note 16**) and use one label per sorted donor/patient sample.
3. Add each single diluted iTRAQ™ label to one single peptide sample and mix well.
4. Incubate for 2 h (*see* **Note 17**).
5. Vacuum dry samples.
6. Dissolve peptides in 50 μL RP buffer B.
7. Purify peptides as described in Subheading 3.4, starting from **steps 4** to **11** (*see* **Note 18**).
8. If necessary, samples can be stored at −20 °C.

3.6 Peptide Fractionation by Basic Reverse Phase (b-RP) Chromatography

1. Pre-activate the Oasis HLB 1 cc (10 mg) Extraction Cartridge columns with 500 μL 100% methanol.
2. Equilibrate the columns with 20 bed volumes 80% ACN in b-RP buffer.
3. Equilibrate the columns 3× with 20 bed volumes b-RP buffer.
4. Solve peptides from Subheading 3.5, **step** 7 in 50 μL 200 mM ammonium formate.
5. Combine eight iTRAQ-labeled samples, mix, and shortly spin down.
6. Apply the sample onto the columns and collect the flow-through fraction (*see* **Note 19**).
7. Wash the columns with 20 bed volumes with b-RP buffer and collect the wash fraction.
8. Elute with 20 bed volumes 3% ACN in b-RP buffer; fraction 1.
9. Elute with 20 bed volumes 6% ACN in b-RP buffer; fraction 2.
10. Elute with 20 bed volumes 9% ACN in b-RP buffer; fraction 3.
11. Elute with 20 bed volumes 12% ACN in b-RP buffer; fraction 4.
12. Elute with 20 bed volumes 15% ACN in b-RP buffer; fraction 5.
13. Elute with 20 bed volumes 80% ACN in b-RP buffer; fraction 6.
14. Vacuum dry samples.
15. Resolve the samples in 13 μL 0.2% TFA and transfer samples into ultracentrifuge tubes.
16. Clear samples by ultracentrifugation at 109.000 × *g* for 20 min.
17. Transfer supernatants into auto sampler HPLC vials and analyze fractions by LC-MS/MS.

3.7 Mass Spectrometry and Protein Identification

Here LC-MS/MS is performed using a Dionex UltiMate 3000 n-RSLC system connected to an Orbitrap™ Fusion Tribid™ mass spectrometer.

1. Load each peptide fraction onto a C18 pre-column.
2. Wash column with 0.1% FA for 3 min at a flow rate of 6 μL/min.
3. Separate peptides on a C18 analytical column (3 mm, Acclaim™ PepMap RSLC, 75 mm × 25 cm, Dionex, Sunnyvale, USA) at flow rate of 350 nL/min via a linear 120 min

gradient from 3% MS buffer B (97% MS buffer A) to 25% MS buffer B (75% MS buffer A).

4. Run linear 15 min gradient from 25% MS buffer B (75% MS buffer A) to 60% MS buffer B (40% MS buffer A).
5. Electro-spray the effluent by the stainless emitter into the mass spectrometer.
6. Control and operate the mass spectrometer in the "top speed" mode, allowing the automatic selection of as many doubly and triply charged peptides as possible in a 3-s time window, and subsequent fragmentation of these peptides.
7. Peptide sequencing based on MS/MS data files is performed via the Proteome Discoverer™ linked to a Mascot™ server using the SwissProt/Uniprot database (*see* **Note 20**).
8. Use the following parameters for Mascot search:
   (a) Maximum missed cleavage site: 1.
   (b) Precursor tolerance: 10 ppm.
   (c) Fragment mass tolerance: 0.05 Da.
   (d) Protease: trypsin.
   (e) Modification: oxidation of M (variable), iTRAQ™ (fixed), methylthio (fixed).
9. Use the following filter in Proteome Discoverer™:
   (a) Peptide confidence: high.
   (b) Search engine rank: 1.
   (c) Peptide score: 30.

4 Notes

1. MR1-5-OP-RU tetramer can also be produced in house according to [17]. MR1-5-OP-RU tetramers and antibodies can be used in different fluorochromes depending on the laser with which the cell sorter is equipped and your own preferences.
2. Proceed with PBMC isolation within 12 h. MAIT cell frequency varies notably in different donors; one can yield between 10.000 and 80.000 MAIT cells from 8 mL blood.
3. From now on work on ice.
4. Antibody concentrations should be optimized for your individual experiment by titration. However, manufacturer's recommendations are a good starting point.
5. Here we have used the FACSAria II flow cytometer (BD Biosciences).

6. To avoid high concentrations of fetal calf serum in your samples, cells should be sorted using PBS and collected in new 15 mL falcon tubes.
7. 1×10^5 human MAITs cells equal approximately 1–1.5 μg total protein. The here described protocol is optimized for 50.000 sorted human MAIT cells per donor/patient.
8. Do not exceed 37 °C to overcome carbamylation of proteins/peptide.
9. Alkylation time might be reduced since both reduction and alkylation will proceed during protein.digestion.
10. Digestion at high urea concentration will allow optimal digestion by Lys-C. Reducing the urea concentration allows additional digestion by trypsin.
11. The pH should be between 4 and 6, and it will be achieved by adding 2–5 μL 10% TFA. Add TFA in steps of 1 μL (mix samples) and check pH by applying 1 μL of the total sample onto pH paper.
12. Maximal speed on any tabletop centrifuge.
13. The capacity of the columns used in this protocol is optimized for the amount of proteins/peptides isolated from the amount of cells used in this protocol. Other columns are available for higher and lower amounts of proteins/peptides.
14. Shortly spin down samples before drying.
15. Before starting peptide iTRAQ™ labeling, one could perform a digestion control of 5% of the sample using short LC-MS/MS runs.
16. iTRAQ™ labels are shipped in 15–20 μL 100% ACN.
17. Too long labeling will result in labeling of other amino acids like histidine.
18. This purification step is included to remove excess iTRAQ™ labels.
19. For maximal peptide recovery (e.g. acidic peptides), keep the flow-through as a separate fraction in the LC-MS/MS.
20. Peptide sequencing based on MS/MS data can be realized by further academic and commercial software products such as MaxQuant (free of charge for academics).

References

1. Vuadens F, Gasparini D, Déon C et al (2002) Identification of specific proteins in different lymphocyte populations by proteomic tools. Proteomics 2:105–111
2. Tan H, Yang K, Li Y et al (2017) Integrative proteomics and phosphoproteomics profiling reveals dynamic signaling networks and bioenergetics pathways underlying T cell activation. Immunity 46:488–503. https://doi.org/10.1016/j.immuni.2017.02.010
3. Bulitta B, Zuschratter W, Bernal I et al (2018) Proteomic definition of human mucosal-associated invariant T cells determines their unique molecular effector phenotype. Eur J Immunol 48:1336–1349. https://doi.org/10.1002/eji.201747398
4. Godfrey DI, Uldrich AP, McCluskey J et al (2015) The burgeoning family of unconventional T cells. Nat Immunol 16:1114–1123. https://doi.org/10.1038/ni.3298
5. Reantragoon R, Corbett AJ, Sakala IG et al (2013) Antigen-loaded MR1 tetramers define T cell receptor heterogeneity in mucosal-associated invariant T cells. J Exp Med 210:2305–2320
6. Dusseaux M, Martin E, Serriari N et al (2011) Human MAIT cells are xenobiotic resistant, tissue-targeted, CD161hi IL-17 secreting T cells. Blood 117:1250–1260. https://doi.org/10.1182/blood-2010-08-303339
7. Sharma PK, Wong EB, Napier RJ et al (2015) High expression of CD26 accurately identifies human bacteria-reactive MR1-restricted MAIT cells. Immunology 145:443–453. https://doi.org/10.1111/imm.12461
8. Cosgrove C, Ussher JE, Rauch A et al (2013) Early and nonreversible decrease of CD161++/MAIT cells in HIV infection. Blood 121:951–961. https://doi.org/10.1182/blood-2012-06-436436
9. Ussher JE, Bilton M, Attwod E et al (2014) CD161++ CD8+ T cells, including the MAIT cell subset, are specifically activated by IL-12+IL-18 in a TCR-independent manner. Eur J Immunol 44:195–203. https://doi.org/10.1002/eji.201343509
10. Gherardin NA, Souter MN, Koay H-F et al (2018) Human blood MAIT cell subsets defined using MR1 tetramers. Immunol Cell Biol 96:507–525. https://doi.org/10.1111/imcb.12021
11. Dias J, Boulouis C, Gorin J-B et al (2018) The CD4-CD8- MAIT cell subpopulation is a functionally distinct subset developmentally related to the main CD8+ MAIT cell pool. Proc Natl Acad Sci U S A 115:E11513–E11522. https://doi.org/10.1073/pnas.1812273115
12. Le Bourhis L, Martin E, Péguillet I et al (2010) Antimicrobial activity of mucosal-associated invariant T cells. Nat Immunol 11:701–708. https://doi.org/10.1038/ni1010-969a
13. Serriari N-E, Eoche M, Lamotte L et al (2014) Innate mucosal-associated invariant T (MAIT) cells are activated in inflammatory bowel diseases. Clin Exp Immunol 176:266–274. https://doi.org/10.1111/cei.12277
14. Walker LJ, Tharmalingam H, Klenerman P (2014) The rise and fall of MAIT cells with age. Scand J Immunol 80:462–463. https://doi.org/10.1111/sji.12237
15. Wiese S, Reidegeld KA, Meyer HE, Warscheid B (2007) Protein labeling by iTRAQ: a new tool for quantitative mass spectrometry in proteome research. Proteomics 7:340–350. https://doi.org/10.1002/pmic.200600422
16. Pottiez G, Wiederin J, Fox HS, Ciborowski P (2012) Comparison of 4-plex to 8-plex iTRAQ quantitative measurements of proteins in human plasma samples. J Proteome Res 11:3774–3781. https://doi.org/10.1021/pr300414z
17. Corbett AJ, Eckle SBG, Birkinshaw RW et al (2014) T-cell activation by transitory neo-antigens derived from distinct microbial pathways. Nature 509:361–365. https://doi.org/10.1038/nature13160

Chapter 15

Generation of MR1-Restricted T Cell Clones by Limiting Dilution Cloning of MR1 Tetramer⁺ Cells

Meghan Cansler, Megan Null, Erin Meermeier, Gwendolyn Swarbrick, David Lewinsohn, and Melanie Harriff

Abstract

Tetramers are a powerful tool for identification of T cell subsets that are restricted by specific antigen presenting molecules and their cognate antigens. The generation of T cell clones from specific T cell subsets allows for further investigation of the phenotype and function of these cells. Here, we describe a method for sorting and cloning of MR1-restricted T cells using the MR1/5-OP-RU tetramer. This protocol can be easily modified to enrich for expansion of specific or unique subsets of MR1-restricted T cell clones from any tissue to further characterize the phenotype and function of those cells.

Key words MAIT cells, MR1-restricted T cells, T cell cloning

1 Introduction

MR1-restricted T cells are a subset of unconventional T cells that recognize microbially derived small molecule metabolites presented on the Class I-like molecule, MR1. These cells were originally detected in the $CD4^{-}CD8^{-}$ T cell compartment of mucosal tissues and defined as mucosal-associated invariant T (MAIT) cells based on this localization and the expression of an invariant T cell receptor (TCR)α chain characterized by TRAV1-2 rearranged with TRAJ33 [1]. As such, MAIT cells were initially identified based on their expression of TRAV1-2, CD161, and CD26 [2–5]. Using *Mycobacterium tuberculosis*-infected dendritic cells as a stimulus in a limiting dilution assay with purified $CD8^{+}$ T cells, our group expanded numerous MAIT cell clones with this phenotypic definition [4]. Subsequent reports have demonstrated that the TCR repertoire is more diverse than originally thought, with expression of additional TRAJ and TCRβ chains, as well as TCRα chains besides TRAV1-2 [6–10]. Many of these findings were made possible following the identification of small molecule metabolite

Helen Kaipe and Isabelle Magalhaes (eds.), *MAIT Cells: Methods and Protocols*, Methods in Molecular Biology, vol. 2098,
https://doi.org/10.1007/978-1-0716-0207-2_15,

ligands and subsequent generation of an MR1 tetramer with one of these ligands, 5-(2-oxopropylideneamino)-6-D-ribitylaminouracil (5-OP-RU) [11, 12]. Indeed, the use of antigen-specific tetramers has uncovered the existence of a broader diversity of MR1-restricted T cells, even beyond TCR diversity. Here, we describe a protocol for sorting and cloning of MR1-restricted T cells with the MR1/5-OP-RU tetramer. The ability to isolate and clone from unique MR1-restricted T cell subsets, such as those that are TRAV1-2^{-} or CD4^{+}, provides the opportunity for further investigation into the functional potential for these cells.

2 Materials

Prepare all media and solutions in a Class II laminar flow biosafety cabinet.

2.1 T Cell Enrichment (Optional, See Note 1)

1. Supplies: Polystyrene FACS tubes, 15- and 50-mL conical tubes, sterile media bottles, bottle-top filters for sterile filtration.
2. Equipment: Benchtop centrifuge with swinging bucket rotor, water bath, hemocytometer, light microscope, 37 °C tissue culture incubator with 5% CO_2.
3. RPMI-FBS media: 500 mL RPMI, 60 mL heat-inactivated fetal bovine serum (FBS) (*see* **Note 2**), 50 μg/mL gentamicin, 4 mM L-glutamine, sterile filtered.
4. DNase I: 100 mg vial of DNase I resuspended in RPMI-HuS media at 3 mg/mL, aliquoted, and stored at −80 °C.
5. RPMI-HuS media: 500 mL RPMI, 55 mL heat-inactivated human serum (HuS) (*see* **Note 3**), 50 μg/mL gentamicin, 4 mM L-glutamine, sterile filtered.
6. EasySep™ Human T cell enrichment kit: Negative Selection (*see* **Note 4**). Kit includes enrichment cocktail and magnetic particles.
7. EasySep™ buffer: Phosphate buffered saline (PBS), 2% FBS, 1 mM ethylenediaminetetraacetic acid (EDTA).
8. EasySep™ Magnetic block.

2.2 Staining and Sorting of T Cells

1. Supplies: 96-well V-bottom plate, 96-well U-bottom plates, 15 mL conical tubes, ice.
2. Equipment: Hemocytometer, light microscope, benchtop centrifuge with swinging bucket rotor, flow sorter (e.g., BD Influx), 37 °C tissue culture incubator with 5% CO_2.
3. PE-conjugated MR1 Tetramer: NIH Tetramer core (*see* **Note 5**).

Table 1
Sample sorting panel to isolate CD4⁻ tetramer⁺ MR1T cells

Marker[a]	Fluorophore	Clone	Manufacturer	Cat. #	Volume/2e6 cells (μL)[b]
CD4[c]	PeCy7	SK3	BD Biosciences	348789	2
CD14	PeCy7	M5E2	Biolegend	301814	2
CD19	PeCy7	HIB19	Biolegend	302216	2
γδTCR[d] (optional)	FITC	5A6. E9	ThermoFisher Scientific	TCR2061	2
Live/dead[e]	Aqua	N/A	ThermoFisher Scientific	L34957	0.25
MR1 tetramer	PE	N/A	NIH Tetramer core	N/A	25[f]

[a]Markers can be added or removed depending on desired MR1T cell population for cloning and sorter capacity. In this case, CD4, CD14, CD19, and γδTCR are all included for "dump channels" to remove these cells from the tetramer⁺ population

[b]These antibody amounts have been titrated with each new lot and for our sorters (BD Influx and FACSAria). As such, this is a recommended starting volume, and all antibodies should be titrated for use at a new sorter facility

[c]The volume of CD4 indicated is for adult samples. CD4 is often dimmer on cord blood and pediatric samples. Increasing the stain amount will improve population separation

[d]This is an optional marker to include in the sort panel. It is not included in the example shown in Fig. 1; however, we have used it in other sorts to gate on the γδTCR⁻ cells after gating on the CD4−CD14−CD19− dump channel

[e]Live/dead is supplied in a desiccated form. It should be resuspended in 50 μL DMSO prior to use

[f]The MR1 tetramer should be titrated prior to use. We generally dilute 1:500 in tetramer buffer, then use 25 μL of the diluted tetramer/2e6 cells

4. Tetramer buffer: PBS with 2% FBS, sterile filtered.
5. Antibodies: *See* Table 1.
6. RPMI-HuS with antibiotics (e.g., 100 units/mL penicillin, 0.1 mg/mL streptomycin, 250 ng/mL Amphotericin B) (*see* **Note 6**).
7. IL-2 (stock concentration 3 μg/mL in RPMI-HuS, final concentration in media for rest 0.5 ng/mL).

2.3 Cloning by Limiting Dilution Assay

1. Supplies: 96-well U-bottom plates, 15- and 50-mL conical tubes, Mr. Frosty™ (*see* **Note 7**).
2. Equipment: Benchtop centrifuge with swinging bucket rotor, water bath, irradiation source (*see* **Note 8**), 37 °C tissue culture incubator with 5% CO_2, inverted light microscope.
3. RPMI-HuS (*see* Subheading 2.1, **item 3**).
4. Lymphoblastoid cell lines (LCL) from two donors (*see* **Note 9**).
5. Peripheral blood mononuclear cells (PBMC) from at least two donors (*see* **Note 10**).
6. Cytokines: *See* Table 2.

Table 2
Cytokines and antibodies used for limiting dilution assay

Cytokine[a]/antibody	Source	Stock concentration[b]	Working concentration
IL-2	Pharmacy[c]	3 μg/mL	2 ng/mL
IL-12	Biolegend	10 μg/mL	0.5 ng/mL
IL-7	Biolegend	3 μg/mL	0.5 ng/mL
IL-15	Biolegend	20 μg/mL	0.5 ng/mL
α-CD3	eBioscience OKT3[d]	1 mg/mL	0.03 μg/mL

[a]IL-2 and α-CD3 are required for cloning. Other cytokines are used when cloning from challenging samples such as cord blood or bronchoalveolar lavage. They are not necessary when cloning from PBMC
[b]All stocks are made in RPMI-HuS and stored in aliquots at −80 °C
[c]IL-2 is obtained from our University in-patient pharmacy under the name Proleukin
[d]Other α-CD3 antibody clones have not performed as well in cloning as OKT3. The functional grade OKT3 from eBioscience is preferred

7. α-CD3 (clone OKT3) (*see* Table 2).
8. Freezing media: 50% FBS, 40% RPMI, 10% dimethylsulfoxide (DMSO), sterile filtered. For samples with fewer or less hearty cells, such as cord blood, pediatric PBMC, or bronchoalveolar lavage, 90% FBS with 10% DMSO can be used.

2.4 Phenotypic and Functional Analysis of Buttons

1. Supplies: 96-well V-bottom plates, 15- and 50-mL conical tubes, ELISPOT Plates (mixed cellulose ester member *see* **Note 11**), syringe, and 0.22 μM syringe filter.
2. Equipment: Benchtop centrifuge with swinging bucket rotor, water bath, light microscope, hemocytometer, ELISPOT reader (*see* **Note 12**), and flow cytometer (*see* **Note 13**).
3. Tetramer buffer (*see* Subheading 2.2, **item 4**).
4. PBMC or MR1 clone control for staining (e.g., ref. 9).
5. Antibodies: *See* Table 3.
6. 1% paraformaldehyde in PBS.
7. ELISPOT Buffers (EB): EB I: Prepare 0.1 M $NaHCO_3$ (4.2 g $NaHCO_3$ in 500 mL diH_2O). Prepare 0.1 M Na_2CO_3 (5.3 g Na_2CO_3 in 500 mL diH_2O). Add the $NaHCO_3$ solution to the Na_2CO_3 solution until the pH is 9.6, then sterile filter. EB II: Add 5 g of bovine serum albumin (BSA) to 1 L PBS. Warm at 37 °C to dissolve the BSA. When BSA is in solution, add 500 μL Tween-20 and sterile filter.
8. PBS, PBS-Tw: 19.2 g of PBS powder, 1 mL Tween20, 2 L diH_2O, stir for 1 h.
9. ELISPOT antibodies: e.g., Anti-human IFNγ coating antibody (clone 1_D1K), anti-human IFNγ ALP-conjugated detection antibody (clone 7-B6-1) (Mabtech) (*see* **Note 14**).

Table 3
Sample phenotyping panel for button analysis

Marker[a]	Fluorophore	Clone	Company	Cat. #	Volume per test (μL)[b]
CD3	BV650	OKT3	Biolegend	317324	0.5
CD4[c]	BV421	OKT4	Biolegend	317434	0.5
CD8	APC-Cy7	SK1	Biolegend	344714	0.025
CD26	FITC	BA5b	Biolegend	302704	0.25
CD161	PE-Cy7	HP-3G10	Biolegend	339918	0.5
TRAV1-2	APC	3C-10	Biolegend	351807	0.25
Live/dead[d]	Aqua	N/A	ThermoFisher Scientific	L34957	0.05
MR1 tetramer[e]	PE	N/A	NIH tetramer core	N/A	25

[a]Markers can be added or removed depending on desired MR1T cell buttons and flow cytometer capacity
[b]These antibody amounts have been titrated with each new lot and for our cytometers (BD LSR, Fortessa, and Symphony). As such, this is a recommended starting volume, and all antibodies should be titrated for use at a new flow cytometry facility. Additionally, these volumes have been reduced for screening buttons. If performing flow cytometry on MR1T cells from rapid expansion flasks, increase each antibody volume by a factor of 4
[c]Volume of CD4 is for adult samples. CD4 is often dimmer on cord blood and pediatric samples. Increasing the stain amount will improve population separation
[d]Live/dead is supplied in a desiccated form. It should be resuspended in 50 μL of DMSO prior to use
[e]The MR1 tetramer should be titrated prior to use. For screening of buttons, we generally dilute the tetramer 1:2000 (due to the low cell numbers in a button), then use 25 μL of this dilution for each sample. If performing flow cytometry on MR1T cells from rapid expansion flasks, dilute the tetramer 1:500 instead

10. 5-Bromo-4-chloro-3-indolyl-phosphate/nitro blue tetrazolium (BCIP-NBT) solution.
11. A549 cell line (ATCC CCL-185) and the A549-MR1$^{-/-}$ cell line [9] (*see* **Note 15**).
12. RPMI-HuS (*see* Subheading 2.1, **item 3**).
13. *M. smegmatis* or other MAIT cell bacterial antigen (*see* **Note 16**).
14. Phytohemagglutinin (PHA): 5 mg/mL stock in RPMI-HuS, 10 μg/mL final concentration.

2.5 Rapid Expansion of MAIT Cell Clones from Buttons

1. Supplies: T-25 tissue culture flasks, 15- and 50-mL conical tubes, Mr. Frosty™.
2. Equipment: Benchtop centrifuge with swinging bucket rotor, water bath, irradiation source, hemocytometer, light microscope.
3. RPMI-HuS (*see* Subheading 2.1, **item 3**).
4. LCL, PBMC (*see* Subheading 2.3, **items 4** and **5**).
5. α-CD3 (*see* Subheading 2.3, **item** 7, Table 2).
6. IL-2 (*see* Table 2).
7. Freezing media (*see* Subheading 2.3, **item 8**).

3 Methods

All tissue culture procedures should be carried out in a Class II laminar flow biosafety cabinet. All incubations should be done in a 37 °C incubator with 5% CO_2.

3.1 T Cell Enrichment (Optional, See Note 1)

1. Begin with cryopreserved samples containing an appropriate number of cells from tissue of choice (e.g., PBMC, single cell suspensions from lung, cord blood, etc.) (*see* **Note 17**). Thaw frozen cell sample(s) in 37 °C water bath and immediately transfer to RPMI-FBS containing 100 μL DNase I. Centrifuge at room temperature for 10 min at 300 × *g*.
2. Resuspend cell pellet in pre-warmed RPMI-HuS containing 100 μL DNase I and count using a hemocytometer.
3. Centrifuge cells again and resuspend at 50e6 cells/mL in Easy-Sep buffer.
4. Transfer sample to polystyrene FACS tube and add Enrichment cocktail at 50 μL/50e6 cells, mix gently with a pipette, and incubate at room temperature for 10 min.
5. Just before the end of the incubation, prepare magnetic particles by vortexing for 30 s. Add 150 μL magnetic particles per 50e6 cells and incubate for 5 min.
6. Add 2.5 mL EasySep buffer to the samples and mix gently with a pipette.
7. Incubate tubes upright in the EasySep magnetic block for 5 min. After incubation, lift magnet and pour contents into a fresh tube (*see* **Note 18**). Proceed with Subheading 3.2 to stain and sort T cells.

3.2 Staining and Sorting of T Cells

1. Determine the number of samples required for staining. Samples should consist of unstained cells, single stain controls for all markers, and cells stained with the full panel (*see* **Note 19**).
2. From the optional T cell enrichment in Subheading 3.1, **step** 7, or from freshly thawed PBMC if T cell enrichment is not necessary, remove a sample for counting, then centrifuge cells at 300 × *g* for 10 min.
3. After centrifugation, resuspend samples in 25 μL of tetramer buffer for every 2e6 cells.
4. Transfer each staining sample (consisting of 2e6 cells) to a 96-well V-bottom plate and centrifuge at 300 × *g* for 5 min.
5. Resuspend each sample in a total of 25 μL tetramer buffer containing the appropriate MR1 tetramer dilution, or no tetramer (for controls) and incubate in the dark for 45 min at room temperature.

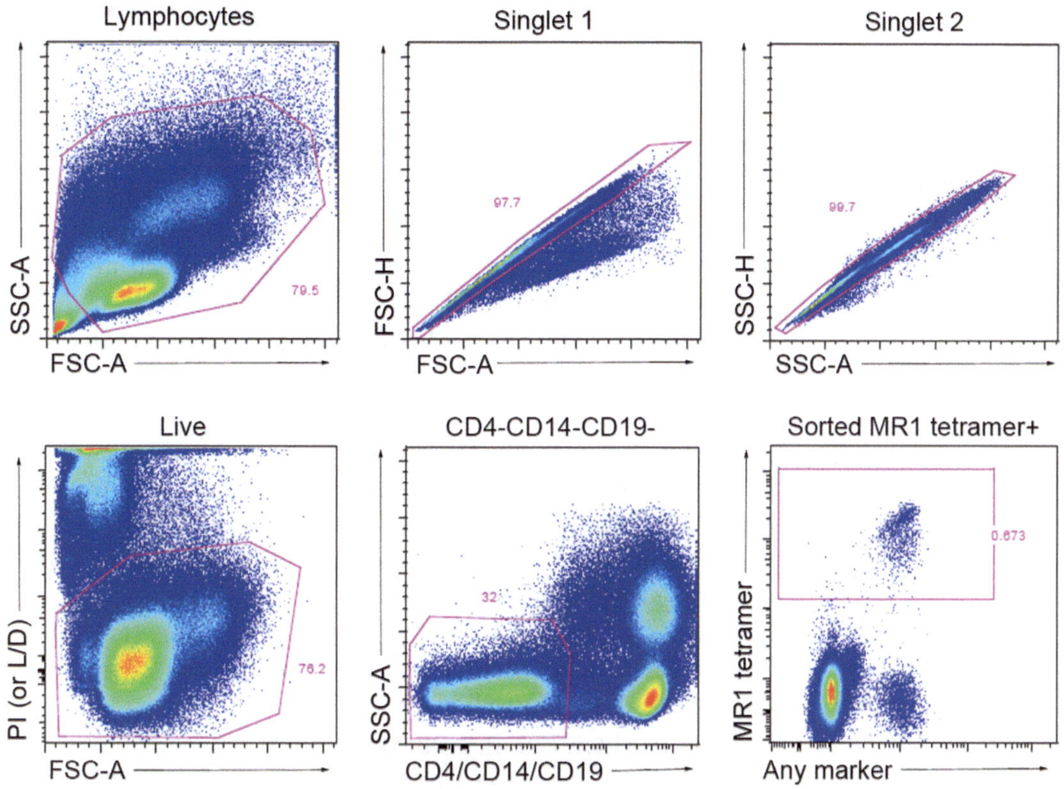

Fig. 1 Sample staining and gating strategy for sort. All samples should be gated for live, single, lymphocytes that do not express CD14 or CD19. From there, optional dump channel markers include γδTCR or CD4. Optional markers for tetramer sorting could include TRAV1-2, CD4, CD8, or any other marker that could define a subset of tetramer$^+$ MR1-restricted T cells

6. While incubating, prepare the antibody master mixes in tetramer buffer (for controls and full panel), with a final volume of 25 μL per sample. *See* Table 1 for an example sorting panel.
7. Add the 25 μL antibody mixture including live/dead directly to the 25 μL tetramer mixture already in the wells and mix gently. Incubate at 4 °C for 30 min, then wash at 4 °C twice with tetramer buffer.
8. Transfer cells on ice to a flow sorter. After using single stained controls to appropriately set up voltage parameters and gates, sort tetramer$^+$ cells of interest into a 15-mL conical tube containing 5 mL of RPMI-HuS with a low dose of IL-2 (0.5 ng/mL). Figure 1 demonstrates an example of sort plots with proper staining and gating.
9. Wash cells twice in RPMI-HuS media (containing PSF, depending on the source of the cells (*see* **Note 6**)).

10. Rest cells overnight in a 96-well U-bottom plate in 200 μL of RPMI-HuS with a low dose of IL-2 (0.5 ng/mL) at 15e6 cells/well (*see* **Note 20**).

3.3 Cloning by Limiting Dilution Assay (LDA)

Use RPMI-HuS for all steps of the LDA.

1. Determine the number of wells that will be used for cloning, based on the estimated number of T cells remaining after sorting and rest. Ideally, T cells should be seeded in 1 plate each of 1, 3, 10, and 30 cells/well (*see* **Note 21**).
2. Centrifuge LCL to wash, resuspend in media, and count. Resuspend at a final concentration of 3e4/50 μL/well (6e5/mL).
3. Thaw and wash allogeneic PBMC, resuspend in media, and count. Resuspend at a final concentration of 1.5e5/well in 50 μL/well (3e6/mL).
4. Irradiate the LCL with a total dose of 6000 cGy and the PBMC with a total dose of 3000 cGy.
5. Combine cytokines (IL-2, IL-12, IL-7, and IL-15) and α-CD3 at the concentrations indicated in Table 2 into a total volume of 50 μL/well (*see* **Note 22**).
6. Add the LCL (50 μL), PBMC (50 μL), and cytokine/α-CD3 (50 μL) mixtures to each well of the 96-well U-bottom plate (s) for a total volume of 150 μL/well.
7. Finally, add the T cells at 1, 3, 10, or 30 T cells/well in 50 μL volumes for a final total volume of 200 μL/well.
8. Cover the plates loosely with foil and incubate for 7–14 days at 37 °C.
9. On day 5, the α-CD3 should be washed from the wells as follows. Visually observe the plate to confirm that the cells are settled at the bottom of the wells. Carefully remove 100 μL of media from each well without disturbing the cells at the bottom of the well. Replace with 100 μL of fresh media supplemented with all cytokines being used. Repeat this wash every 2–3 days until buttons are screened.
10. At day 7, check the plates for growth of buttons by visual inspection of the wells (*see* Fig. 2). If buttons are not observed check the plates again at days 10–14, until buttons are clearly observed. Plates with fewer than 30 buttons are preferred as these are more likely to be clonal.
11. Visually screen a second time using an inverted microscope and select the best buttons to carry forward. The best buttons will be nearly perfectly round, with a well-defined edge. The cells should be densely cluttered and free from visible contamination with other cell types.

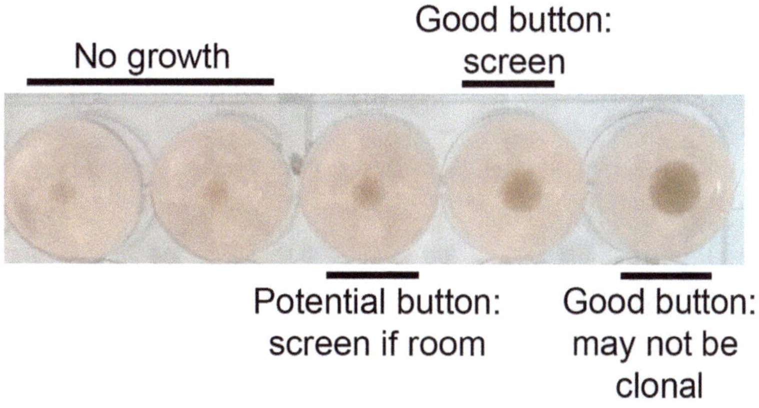

Fig. 2 Analysis of button growth. Shown is the view from the bottom of a cloning plate without magnification. The majority of wells should show no growth. More than 30 buttons on a single plate could indicate that too many cells/well were used for cloning and the buttons may not be clonal. Prioritize buttons that have intermediate growth as shown above, adding in buttons with more or less growth, depending on how many will be screened. When viewed using an inverted microscope, good buttons will be nearly perfectly round, with a well-defined edge. The cells should be densely cluttered and free from visible contamination with other cell types

12. Transfer all buttons to be tested to a new 96-well U-bottom plate. Due to expected evaporation, the original 200 μL volume will be closer to 170 μL.
13. Gently mix and remove 45 μL from each well and transfer to a new 96-well U-bottom plate to stain as described in Subheading 3.4.
14. 30 μL of the remaining button will be used for ELISPOT assay as described in Subheading 3.5.
15. Bring the volume of the remaining button up to a total of 200 μL of RPMI-HuS with a low dose of IL-2 (0.5 ng/mL) and incubate overnight. If the button has the appropriate phenotype and function, this will be used for rapid expansion as described in Subheading 3.6 (*see* **Note 23**).

3.4 Flow Cytometry Staining of Buttons

1. Staining volumes for screening are reduced to account for the reduction in cell numbers from those described in Subheading 3.2. *See* Table 3 for an example staining panel for screening the buttons. As with sorting, use an existing MAIT cell clone or PBMC sample for single stain controls and as a control for the full staining panel.
2. Centrifuge cells transferred to the 96-well U-bottom plate in Subheading 3.3, **step 13** and wash once with tetramer buffer. Resuspend the cells in 25 μL of total volume tetramer buffer that contains MR1 tetramer. Incubate at room temperature for 45 min.

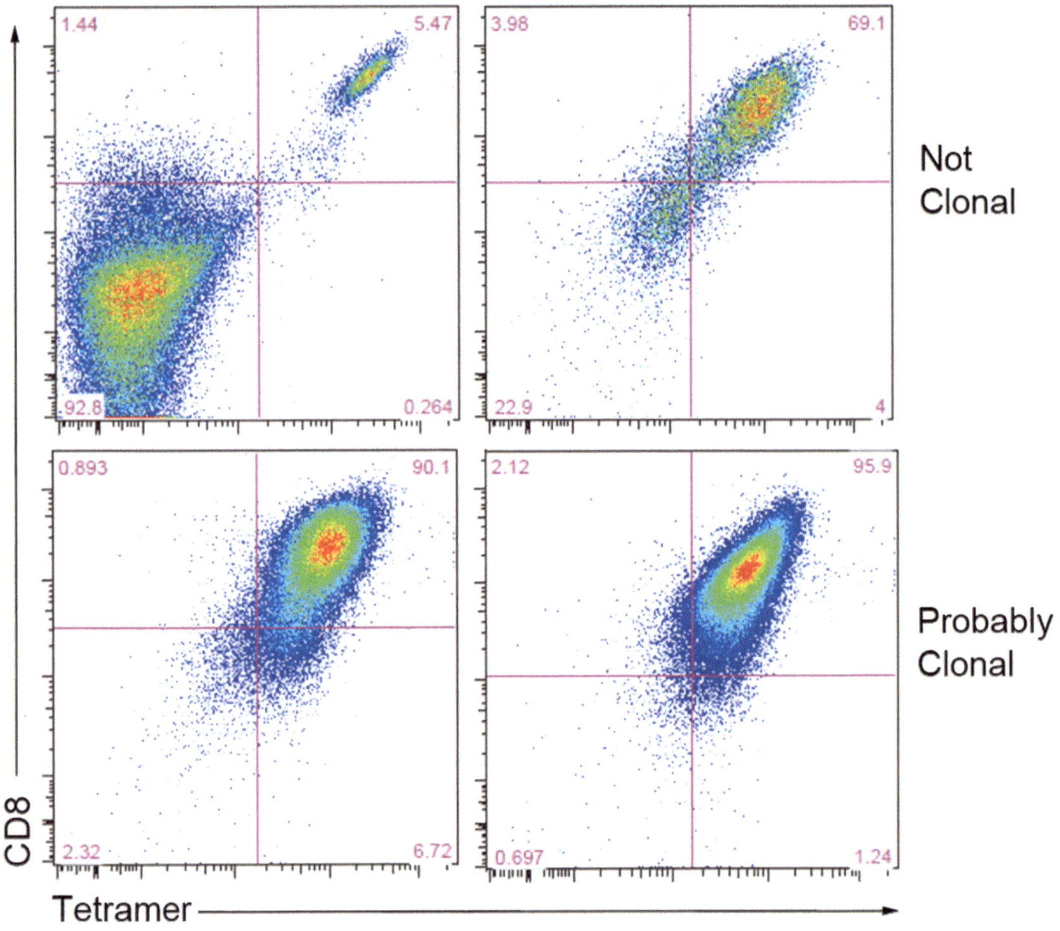

Fig. 3 Flow cytometry analysis of buttons. Initially, cells should be gated on live, singlet, CD3+ lymphocytes. Of this population, cells should be further analyzed for MR1 tetramer staining and other markers of interest, depending on the original sort parameters. Shown here are cells from buttons that were initially sorted on CD8 and tetramer staining. If the whole population does not express these markers, it is not clonal, or not the cell subset that you were trying to expand. Populations that appear clonal should be further verified by TCR sequencing after rapid expansion

3. During the incubation, prepare surface stains and live/dead master mix in 25 μL tetramer buffer per well. Staining panel should include MR1 tetramer, TRAV1-2, CD3, CD8, CD4, and live/dead, as well as CD26, CD161, or other markers as desired.
4. After 45 min, add 25 μL of the antibody master mix to each well on top of the 25 μL tetramer already in the well. Incubate an additional 30 min at 4 °C. Centrifuge the plate and wash the pellet twice with tetramer buffer.
5. Resuspend pellets in 1% PFA in PBS overnight to fix.

6. Acquire samples on a flow cytometer within 24 h of staining. Figure 3 shows examples of MAIT cell buttons likely to be clonal versus those that are not.

3.5 Functional Testing of Buttons by ELISPOT Assay

1. On the day before the buttons are to be stained (Subheading 3.4), coat 3 ELISPOT plates as follows: Per plate, add 50 μL α-IFNγ coating antibody to 5 mL EB1. Add 50 μL antibody mixture to each well of the ELISPOT plate and tap gently to completely coat the entire well. Incubate overnight at 4 °C. If more than 95 buttons will be screened, additional plates will be required.
2. Remove the coating antibody from the plates by flicking. Take care to limit the exposure of the open plate outside of the biosafety cabinet. Add 200 μL PBS to each well and incubate for 15 min at room temperature. Repeat for a total of three PBS washes.
3. Add 100 μL of RPMI-HuS to each well and incubate at room temperature for 1 h to block.
4. While blocking, prepare the antigen presenting cells (APC). Harvest A549 and A549-MR1$^{-/-}$ cells from flasks (*see* **Note 24**). Resuspend pellet in RPMI-HuS and count cells. Add RPMI-HuS for a final concentration of 1e5 cells/mL.
5. After the plates have blocked for 1 h, flick the media from the wells. Then add APC to the ELISPOT wells at 100 μL (1e4 cell) per well, such that there are two plates with A549 cells and one plate with A549-MR1$^{-/-}$ cells. Figure 4 shows a sample ELISPOT plate set up.
6. Dilute bacterial antigen in RPMI-HuS and add 50 μL antigen mixture to the wells of one A549 plate and one A549-MR1$^{-/-}$ plate (*see* **Note 16**).
7. Incubate the plates for 1 h at 37 °C.
8. Gently mix the buttons and pipet 10 μL of each button into the appropriate wells (using 30 μL total of each button). Incubate for 18 h at 37 °C.
9. The following day prepare the detection antibody by adding 10 μL of anti-IFNγ-ALP to 10 mL of EBII.
10. Remove all media and cells from wells by flicking the plate into a biohazard waste bag.
11. Wash plate six times with PBS-Tw. Tap out excess wash onto paper towels prior to adding secondary antibody.
12. Add 100 μL of EBII/antibody mixture to each well and incubate at room temperature for 2 h.
13. Remove EBII/antibody mixture from plate by flicking.

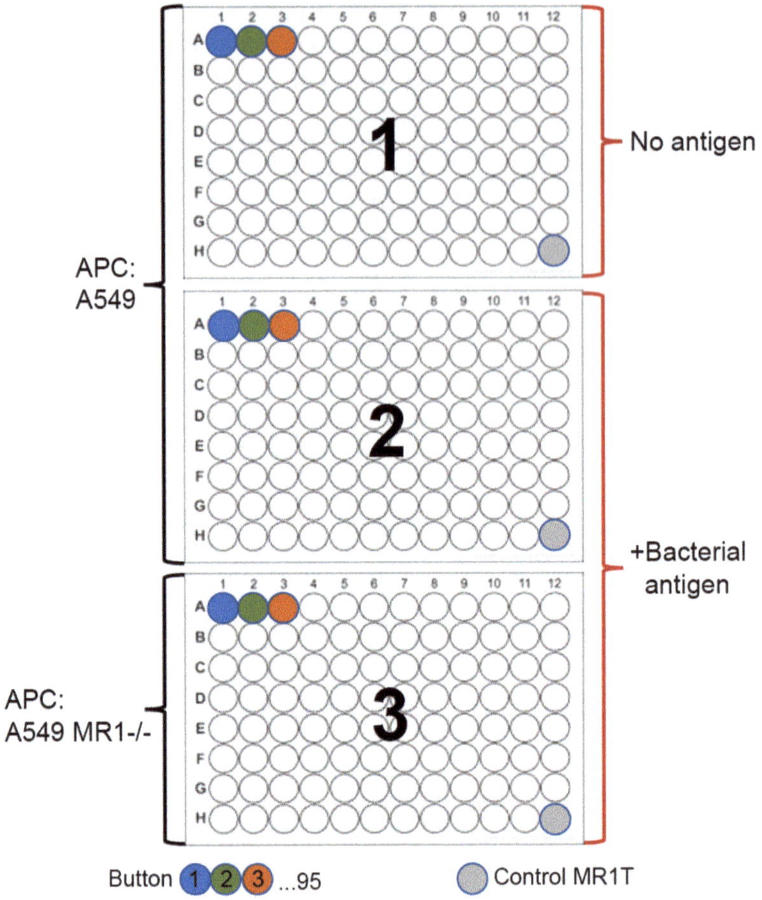

Fig. 4 Layout for ELISPOT screening of buttons. Responses are only expected in plate 2 (center). Buttons that exhibit this phenotype are good candidates for further expansion. Responses in plate 1 (top) would indicate background responses from cells in the button. No responses in plate 2 (center) would indicate that the cells from the button do not recognize bacterial antigen. Responses in plate 3 (bottom) would indicate that the cells from the button are not responding to the bacterial antigen in an MR1-dependent manner. Depending on the desired outcome of the cloning, these may not be good candidates for expansion

14. Wash plate six times with PBS-Tw. Incubate with sixth wash for 15 min.
15. Wash plate three times with PBS, taking care to fill each well completely with PBS to remove excess PBS-Tw buffer.
16. During the 15-min incubation, prepare 10 mL BCIP/NBT solution by passing it through a 0.22-μm filter. Ensure that BCIP/NBT solution comes to room temperature before adding to the plate.

17. Add 100 μL BCIP/NBT solution to each well and incubate in the dark for 10–15 min until purplish-blue spots appear in the wells. At this time, flick BCIP/NBT solution from wells and wash well with deionized water. Remove ELISPOT plate backing during washing.
18. Allow plate to dry and assess button responses with an ELISPOT reader program.

3.6 Rapid Expansion of MAIT Clones from Buttons

1. After phenotypic analysis by flow cytometry and functional analysis by ELISPOT (or other assay), select desired buttons for rapid expansion (*see* **Note 23**).
2. Prepare LCL and PBMC in RPMI-HuS as described in Subheading 3.3, **steps 1** and **2**. Each rapid expansion flask will require 5e6 LCL and 25e6 PBMC in 5 mL each.
3. In an upright T25 flask combine the following: LCL (5e6 in 5 mL), PBMC (25e6 in 5 mL), 30 ng/mL α-CD3, the remaining T cell button, and RPMI-HuS to bring the total volume to 30 mL. Incubate overnight at 37 °C. Cells should settle at the bottom of the flask.
4. Add IL-2 to a final concentration of 2 ng/mL after 24 h (Day "1" of the expansion).
5. Day 5: Wash α-CD3 from the cells as follows. Mix cells and transfer to a conical to centrifuge. Remove all excess media containing α-CD3 from the flasks by aspiration during the centrifugation. Resuspend the cell pellet in 30 mL fresh RPMI-HuS with 2 ng/mL IL-2 and return to the flask.
6. Incubate an additional 5 days, adding 2 ng/mL IL-2 every 2–3 days.
7. Culture can be tested for phenotype and function as described in Subheadings 3.3 and 3.4 starting on Day 10. PHA should be used as a control for T cell responses when testing rapid expansions.
8. Fresh IL-2 should continue to be added every 2–3 days. If pH of media drops, remove and replace 10–15 mL of media from the top of the culture with fresh media (do not disturb cells at bottom of flask). Cells will be functional from Day 10–24, depending on the clone (*see* **Note 25**).
9. Generate a frozen stock of the rapid expansion between Day 10–14. Pellet and resuspend 1e6 cells for every 1 mL freezing media. Place cells in a −80 °C freezer using a Mr. Frosty™, then transfer cells to a liquid nitrogen freezer for long-term storage.
10. Flow cytometry can be used as previously described to determine whether a rapid expansion is likely to be clonal.

11. MAIT cells can be re-expanded by repeating the **steps 2** and **8** by using 5e5 freshly expanded cells (ideally during days 10–14), or by using 5e5 of the cryopreserved T cell rapid expansion (**step 9**).

4 Notes

1. The EasySep T cell enrichment is optional. It can help clean up complex samples like those from the lung (e.g., bronchoalveolar lavage) and reduce sort time. It is not necessary for cloning from peripheral or cord blood samples.
2. In our experience, FBS can cause background in ELISPOT and flow cytometry assays. As a result, we pretest all lots of FBS prior to use for cloning and downstream assays and select those with minimal background. FBS should be heat inactivated by incubation at 56 °C for 45 min after thawing.
3. Our human serum is prepared by drawing 400 cc of blood from research donors. The blood is allowed to clot for 2 h, then the serum is removed and heat inactivated at 56 °C for 45 min. The serum from the donors is pooled together (4 pools of approximately 7 donors each), divided into 30 mL aliquots and frozen at −80 °C. As with FBS, pools are tested for background prior to use. Serum from at least 2 pools is used to make RPMI-HuS. Commercial sources of human serum are available and should be tested prior to use to assess T cell expansion and background activation.
4. Miltenyi T cell negative selection kit and columns can also be used (Pan T Cell Isolation Kit, human).
5. Brighter conjugates such as PE or BV421 are preferable for tetramer staining. A titration should be performed on each new stock of tetramer received from the NIH tetramer core to determine the best dilution for staining. We typically use a 1:500 dilution of tetramer for sorting. The MR1-6FP control tetramer (also available from the NIH Tetramer Core) should be used on a separate sample prior to sorting to verify where positive tetramer staining appears relative to the control to determine appropriate gates for sorting.
6. Antibiotics/antifungals are not necessary when sorting from PBMC or cord blood samples, but should be used if sorting from a sample that could result in contamination (e.g., bronchoalveolar lavage).
7. A Mr. Frosty™ is designed to cool cells by −1 °C/min for preservation. Any other mechanism that achieves this optimal cooling rate for cell cryopreservation will also work.

8. Irradiation is used to stop replication of the LCL and PBMC feeder cells so that they will not expand during the course of the T cell expansions. We have used both cesium irradiation and X-ray irradiation with equal success. It is important to ensure that samples are in the optimal location in the irradiation source for full dosing to prevent replication. Prior to cloning, proper irradiation methods should be confirmed by culturing irradiated cells and verifying that there is no growth.
9. Human lymphoblastoid cell lines (LCL) can be established by infecting resting B cells from PBMC in vitro with Epstein Barr Virus (EBV) [13, 14]. LCL should be cultured in RPMI-FBS, and maintained at a cell density of 1e5–1e6 cells/mL. LCL are available for purchase commercially.
10. Peripheral blood mononuclear cells (PBMC) are isolated from whole blood or apheresis product by standard protocols (e.g., density gradient centrifugation). PBMC are available for purchase commercially.
11. PVDF membranes can also be used but require additional steps to activate the membrane. We have had the best success in our assays with the MultiScreen$_{\mathrm{HTS}}$ plates; however, any mixed cellulose or PVDF membrane and plate that is sterile can be used for this assay.
12. This protocol describes the ELISPOT method to assess the functional capacity of putative MAIT cell clones (buttons), which requires an ELISPOT reader and software. Testing for function of the buttons could also be achieved through different flow cytometry-based methods if an ELISPOT reader is not available (e.g., refs. 9, 10).
13. The flow cytometer should have the capacity to detect a minimum of 6 parameters to analyze live/dead, CD3, CD4, CD8, TRAV1-2, and the MR1 tetramer to confirm that the expanded button is a MAIT cell. Examples include the BD LSR, Fortessa, or Symphony.
14. Examples are given for IFNγ coating and secondary antibodies and the related developing reagent. ELISPOT assays can also be done to analyze markers such as TNFα or Granzyme B. There is also the option to use an HRP-conjugated secondary or to use biotin-conjugated secondary antibodies followed by streptavidin-conjugated HRP or ALP. See the Millipore or Mabtech websites for alternative ELISPOT antibodies.
15. As an alternate to the A549-MR1$^{-/-}$ cell line, blocking with the α-MR1 26.5 antibody clone can be used to demonstrate MR1 restriction as described by Gold et al. [4].
16. *E. coli* or another microbial antigen can also be used as an antigen for the ELISPOT assay. A functional titration should

be done ahead of time for any bacterial antigen stock to determine the appropriate concentration to use in the ELISPOT assay.

17. The minimum number of cells to use will depend on the expected MR1T cell frequency. We have achieved successful clones starting with as few as 3.5e6 live PBMC per cloning sample.
18. Ensure that liquid contents are retained following magnetic bead selection. When inverting the magnet, do so in one continuous motion, and pipet drips from the tube lip to maximize yield.
19. Markers used in sorting panel will be determined by the desired cloning outcome. For example, TRAV1-2 should be included in the panel if TRAV1-2^+ or TRAV1-2^- clones are specifically desired.
20. Cells should be placed in wells near center of plate to minimize evaporation. Assume that approximately half of the cells will die overnight.
21. The number of cells per well and the number of plates may need to be adjusted depending on the starting sample. For example, if no or few buttons grow in plates that are seeded with 1 or 3 cells/well, it may be wise to increase plates with higher numbers of cells per well. However, too many cells per well can often result in buttons that are not clonal. If the number of cells obtained after sorting are limiting, we would recommend starting with plates containing 3 and 10 cells/well, with incomplete plates of both cell numbers if necessary.
22. IL-2 is required for the LDA, while the other cytokines are recommended, particularly for samples that are more challenging to clone from (e.g., bronchoalveolar lavage or cord blood). Ensure that concentration calculations are based on the final volume of 200 μL/well.
23. Ideally, the flow cytometry and ELISPOT assay will be performed on the day the buttons are ready, so that rapid expansions can be done the following day. Cells from buttons can be cryopreserved if there is not enough time to do both screening methods, or if the buttons will not be screened right away (e.g., buttons from plates where many buttons expanded can be cryopreserved as a backup). It should be noted, however, that the number of cells is relatively small, and will be further reduced by cryopreservation or longer incubations, making subsequent screening and rapid expansion more challenging.
24. A549 cells should be cultured and harvested based on the methods described by the ATCC CCL-185.

25. It is ideal to initially test and cryopreserve expanded clones during day 10–14 of the expansion protocol; however, most MR1T cell clones will maintain their functional capacity out to day 21, and in some cases longer.

References

1. Porcelli S, Yockey CE, Brenner MB, Balk SP (1993) Analysis of T cell antigen receptor (TCR) expression by human peripheral blood CD4-8-alpha/beta T cells demonstrates preferential use of several V beta genes and an invariant TCR alpha chain. J Exp Med 178:1–16
2. Treiner F, Treiner SH, Park H, Carcia C, Lemonnier F, de la Salle H, Bendelac A, Bonneville M, Lantz O (2003) Selection of evolutionarily conserved mucosal-associated invariant T cells by MR1. Nature 422:1018
3. Le Bourhis L, Martin E, Peguillet I, Guihot A, Froux N, Core M, Levy E, Dusseaux M, Meyssonnier V, Premel V, Ngo C, Riteau B, Duban L, Robert D, Huang S, Rottman M, Soudais C, Lantz O (2010) Antimicrobial activity of mucosal-associated invariant T cells. Nat Immunol 11:701–708
4. Gold MC, Cerri S, Smyk-Pearson S, Cansler ME, Vogt TM, Delepine J, Winata E, Swarbrick GM, Chua WJ, Yu YY, Lantz O, Cook MS, Null MD, Jacoby DB, Harriff MJ, Lewinsohn DA, Hansen TH, Lewinsohn DM (2010) Human mucosal associated invariant T cells detect bacterially infected cells. PLoS Biol 8: e1000407
5. Sharma PK, Wong EB, Napier RJ, Bishai WR, Ndung'u T, Kasprowicz VO, Lewinsohn DA, Lewinsohn DM, Gold MC (2015) High expression of CD26 accurately identifies human bacteria-reactive MR1-restricted MAIT cells. Immunology 145:443–453
6. Reantragoon RA, Corbett AJ, Sakala IG, Gherardin NA, Furness JB, Chen Z, Eckle SB, Uldrich AP, Birkinshaw RW, Patel O, Kostenko L, Meehan B, Kedzierska K, Liu L, Fairlie DP, Hansen TH, Godfrey DI, Rossjohn J, McCluskey J, Kjer-Nielsen L (2013) Antigen-loaded MR1 tetramers define T cell receptor heterogeneity in mucosal-associated invariant T cells. J Exp Med 210:2305–2320
7. Lepore M, Kalinichenko A, Colone A, Paleja B, Singhal B, Tschumi A, Lee B, Poidinger M, Zolezzi F, Quagliata L, Sander P, Newell E, Bertoletti A, Terracciano L, De Libero G, Mori L (2014) Parallel T-cell cloning and deep sequencing of human MAIT cells reveal stable oligoclonal TCRβ repertoire. Nat Commun 5:3866
8. Gold MC, McLaren JE, Reistetter JA, Smyk-Pearson S, Ladell K, Swarbrick GM, Yu YY, Hansen TH, Lund O, Nielsen M, Gerritsen B, Kesmir C, Miles JJ, Lewinsohn DA, Price DA, Lewinsohn DM (2014) MR1-restricted MAIT cells display ligand discrimination and pathogen selectivity through distinct T cell receptor usage. J Exp Med 211:1601–1610
9. Meermeier EW, Laugel BF, Sewell AK, Corbett AJ, Rossjohn J, McCluskey J, Harriff MJ, Franks T, Gold MC, Lewinsohn DM (2016) Human TRAV1-2-negative MR1-restricted T cells detect *S. pyogenes* and alternatives to MAIT riboflavin-based antigens. Nat Commun 7:12506
10. Gherardin NA, Keller AN, Woolley RE, Le Nours J, Ritchie DS, Neeson PJ, Birkinshaw RW, Eckle SBG, Waddington JN, Liu L, Fairlie DP, Uldrich AP, Pellicci DG, McCluskey J, Godfrey D, Rossjohn J (2016) Diversity of T cell restricted by the MHC class I-related molecule MR1 facilitates differential antigen recognition. Immunity 44:32–45
11. Kjer-Nielsen L, Patel O, Corbett AJ, Le Nours J, Meehan B, Liu L, Bhati M, Chen Z, Kostenko L, Reantragoon R, Williamson NA, Purcell AW, Dudek NL, McConville MJ, O'Hair RAJ, Khairallah GN, Godfrey DI, Fairlie DP, Rossjohn J, McCluskey J (2012) MR1 presents microbial vitamin B metabolites to MAIT cells. Nature 491:717–723
12. Corbett AJ, Eckle SBG, Birkinshaw RW, Liu L, Patel O, Mahony J, Chen Z, Reantragoon R, Meehan B, Cao H, Williamson NA, Strugness RA, Van Sinderen D, Mak JYW, Fairlie DP, Kjer-Nielsen L, Rossjohn J, McCluskey J (2014) T-cell activation by transitory neo-antigens derived from distinct microbial pathways. Nature 509:361–365
13. Brodie SJ, Lewinsohn DA, Patterson BK, Jiyamapa D, Krieger J, Corey L, Greenberg PD, Riddell SR (1999) In vivo migration and function of transferred HIV-1-specific cytotoxic T cells. Nat Med 5:34–41
14. Lewinsohn DM, Alderson MR, Briden AL, Riddell SR, Reed SG, Grabstein KH (1998) Characterization of human CD8+ T cell reactive with *Mycobacterium tuberculosis*-infected antigen presenting cells. J Exp Med 18:1633–1640

Chapter 16

Reprogramming of MAIT Cells to Pluripotency and Redifferentiation

Hiroshi Wakao

Abstract

Reprogramming differentiated cells into induced pluripotent stem cells (iPSCs) consists in dedifferentiation of the cells into the pluripotent state, i.e., stem cells. Since T cells play a pivotal role in our immune system, T cell reprogramming into iPSCs and subsequent redifferentiation of iPSCs toward the original cells hold a great promise for future cell therapy and for further exploring the biology of such T cells. Mucosal-associated invariant T (MAIT) cells are an innate-like T cells linking innate immunity to adaptive immunity, and believed to be implicated in host protection to infection, in inflammation, and in immune homeostasis, which makes them an attractive target for the clinical intervention. In this chapter, we will outline the protocol for reprogramming MAIT cells to pluripotency with Sendai virus vector and redifferentiation. This technique will allow expansion of MAIT cells for cell therapy against the intractable infectious diseases such as HIV/Tuberculosis or cancer.

Key words Reprogramming, Mucosal-associated invariant T (MAIT) cells, Induced pluripotent stem cells (iPSCs), Redifferentiation, Sendai virus, Feeder cells

1 Introduction

Mucosal-associated invariant T (MAIT) cells are innate-like T cells highly abundant in human, being implicated in many diseases, and thus have been a target of intensive scrutiny. Although recent works have begun to shed light on their roles in health and disease, the fact that human MAIT cells are poor at proliferation has hampered to further MAIT cell biology. Furthermore, that the frequency of MAIT cells declines in many disease settings suggests the provision of exogenous MAIT cells holds a great promise for future therapeutic application such as cell therapy and/or regenerative medicine. Advent of induced pluripotent stem cells (iPSCs) has made it possible to reprogram cells into the pluripotent state. A selective differentiation of the desired cell types from iPSCs makes it possible to procure unlimited number of cells relevant to cell therapy and/or regenerative medicine. Given the paucity of human MAIT

Helen Kaipe and Isabelle Magalhaes (eds.), *MAIT Cells: Methods and Protocols*, Methods in Molecular Biology, vol. 2098, https://doi.org/10.1007/978-1-0716-0207-2_16,

cells in proliferation, MAIT cells are reprogrammed into iPSCs so that redifferentiation into MAIT cells (hereafter referred as reMAIT cells) can be achieved. We hereby present a protocol that enables MAIT cell reprogramming into iPSCs and redifferentiation of MAIT cells from the iPSCs. Such a procedure allows amplification of MAIT cells in vitro. However, one has to keep in mind that reMAIT cells are not identical to MAIT cells present in vivo since they are different in transcription and lack expression of some markers relevant to human MAIT cells [1].

2 Materials

2.1 Purification of Cord Blood MAIT Cells

1. Human cord blood.
2. Biotin-labeled anti-human V alpha 7.2 (3C10 clone) antibody.
3. Phycoerythrin (PE)-conjugated Streptavidin magnetic microbeads.
4. FcR blocking reagent (human).
5. LS and MS column.
6. Magnetic separator for MS and LS columns.
7. Phosphate buffered saline (PBS).
8. MACS buffer: PBS containing 0.5% (w/v) BSA and 2 mM EDTA.
9. Ficoll.
10. Hemocytometer (cell counting chamber).
11. Flow cytometer.
12. Centrifuge.

2.2 MAIT Cell Reprogramming

1. Sendai virus harboring reprogramming factors (KOSM302L, provided by Dr. M. Nakanishi, National Institute of Advanced Industrial Science and Technology, Tsukuba, Japan) [2, 3].
2. HBSS.
3. Human ES medium: DMEM/F12 1:1 culture medium containing 20% Knockout Serum Replacement (KSR), 0.1 mM MEM non-essential amino acid solution, 2 mM L-glutamine, 0.1 mM 2-mercaptoethanol, and 4 ng/mL of basic fibroblast growth factor (bFGF).
4. Mitomycin C (MMC) (1 mg/mL solution).
5. 6- and 96-well plates.
6. MMC-treated mouse embryonic fibroblast (MEF) (*see* **Note 1**) in 6-well plate.

2.3 iPSC Colony Formation and Colony Pickup

1. 23-gauge needle.
2. Microscope.
3. P20 and P200 Pipetman.
4. Trypsin/EDTA (TE) solution: 0.25% (w/v) trypsin/ 1 mM EDTA.
5. PBS.
6. Human ES medium (*see* Subheading 2.2, **item 3**).
7. 24- and 96-well plates.
8. 24-well plate filled with MMC-treated MEF.
9. DMEM containing 10% (v/v) FBS.

2.4 iPSC Stock and Characterization

1. Human ES/iPSC colony-detaching solution (any commercially available one, for example Dissociation solution for human ES/iPS cells).
2. Primate iPSC stock solution (ESC-freezing medium, any commercially available one, for example, Stem Cell Keep).
3. Liquid nitrogen, freezing tube (1.5 mL), and liquid nitrogen tank.
4. Human ES medium (*see* Subheading 2.2, **item 3**).
5. DMEM containing 10% (v/v) FBS.
6. MMC-treated MEF (Ø = 60 mm dish).
7. P1000 Pipetman.
8. Vα19 primer (Forward) 5′-GGTGCCATTGTCCAGATCAAC TGC-3′.
9. Jα33 (Reverse) 5′-CTTTATAATTAGCTTGGTCCCAGC-3′.
10. PCR machine and reagents.
11. Restriction enzyme *SacI*.
12. Antibodies against human alkaline phosphatase, SSEA4, Oct-3/4, Nanog, TRA-1-60, and TRA-1-81.
13. Micro glass plate (Ø = 10 mm).
14. 24-well plate.
15. DMEM containing 10% (v/v) FBS.
16. 4% paraformaldehyde (w/v) in PBS.
17. 1% BSA in PBS.
18. PBS containing 0.1% (v/v) Triton X-100, 1% (w/v) BSA, and 10% FBS.
19. PBS containing 1% (w/v) BSA and 10% FBS.
20. Normal donkey serum.
21. TRIC-conjugated donkey anti-goat IgG antibody.

22. Rhodamine Red-X-conjugated donkey anti-mouse IgM.
23. Mounting medium containing DAPI.
24. Fluorescent microscopy.
25. Immunocompromised mice (NOD.CB17-*prkdcscid/J*).
26. Matrigel.
27. 4% (w/v) paraformaldehyde solution.
28. Xylene.
29. Ethanol.
30. Milli Q water.
31. Pan-cytokeratin antibody (AE1/AE3).
32. Desmin antibody (D33).
33. TBS (25 mM Tris–HCl, 2.7 mM KCl, and 137 mM NaCl).
34. Triton X-100.
35. Horseradish peroxidase-conjugated anti-mouse IgG antibody.
36. DAB solution: DAB 5 mg/10 mL of TBS supplemented with 1 μL of 30% H_2O_2.
37. Hematoxylin/eosin.
38. Paraffin.
39. Microtome.
40. The wash buffer (for telomerase assay): 10 mM Hepes-KOH (pH 7.5), 1.5 mM $MgCl_2$, 10 mM KCI, and 1 mM dithiothreitol.
41. Buffer 1 (for telomerase assay): 10 mM Tris–HCl (pH 7.5), 1 mM $MgCl_2$, 1 mM EGTA, 0.1 mM phenyl-methyl sulfonyl fluoride, 5 mM of β-mercaptoethanol, 0.5% CHAPS, and10% glycerol.
42. The reaction mixture 1 (for telomerase assay): 20 mM Tris–HCl (pH 8.3), 1.5 mM MgCl2, 70 mM KCl, 0.05% Tween-20, 1 mM EGTA, 50 μM dNTP, 100 ng TS oligonucleotide (5′-AATCCGTCGAGCAGAGTT-3′), and 0.02 pM TSNT internal standard (5′-AATCCGTCGAGCAGAGTTAAAAGG CCGAGAAGCGT-3′).
43. Primer set 1 (for telomerase assay) ACX: 5′-GCGCGGCTTA CCCTTACCCTTACCCTAACC-3′, NT: 5′-ATCGCTTCT CGGCCTTTT-3′.
44. Blend Taq DNA polymerase.
45. 10× TBE: Tris base 121.1 g, boric acid 61.8 g, EDTA 7.4 g, dissolved in 1 L of Milli Q water.
46. 10% polyacrylamide gel (10 cm × 10 cm) for mini gel electrophoresis.
47. Power supply and gel electrophoresis apparatus.

48. SYBR green solution (1×).
49. Transilluminator to visualize the gel (for SYBR green).
50. Wizard SV genomic DNA purification kit.
51. MethylCode Bisulfate Conversion Kit.
52. Primer sequences (for DNA methylation analysis): For unconverted OCT3/4: Nested forward primer 5′-GAGGACAGGAA TTCAAGACCAGCCTGGGTA-3′ Forward primer 5′-GAGG CTGGAGCAGAAGGATTGCTTTGGCCC-3′ Reverse primer 5′-CCCCCCTGGCCCATCACCTCCACCACCTGG-3′ NAN OG: Forward primer 5′-TGGCCAGGCTGGTTTCAAACTCC TG-3′ Reverse primer 5′-GACCCACCCTTGTGAATTCTCA GTTA-3′ For converted OCT3/4: Nested forward primer 5′-GAGGATAGG AATTTAAGATTAGTTTGGGTA-3′ Forward primer 5′-GAGGTTGGAGTAGAAGGATTGTTTTGG TTT-3′ Reverse primer 5′-CCCCCCTAACCCATCACCT CCACCACCTAA-3′ NANOG: Forward primer 5′-TGGTT AGGTTGGTTTTAAATTTTTG-3′ Reverse primer 5′-AACCC ACCCTTATAAATTCTCAATTA-3′.
53. Plasmid pMD20.
54. PRISM 3730 DNA sequencer.

2.5 Preparation of Lymphoid Progenitor Cells

1. OP9 cells (Riken Cell Bank).
2. DMEM containing 10% FBS.
3. Differentiation medium 1: αMEM supplemented with 10% fetal bovine serum (FBS) plus 100 μM 1-thioglycerol.
4. Gelatin.
5. Hemocytometer.
6. TE.
7. Human ES/iPSC colony-detaching solution.
8. Collagenase IV.
9. Trypsin.
10. PBS-5% (w/v) BSA.
11. Cell strainer (100 μm).
12. CD34 MultiSort kit.
13. Phycoerythrin (PE)-CD34 (4H11).
14. Fluorescein isothiocyanate (FITC)-CD43 (MEM59).
15. Anti-FITC bead.
16. MS (LS) column.
17. Colony digestion solutions: αMEM containing 1 mg/mL Collagenase IV and 0.05% (w/v) trypsin-0.5 mM EDTA.

2.6 In Vitro Differentiation of reMAIT Cells

1. OP9 cells ectopically expressing delta-like 1 (OP9/dkl1).
2. αMEM containing 20% FBS.
3. Human fetal liver tyrosine kinase 3 ligand (Flt3L).
4. Human stem cell factor (SCF).
5. Human interleukin-7 (IL-7).
6. 24-well plate filled with OP9/dlk1.
7. Biotin-3C10.
8. Streptavidin/PE-Cy7.
9. Pacific Blue-TCRβ (IP26).
10. PE-IL-18Rα (H44).
11. APC-CD161 (DX12).
12. 96-well plate.
13. FACS buffer (PBS containing 2% FBS, 1 mM EDTA, and 0.1% NaN_3).
14. 7-AAD (10 μg/mL).
15. 8-color flow cytometer.
16. BAMBANKER (cell freezing medium).
17. Deep freezer (−80 °C).
18. Liquid nitrogen tank.

3 Methods

For reprogramming MAIT cells, MAIT cells are first enriched from the cord blood with the anti-human V alpha 7.2 antibody and MACS column.

3.1 Purification of Cord Blood MAIT Cells

Mononuclear cell preparation from cord blood.

1. Cord blood is diluted with an equal volume of PBS, overlaid on 15 mL Ficoll in 50 mL plastic tubes, and centrifuged at 400 × *g* for 20 min (no brake) at room temperature.
2. After centrifugation, the layer containing mononuclear cells (buffy coat) is collected with a glass pipet and transferred to a new 50 mL tube and washed twice with 50 mL of PBS (600 × *g* for 10 min). Suspend the pellet with 5–10 mL of PBS.
3. Count the mononuclear cells, then centrifuge at 400 × *g* for 5 min. Remove the supernatant as much as possible, add FcR blocking reagent (10 μL/1.0 × 10^7 mononuclear cells) and Biotin-labeled 3C10 anti-human V alpha 7.2 antibody (use 2.0 μg/1.0 × 10^8 mononuclear cells) in the same 50 mL tube (staining performed at 1 × 10^8 cells/50 μL). Adjust the volume to 100 μL with MACS buffer. Mix well and incubate the mixture at +4 °C for 15 min.

4. Wash the cells with 5 mL of MACS buffer and centrifuge at 400 × *g* for 6 min.
5. Resuspend the pellet in 900 μL MACS buffer, then add 100 μL of streptavidin-PE microbeads, and incubate at +4 °C for 15 min.
6. After washing with 5 mL of MACS buffer, pellet the cells by centrifugation at 400 × *g* for 6 min, and resuspend the cells in 0.5 mL MACS buffer.
7. The cells are loaded onto the LS column attached to a magnetic separator (the column should be previously equilibrated with 2 mL of MACS buffer). Once all cells enter the column, wash the column with 2 mL of MACS buffer three times.
8. Remove the column from a magnetic separator, and flush out with 5 mL of MACS buffer by applying the plunger. Count the cell number with a hemocytometer.
9. The purity of the enriched MAIT cells, from here on referred to as $3C10^+$ cells, should be checked with flow cytometric analysis (PE+ fraction) using 50 μL of the eluate. Generally, this procedure gives ~5–20% of PE-positive cells as $3C10^+$ fractions and requires further purification with MS column. Since the percentage of $3C10^+$ cells varies from one cord blood to another, it is not possible to predict the purity at this stage.

   Second round purification with MS column.
10. After centrifugation (400 × *g* for 5 min) of the $3C10^+$ fraction isolated from the LS column, the cell pellet is resuspended in 500 μL of MACS buffer and loaded onto the MS column. The procedure for MS column is essentially same as that for LS column except that 0.5 mL and 1 mL of MACS buffer are used for washing and flushing out, respectively.
11. The purity of $3C10^+$ cells should be checked as described above (*see* **Note 2**).

3.2 Reprogramming of $3C10^+$ Cells (Considered to Be MAIT Cells) with KOSM302L

We use the sendai virus vector encoding the Yamanaka factors (Klf4, Oct3/4, Sox2, and c-Myc, KOSM302L) for reprogramming MAIT cells. Use of such a vector allows generation of iPSCs free from the genomic scar and minimizes the occurrence of tumors, thus is relevant to future therapeutic applications such as cell therapy and/or regenerative medicine. Furthermore, KOSM302L is superior to other virus in that virus *per se* is easy to be deleted with RNAi after reprogramming [1].

1. KOSM302L is incubated with $3C10^+$ cells (the purity >75%) at Multiplicity of Infection = 2.5 for 2 h at room temperature in HBSS in a 96-well plate (total volume <50 μL).
2. After incubation the mixture is centrifuged at 300 × *g* for 3 min to remove the virus and the pellet is suspended in 100 μL

human ES medium. The resultant cells are transferred to 6-well plates (1×10^4 to 1×10^5 cells/well) filled with MMC-treated MEF in 2 mL of human ES medium, and cultured at 37 °C in 5% CO_2.

3.3 iPSC Colony Pickup

1. Change the culture medium every other day.
2. iPSC or Embryonic stem cell (ESC)-like colonies appear on 12–14 days after transduction.
3. Wash with PBS once and fill the dish with 2 mL of PBS. With a 23-gauge needle, cut the MEF around the colony so that the colony is detached from MEF under the microscope (10–40×).
4. Aspirate the colony with P-20 Pipetman and transfer to a well of a 96-well plate filled with 50 μL of TE. Incubate the 96-well plate at 37 °C for 20 min to dissolve into single cells (*see* **Note 3**).
5. Neutralize cells with 100 μL of DMEM containing 10% (v/v) FBS, and pipet up and down in the well to obtain a single-cell suspension (20–30 times with P200 Pipetman).
6. Harvest the cells by centrifugation at 400 × *g* for 6 min, then resuspend the pellet in 2 mL of human ES medium and seed onto MMC-treated MEF in 24-well plate.
7. After iPSC colony formation (takes 4–6 days), the cells are either stocked or subject to other experiments.

3.4 iPSC Stock and Characterization

iPSCs should be stocked for further analysis and/or for future use such as characterization of the stemness and redifferentiation. While the achievement of reprogramming can be assessed by expression of the markers relevant to pluripotency such as alkaline phosphatase, SSEA4, Oct-3/4, Nanog, TRA-1-60, and TRA-1-81 as will be described in the following section, complementary assays are often required to confirm the authentic pluripotency of iPSCs generated in house.

3.4.1 Harvesting and Freezing iPSCs

1. When iPSC colony covers 20–30% of the well surface, wash the well with 1 mL PBS, add 150 μL Human ES/iPSC colony-detaching solution and incubate for 2–4 min at 37 °C until MEF are detached from the dish (iPSC colony should remain intact). iPSCs in 24-well plate are then neutralized with 1 mL DMEM containing 10% (v/v) FBS by pipetting five to six times with P1000 Pipetman and centrifuged at 400 × *g* for 5 min. The pellet is suspended in 1 mL of human ES medium (pipet less than four times), and seeded onto MMC-treated MEF (Ø = 60 mm dish) to let iPSCs expand in 4 mL human ES medium.

2. When iPSCs covers ~50–60% of the surface, wash the cells with PBS, then treat cells with 1 mL Human ES/iPSC colony-detaching solution as described above.
3. Incubate at 37 °C until the iPSC colonies completely detach from the dish (iPSC colonies form the hemi-sphere). Neutralize with DMEM containing 10% (v/v) FBS and centrifuge at 400 × *g* for 5 min.
4. Resuspend the pellet in 200 μL ESC-freezing medium, transfer to a freezing tube, and immediately put in liquid N_2 (*see* **Note 4**).

3.4.2 PCR to Confirm the Origin of iPSCs

Since $3C10^+$ fraction contains cells other than MAIT cells even after the second round of purification, it is highly recommended to confirm the resulting iPSCs indeed stemmed from MAIT cells. For this purpose, genomic DNA is prepared by any convenient methods (for example, using the spin-column) and subjected to PCR to amplify the rearranged fragment Vα19-Jα33, the TCR α chain specific for human MAIT cells with a primer set (Vα19(forward) and Jα33(reverse)) (see Subheading 2.4).

1. Representative PCR conditions: 98 °C 30 s, 36 rounds (98 °C 10 s 64 °C 10 s 72 °C 10 s), and 72 °C 20 s (*see* **Note 5**).
2. The PCR product size is 282 bp and the resultant product can be digested with the restriction enzyme *SacI*, conferring 191 and 91 bp DNA [1]. If desired, a 282-bp product can be subject to DNA sequencing with the primers described in Subheading 2.4, **items 8** and **9**. We have established more than 50 MAIT cell-derived iPSCs (MAIT-iPSCs) from three donors [1].

3.4.3 Pluripotency Marker Expression in MAIT-iPSCs

Once reprogramming has successfully been done, iPSCs express the markers specific for pluripotent cells such as alkaline phosphatase, SSEA4, Oct-3/4, Nanog, TRA-1-60, and TRA-1-81. These markers can be detected with immunofluorescence microscopy.

1. Put a micro glass plate in a 24-well plate, seed MMC-treated MEF at 5 × 10^4/cm^2 in 2 mL DMEM containing 10% FBS. On next day, remove the medium, wash with 1 mL PBS, and seed MAIT-iPSCs in a 24-well plate in 2 mL human ES medium. Change the medium every other day. Wait until iPSC colonies covers 20–30% of the glass surface in the 24-well plate.
2. Wash the cells twice with PBS (1 mL/well of a 24-well plate).
3. Fix the cells with 4% paraformaldehyde (w/v) in PBS for 20 min at room temperature.
4. Wash the cells three times with 1% BSA in PBS for 5 min (0.5 mL/well of a 24-well plate).

5. Permeabilize and block the cells with PBS containing 0.1% (v/v) Triton X-100, 1% (w/v) BSA, and 10% FBS (for oct3/4 and nanog stainings) or PBS containing 1% (w/v) BSA and 10% FBS (for SSEA4, alkaline phosphatase, TRA-1-60, and TRA-1-81 stainings) at room temperature for 45 min (0.5 mL/well of a 24-well plate).
6. During the blocking, dilute the reconstituted antibody in PBS containing 0.3% Triton X-100, 1% BSA, and 10% normal donkey serum to a final concentration of 10 μg/mL for the SSEA4, alkaline phosphatase, oct3/4, and nanog antibodies, and of 2 μg/mL for the TRA-1-60 and TRA-1-81 antibodies.
7. After blocking, incubate the cells with the diluted antibodies (300 μL/well of a 24-well plate) for overnight at 2–8 °C.
8. Next day wash the cells three times with 1% BSA in PBS for 5 min (0.5 mL/well of a 24-well plate).
9. Dilute the appropriate secondary antibody at 1:200 in PBS containing 1% BSA (For oct3/4 and nanog, use TRIC conjugated donkey anti-goat IgG antibody; for alkaline phosphatase, SSEA4, TRA-1-60, and TRA-1-81, use Rhodamine Red-X-conjugated donkey anti-mouse IgM).
10. Incubate the cells with diluted secondary antibody in the dark for 60 min at room temperature (300 μL/well of a 24-well plate).
11. Wash the cells three times with PBS containing 1% BSA for 5 min (500 μL/well of a 24-well plate).
12. Cover the cells with a slide glass and add a drop of mounting medium containing DAPI. Cells are now ready for fluorescent microscopy. Representative images are shown in Fig. 1. (Reprinted from Cell Stem Cell, 12, Wakao et al./Reprogramming of MAIT cells to pluripotency and redifferentiation, 546–558. Copyright (2013), with permission from Elsevier.)

3.4.4 Teratoma Formation

Given that iPSCs are pluripotent, they should form teratoma in vivo. This can be examined by injecting MAIT-iPSCs into immunocompromised mice.

1. Eight to ten million MAIT-iPSC ($8–10 \times 10^6$) are subcutaneously injected at the neck together with 60 μL Matrigel into 8- to 10-week-old female NOD/scid mice (NOD.CB17-*prkdcscid/J*).
2. At 10–14 weeks post-injection, mice are euthanized under CO_2 and teratomas are dissected.
3. Teratomas are fixed in 4% (w/v) paraformaldehyde solution for 24 h at room temperature, and paraffin-embedded according to the standard protocol. In brief, fixed tissues are dehydrated

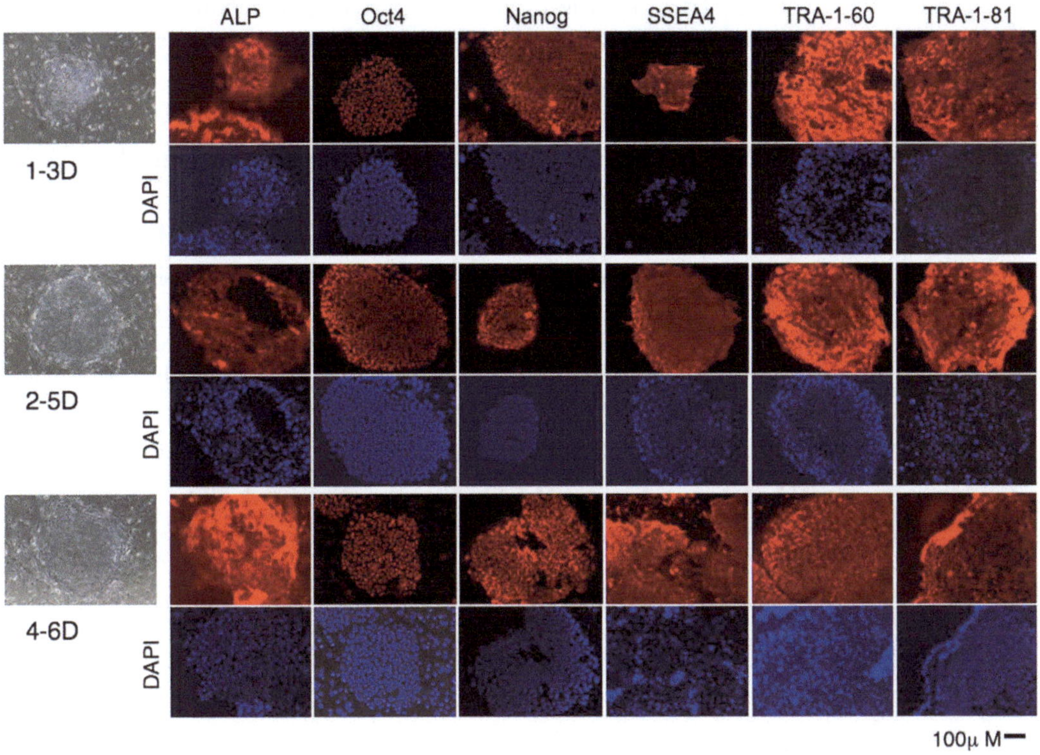

Fig. 1 Pluripotency marker expression in MAIT-iPSCs. MAIT-iPSC colonies exhibit similar morphology to human ESCs and express pluripotency markers, alkaline phosphatase (ALP), Oct3/4, Nanog, SSEA4, TRA-1-60, and TRA-1-81, as evidenced by immunofluorescence staining. Nuclei are stained with DAPI. Scale bar, 100 μM. 1-3D, 2-5D, and 4-6D represent MAIT-iPSC clones. (Reproduced from [1] with permission from Elsevier)

with increasing concentrations of ethanol (70–100%), cleared with xylene at room temperature, and waxed at 58 °C. Sample slices (5–10 μM thickness) are prepared with a microtome on glass slides.

4. Samples are air-dried for 30 min and baked in incubator at 45 °C for 16 h.

5. Deparaffinize the samples in a glass slide staining rack with xylene twice (10 min × 2) in a staining dish. Hydrate the samples with 100% ethanol twice (3 min × 2), 95% ethanol (v/v), and 80% (v/v) ethanol, respectively, for 1 min. Then sample is rinsed with Milli Q water.

6. The samples are immunostained with anti-pan-cytokeratin (an endoderm marker, 1.0 μg/mL) and anti-desmin (a mesoderm marker, 1.0 μg/mL) antibodies in TBS with 1% (w/v) BSA, and incubate for 16 h at 4 °C.

7. Rinse with TBS containing 0.025% (v/v) Triton X-100 twice (5 min × 2) under gentle agitation.

8. To block the excessive superoxide, incubate the sample with 0.3% (v/v) H_2O_2 in TBS for 15 min at room temperature.
9. Add horseradish peroxidase-conjugated anti-mouse IgG as the secondary antibody (1 μg/mL).
10. The reaction is developed using diaminobenzidine (DAB) solution for 10 min at room temperature.
11. Stop the color development by washing with Milli Q water.
12. Samples are then stained with hematoxylin/eosin. Representative results are shown in Fig. 2. (Reprinted from Cell Stem Cell, 12, Wakao et al./Reprogramming of MAIT cells to pluripotency and redifferentiation, 546–558. Copyright (2013), with permission from Elsevier.)

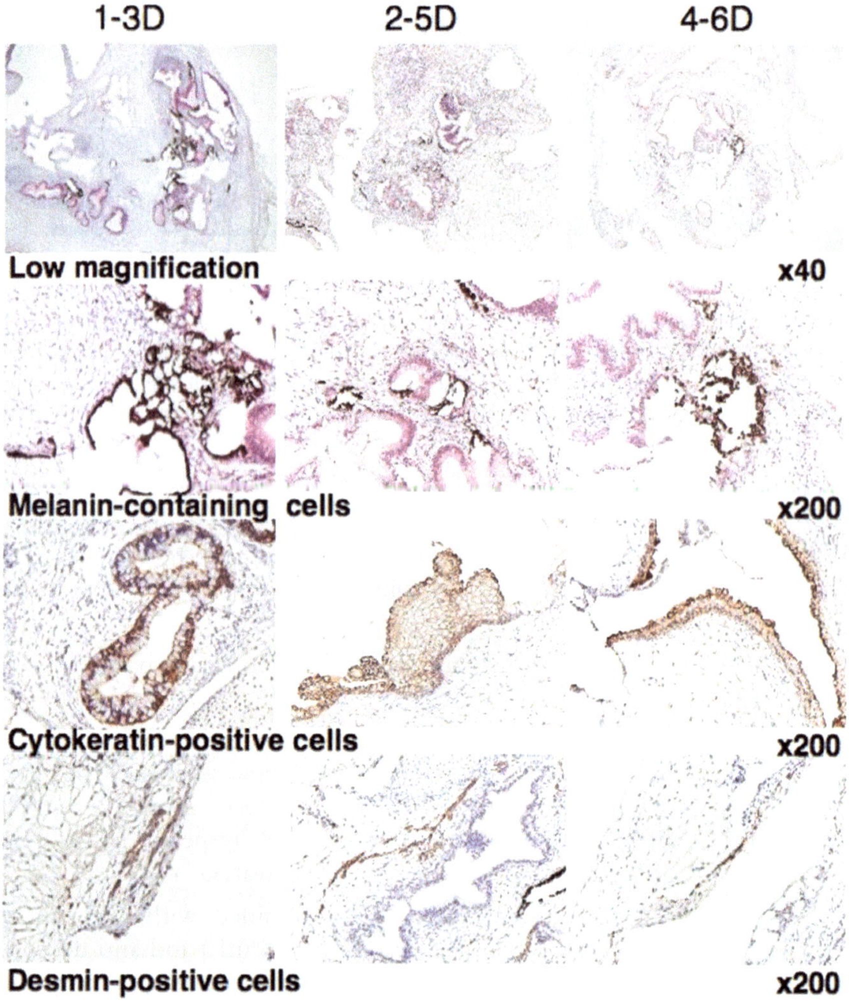

Fig. 2 In vivo pluripotency of MAIT-iPSCs. Teratoma formation 10–12 weeks after transplantation of 1-3D, 2-5D, and 4-6D in NOD/SCID mice. Histological sections of identified cells after HE and DAB staining representing all three germ layers: ectoderm (melanin-containing cells), endoderm (cytokeratin-positive cells), and mesoderm (desmin-positive cells). Low- and high-magnification images are shown. (Reproduced from [1] with permission from Elsevier)

3.4.5 Telomerase Activity

Telomerase activity is a hallmark of immortalized cells such as iPSCs and cancer cells. Telomerase activity can be detected by PCR and visualized on polyacrylamide gel [4].

1. Cell lysates containing the telomerase are prepared as follows. iPSCs (30–40% confluent in 60 mm dish) are washed once in PBS, pelleted at 10,000 × *g* for 1 min, resuspended in the ice-cold wash buffer, and pelleted again, and resuspended in 20 μL of buffer 1. The suspension is incubated for 30 min on ice and then centrifuged at 15,000 × *g* for 30 min at 4 °C. The supernatants are aliquoted and snap frozen in dry ice and kept at −70 °C.
2. Quickly thaw the lysate in a water bath at 37 °C, and use 5–10 μL in the reaction mixture 1 in a total reaction volume of 20 μL, and incubate for 30 min at room temperature.
3. Aliquot 5 μL and heat at 85 °C for 10 min before PCR reaction. This serves as a negative control.
4. Add the primer set 1 (10 μM for each) and 1.25 U of Blend Taq DNA polymerase to the reaction mixture.
5. Run PCR as follows: 94 °C for 3 min and 32 rounds of (94 °C 25 s, 50 °C 25 s, and 72 °C 60 s).
6. Separate PCR products on polyacrylamide gel in 0.5 × Tris-Borate-EDTA (TBE). The ladder can be seen with SYBR green solution. In our hand, the cell lysate equivalent to 2000 cells is sufficient to visualize the ladder (we failed to see the ladder when we used the cell lysate equivalent to 10,000 cells). Representative results are shown in Fig. 3. (Reprinted from Cell Stem Cell, 12, Wakao et al./Reprogramming of MAIT cells to pluripotency and redifferentiation, 546–558. Copyright (2013), with permission from Elsevier.)

3.4.6 Bisulfate Sequencing

Upon cell reprogramming *OCT3/4* and *NANOG* promoters are demethylated. Methylation status at cytosine residues in CpG can be interrogated by PCR. While non-methylated cytosines are converted to uracil, methylated cytosines are resistant to the conversion upon bisulfate modification. The unconverted and converted regions can be amplified with the unconverted and converted region specific primer sets, respectively.

1. Isolate genomic DNA from CB-MAIT cells, human ES cells or iPSCs, and MAIT-iPSCs (60 mm dish, 30–40% confluent) with the Wizard SV genomic DNA purification kit.
2. Measure the DNA concentration with a spectrometer.
3. Genomic DNA (200 ng) is subjected to bisulfate conversion with the MethylCode Bisulfate Conversion Kit.

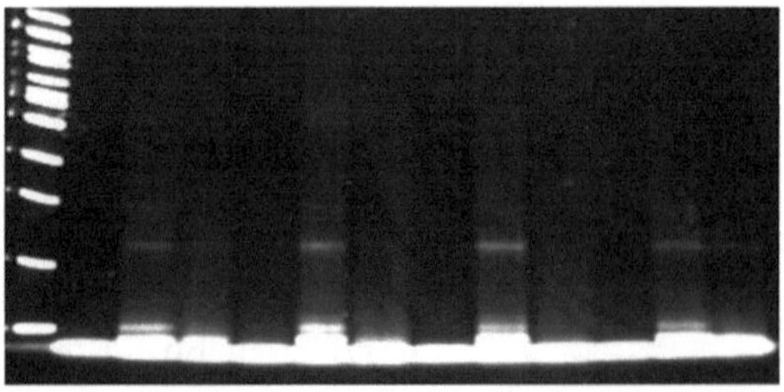

Fig. 3 Telomerase activity in MAIT-iPSCs. M; DNA molecular weight marker, lanes 1–3; khES3, lanes 4–6; 1-3D, lanes 7–9; 2-5D, lanes 10–12; 4-6D. Heat-inactivated extracts of 10,000 cells (lanes 1, 4, 7, and 10). Extracts of 2000 cells (lanes 2, 5, 8, and 11). Extracts of 10,000 cells (lanes 3, 6, 9, and 12) (Reproduced from [1] with permission from Elsevier)

4. Converted DNA (50 ng) is used as a template for conventional nested PCR to amplify 467 and 336 bp of the human *OCT3/4* and *NANOG* promoters, respectively [5].
5. To amplify converted *OCT3/4*, run nested PCR as follows: 94 °C for 2 min and 22 rounds of (94 °C 30 s, 60 °C 30 s, and 72 °C 40 s) 72 °C 10 s (primer set: nested forward primer and reverse primer). Take 0.5 μL of the reaction mixture, run PCR as follows: 94 °C 2 min and 36 rounds of (94 °C 30 s 60 °C 30 s 72 °C 40 s) 72 °C 10 s (primer set: forward primer and reverse primer). Similarly, run nested PCR to amplify unconverted *OCT3/4* as follows: 94 °C 2 min and 22 rounds of (94 °C 30 s, 60 °C 30 s, and 72 °C 40 s) 72 °C 10 s (primer set: nested forward primer and reverse primer). Take 0.5 μL of the reaction mixture, run PCR as follows: 94 °C for 2 min and 36 rounds of (94 °C 30 s, 60 °C 30 s, and 72 °C 40 s) 72 °C 10 s (primer set: forward primer and reverse primer).

   For *NANOG*, run PCR as follows: 94 °C for 2 min and 36 rounds of (94 °C 30 s, 55 °C 30 s, and 72 °C 40 s) 72 °C 10 s for amplification of unconverted and converted regions.
6. The purified PCR products are cloned into the pMD20 vector.
7. Ten randomly picked clones are subject to DNA sequencing.
8. Depict the figure so that the most 5′-cytosine residue is put at the top and the most 3′- cytosine residue is located at the bottom per clone (there are 12 cytosines for *OCT3/4* and 8 for *NANOG* in the amplified region). Demethylated cytosines are marked with an open circle, while methylated ones with a filled circle. Put 10 clones together side by side (*see* **Note 6**). Representative results are shown in Fig. 4. (Reprinted from Cell Stem Cell, 12, Wakao et al./Reprogramming of MAIT cells to pluripotency and redifferentiation, 546–558. Copyright (2013), with permission from Elsevier.)

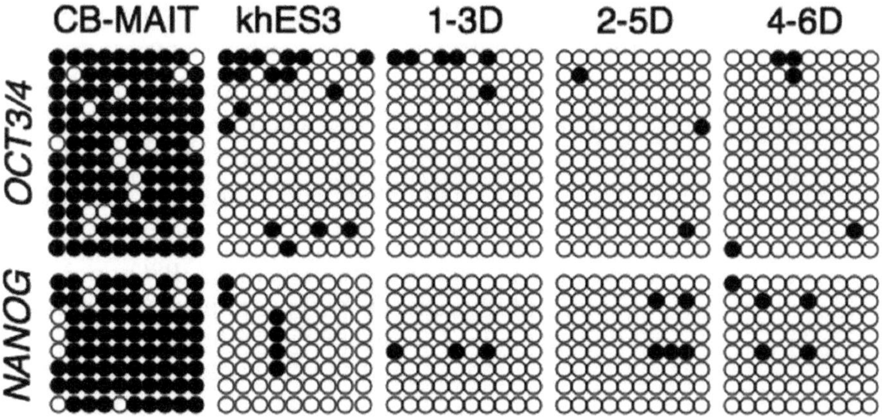

Fig. 4 Methylation analysis of *OCT3/4* and *NANOG* promoters in MAIT cells, human ESCs, and MAIT-iPSCs (1-3D, 2-5D, and 4-6D). Genomic DNA from cord blood (CB)-MAIT cells, human ESCs (khES3), 1-3D, 2-5D, and 4-6D was subjected to bisulfite modification. The promoter regions of *OCT3/4* and *NANOG* were amplified by PCR with the primer sets described in the text, and methylation states were analyzed. Open and filled circles show non-methylated and methylated CpG dinucleotides, respectively (Reproduced from [1] with permission from Elsevier)

Besides the above techniques, in vitro differentiation into three layers (ectoderm, mesoderm, and endoderm) allows to show the pluripotency of iPSCs. Also, expression of pluripotent cell-specific transcripts such as *Oct3/4*, *Nanog* can be examined by Revere Transcription-PCR. The protocols are found elsewhere (for example, [1]).

3.5 Preparation of Lymphoid Progenitor Cells

MAIT-iPSCs harbor a rearranged *Vα7.2-Jα33* specific for human MAIT cells.

This genomic configuration is a prerequisite for quasi-exclusive differentiation of MAIT cells from iPSCs (reMAIT cells) under the T cell permissive culture conditions. The present protocol essentially consists of two steps to induce $3C10^+$ ($V\alpha 7.2\text{-}J\alpha 33^+$) $TCR\alpha\beta^+IL\text{-}18R\alpha^+CD161^+$ cells, as reMAIT cells, more precisely in vitro-generated immature MAIT cells, but not immature MAIT cells present in vivo. The first step consists in inducing lymphoid precursors such as $CD34^+CD43^+$ cells from MAIT-iPSCs [6]. The second step is differentiation of such precursors to reMAIT cells as defined $3C10^+$ $TCR\alpha\beta^+IL\text{-}18R\alpha^+CD161^+$ cells (*see* **Note 7**).

3.5.1 Seeding MAIT-iPSCs on OP9 Cells

1. Prepare five dishes (Ø = 60 mm) of iPSCs as described in Subheading 3.4.1, **step 2**.
2. Wash with 3 mL PBS once, add 1 mL of Human ES/iPSC colony-detaching solution, and incubate for 2–4 min at 37 °C until MEF are detached from the dish.
3. Add 1 mL DMEM containing 10% FBS and pipet up and down six to eight times with P1000 Pipetman.

4. Take 50 μL from the mixture, transfer them in an Eppendorf tube, and centrifuge at 400 × *g* for 4 min.
5. Remove the supernatant, resuspend in 100 μL TE, incubate for 15 min at 37 °C to obtain a single-cell suspension and count the cell number with a hemocytometer.
6. Calculate the MAIT-iPSC cell number in 3 from the cell count with TE treatment.
7. Seed 1.0 × ~10^6 fragmented MAIT-iPSC on 3–6 days post confluent OP9 cells (Ø = 10 cm culture dish) in the differentiation medium 1. Prepare ten dishes for the seeding (total 1.0 × ~10^7 fragmented MAIT-iPSCs) (*see* **Notes 8–10**).

3.5.2 Cell Culture and Harvest

1. Half of the medium was changed 4 days later and then on days 6, 8, and 10.
2. On day 11, the differentiated cells were harvested by mild enzymatic digestion followed by mild trypsin treatment as follows.
3. Single-cell suspension is prepared by treatment of the MAIT-iPSCs/OP9 cocultures with 1 mL of collagenase IV (1 mg/mL in αMEM per dish) for 20 min at 37 °C.
4. Remove the collagenase solution and add 2 mL of 0.05%(w/v) trypsin-0.5 mMEDTA for 15 min at 37 °C [7].
5. Pipet 40–50 times with P-1000 Pipetman to obtain a single-cell suspension.
6. Add 5 mL of PBS-5% (w/v) BSA per dish and pass through the cell strainer (100 μm).
7. Count the cells and subject to magnetic enrichment of the precursor cells.

3.5.3 Enrichment of Lymphoid Precursor Cells

Although a previous study has shown that T cells stem from $CD34^{high}$ $CD43^{med}$ cells [6], this technique requires a cell sorter that is often not available in every laboratory. We therefore established a more convenient method to enrich the precursor cells with MACS column.

1. After enzymatic digestion, cells are subjected to CD34 bead selection (CD34 MultiSort kit). During magnetic labeling, PE-labeled anti-human CD34 monoclonal antibody is added.
2. Perform positive selection with LS column as described in Subheading 3.1.
3. After the LS column, the positive fractions were stained with FITC-labeled anti-human CD43, followed by magnetic labeling with anti-FITC beads.

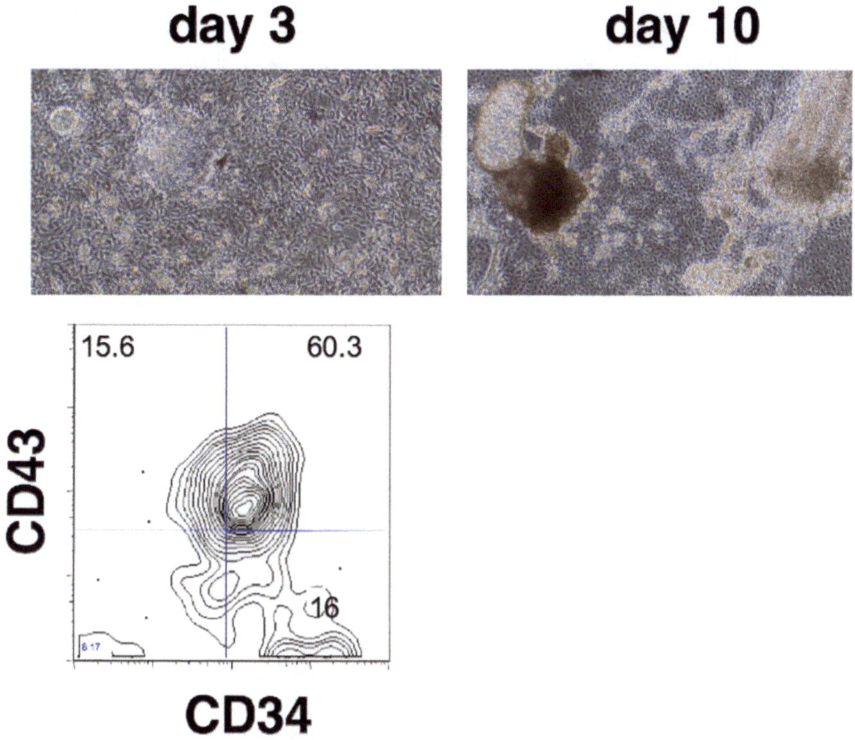

Fig. 5 Differentiation of MAIT-iPSCs on OP9 and enrichment of the precursor cells. Picture shows day 3 (upper left panel) and day 10 culture (upper right panel) of MAIT-iPSC on OP9. Lower panel shows the flowcytometry for $CD34^+CD43^+$ and $CD34^+$ cells (reMAIT cell precursors) after the MACS column

4. CD43-$FITC^+$ cells are subjected to an MS column, and PE^+/$FITC^+$ cells and PE^+ cells are collected (fractions corresponding to $CD34^+CD43^+$ and $CD34^+$ cells, respectively).
5. Count $CD34^+CD43^+$ and $CD34^+$ cells. After the second column, more than 90% of the cells should be either $CD34^+$ or $CD34^+CD43^+$ when analyzed by flow cytometer. Representative results are shown in Fig. 5.

3.6 In Vitro Differentiation of reMAIT Cells

$CD34^+CD43^+$ and $CD34^+$ cells are suspended in αMEM containing 20% FBS, human stem cell factor (SCF, 5 ng/mL), interleukin 7 (IL-7, 5 ng/mL), and fetal liver tyrosine kinase 3 (Flt3) ligand (5 ng/mL) and seeded onto 24-well plates filled with the confluent OP9 cells ectopically expressing delta-like ligand 1 (OP9/dlk1) [8]. Half of the medium is changed every 4 days, and culture continued for up to 30 days at 37 °C 5% CO_2 (*see* **Note 11**).

As described above, MAIT cells are defined as $3C10^+TCR\alpha\beta^+IL\text{-}18R\alpha^+CD161^+$ cells. One can follow the status of differentiation with a flow cytometer at different time points.

1. Stain cells (1.0×10^4) with biotin-3C10 (use 0.2 ng/1.0×10^4 cells) for 30 min on ice in 96-well plate, wash the cells with

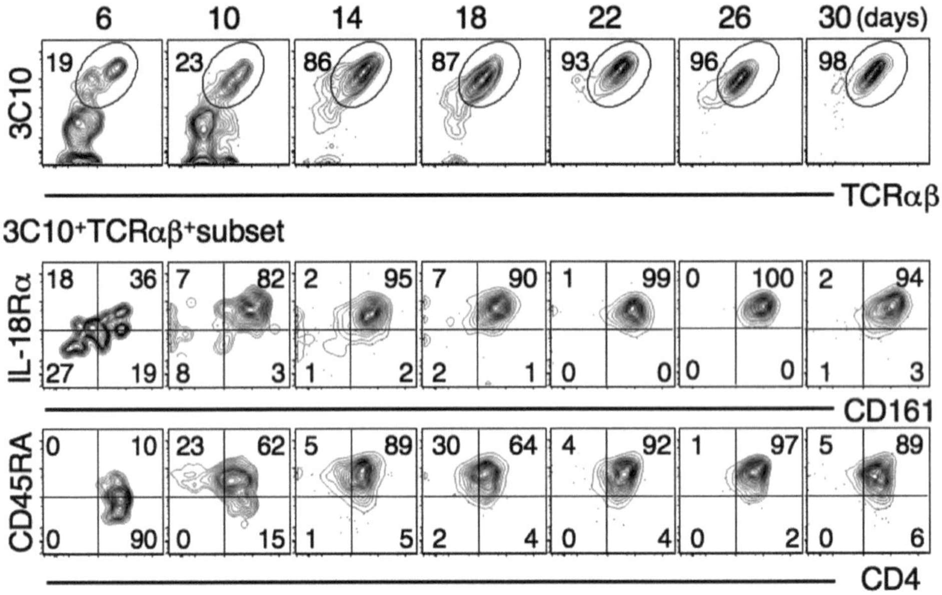

Fig. 6 Time course of reMAIT cell generation from MAIT-iPSCs. FACS analysis of the lymphocytes harvested at the indicated times on OP9/DL1. The numbers show the percentages of $3C10^{+}TCR\alpha\beta^{+}$cells (upper panel). The expression profile of IL-18Rα, CD161, CD45RA, and CD4 in the $3C10^{+}TCR\alpha\beta^{+}$-gated populations is shown (middle and lower panels) (Reproduced from [1] with permission from Elsevier)

150 μL of FACS buffer, centrifuged at 500 × *g* for 5 min, and discard the supernatant.

2. Add appropriate amounts of Pacific Blue-TCRβ, PE-IL-18Rα, and APC-CD161 antibodies to stain 1.0×10^4 cells (final volume <20 μL) for 20 min on ice.
3. Add 40 μL of streptavidin/PE-Cy7 (diluted with FACS buffer to 1/300) and continue to incubate for 30 min on ice.
4. Wash with 150 μL of FACS buffer, centrifuged at 500 × *g* for 5 min, and discard the supernatant.
5. Resuspend the pellet with 100 μL of FACS buffer containing 7-AAD (to exclude dead cells). Samples are now ready for Flow cytometric analysis.

Representative results are shown in Fig. 6 (*see* **Note 12**). (Reprinted from Cell Stem Cell, 12, Wakao et al./Reprogramming of MAIT cells to pluripotency and redifferentiation, 546–558. Copyright (2013), with permission from Elsevier.)

In vitro-generated reMAIT cells are now ready for experiments. In case where immediate use is not envisioned, the cells may be stocked although the survival is severely compromised (less than 10%). The cells are stocked with BAMBANKER (cell freezing medium) and kept in the liquid N_2.

1. Count the in vitro-generated reMAIT cells, transfer to 15 mL Falcon tube, centrifuge at 400 × *g* for 4 min, and discard the supernatant.
2. Add BAMBANKER (100 μL/1 × 10^5 cells), pipet gently for several times, transfer to freezing tube, and put them in the deep freezer (−80 °C).
3. On the next day transfer the tube to liquid N_2 tank.

4 Notes

1. MEF should be treated with MMC before using as an iPSC-feeder cell. We seed one vial of the primary MEF (Oriental Yeast) into five dishes (10 cm plastic dish) in DMEM containing 10% (v/v) FBS and cultured for 3 days. When cells are confluent, they are treated with TE (see below) and expanded in 30 dishes (10 cm plastic dish) for 4–5 days. When cells become confluent, they are treated with DMEM containing 10% (v/v) FBS and MMC (10 μg/mL) for 2 h 15 min at 37 °C, then washed three times with PBS. Cells are dissociated with TE and stored in BAMBANKER cell stock solution (1.0–5.0 × 10^6 cells/vial). The cell density of MEF for a routine use is 3–5 × 10^4/cm^2. For iPSC culture, plastic dishes should be treated with PBS containing 0.1% (w/v) gelatin for 1 h before seeding MEF.
2. In our hand, after the second column, the purity of PE-positive cells in $3C10^+$ fractions increased to 30–95%. We have obtained 50,000–250,000 of the cells for reprogramming with the purity more than 75%. If reprogramming is not anticipated on the same day, $3C10^+$ fractions can be frozen for later use. The cells should be stored in the liquid nitrogen with the cell stock solutions such as BAMBANKER.
3. Pick up as many colonies as possible within 30 min.
4. It is a good practice to minimize the number of pipetting (less than 6 times) to keep iPSC colony as large as possible. Excess pipetting results in poor recovery of iPSC when thawed. iPSCs are kept in the liquid N_2 just before thawing. Remove the cell vial from the liquid N_2, and immediately add prewarmed (37 °C) human ES medium. Pipet quickly several times and centrifuged at 400 × *g* for 5 min. iPSCs are ready for seeding.
5. Annealing temperature is critical for successful PCR. We found Tm = 64 °C is good for Phusion Taq polymerase.
6. The Taq polymerase is important for successful PCR. We have used Platinum Taq polymerase for the above reaction.

7. At present, MAIT cells are defined as 5-OP-RU-loaded human MR1 tetramer^{+}TCR$\alpha\beta^{+}$ CD161^{+} cells [9]. Users are thus encouraged to modify the protocol accordingly.
8. OP9 should be seeded in 10-cm plastic dish coated with 0.1% gelatin (w/v) for one overnight. If not, OP9 cells may detach from the dish during differentiation.
9. Fragmented iPSCs can be obtained by pipetting 10–15 times with P-1000 Pipetman. Most of the fragments should be composed of ~50–100 cells. More or less fragmented MAIT-iPSCs would give less CD34^{+}CD43^{+} cells.
10. FBS lot has a significant impact on the success of differentiation from MAIT-iPSCs. It is of importance to keep in mind that an FBS lot suitable for OP9 is often not optimum for differentiation of MAIT-iPSCs into the precursor cells (CD34^{+}CD43^{+} cells) on OP9. The author thus recommends users to check an array of FBS to find out suitable ones for the proper experiments.
11. The cell number after the MS column remains unchanged after differentiation on OP9/dlk1, i.e., no significant cell proliferation is observed from the precursors, although the percentage of 3C10^{+}CD161^{+}IL-18Rα^{+} cells increases along differentiation.
12. Culture for more than 30 days should be avoided, as in vitro-generated immature MAIT cells tend to die. The number of in vitro-generated immature MAIT cells varies from several thousand to 1 million starting from 10 dishes of 10 cm plastic dish (starting from 1.0×10^6 MAIT-iPSCs per dish).

References

1. Wakao H, Yoshikiyo K, Koshimizu U, Furukawa T, Enomoto K, Matsunaga T, Tanaka T, Yasutomi Y, Yamada T, Minakami H, Tanaka J, Oda A, Sasaki T, Wakao R, Lantz O, Udagawa T, Sekiya Y, Higuchi K, Harada N, Nishimura K, Ohtaka M, Nakanishi M, Fujita H (2013) Expansion of functional human mucosal-associated invariant T cells via reprogramming to pluripotency and redifferentiation. Cell Stem Cell 12:546–558
2. Nishimura K, Sano M, Ohtaka M, Furuta B, Umemura Y, Nakajima Y, Ikehara Y, Kobayashi T, Segawa H, Takayasu S, Sato H, Motomura K, Uchida E, Kanayasu-Toyoda T, Asashima M, Nakauchi H, Yamaguchi T, Nakanishi M (2011) Development of defective and persistent Sendai virus vector: a unique gene delivery/expression system ideal for cell reprogramming. J Biol Chem 286:4760–4771
3. Nishimura T, Kaneko S, Kawana-Tachikawa A, Tajima Y, Goto H, Zhu D, Nakayama-Hosoya K, Iriguchi S, Uemura Y, Shimizu T, Takayama N, Yamada D, Nishimura K, Ohtaka M, Watanabe N, Takahashi S, Iwamoto A, Koseki H, Nakanishi M, Eto K, Nakauchi H (2013) Generation of rejuvenated antigen-specific T cells by reprogramming to pluripotency and redifferentiation. Cell Stem Cell 12:114–126
4. Kim NW, Piatyszek MA, Prowse KR, Harley CB, West MD, Ho PL, Coviello GM, Wright WE, Weinrich SL, Shay JW (1994) Specific association of human telomerase activity with immortal cells and cancer. Science 266:2011–2015
5. Takahashi K, Yamanaka S (2006) Induction of pluripotent stem cells from mouse embryonic and adult fibroblast cultures by defined factors. Cell 126:663–676

6. Timmermans F, Velghe I, Vanwalleghem L, De Smedt M, Van Coppernolle S, Taghon T, Moore HD, Leclercq G, Langerak AW, Kerre T, Plum J, Vandekerckhove B (2009) Generation of T cells from human embryonic stem cell-derived hematopoietic zones. J Immunol 182:6879–6888
7. Vodyanik MA, Bork JA, Thomson JA, Slukvin II (2005) Human embryonic stem cell-derived CD34+ cells: efficient production in the coculture with OP9 stromal cells and analysis of lymphohematopoietic potential. Blood 105:617–626
8. Wakao H, Wakao R, Sakata S, Iwabuchi K, Oda A, Fujita H (2008) In vitro induction of natural killer T cells from embryonic stem cells prepared using somatic cell nuclear transfer. FASEB J 22:2223–2231
9. Corbett AJ, Eckle SB, Birkinshaw RW, Liu L, Patel O, Mahony J, Chen Z, Reantragoon R, Meehan B, Cao H, Williamson NA, Strugnell RA, Van Sinderen D, Mak JY, Fairlie DP, Kjer-Nielsen L, Rossjohn J, McCluskey J (2014) T-cell activation by transitory neo-antigens derived from distinct microbial pathways. Nature 509:361–365

Part IV

MAIT Cells in Murine Models

Chapter 17

Study of MAIT Cell Activation in Viral Infections In Vivo

Timothy S. C. Hinks, Bonnie van Wilgenburg, Huimeng Wang, Liyen Loh, Marios Koutsakos, Katherine Kedzierska, Alexandra J. Corbett, and Zhenjun Chen

Abstract

MAIT cells are abundant, highly evolutionarily conserved innate-like lymphocytes expressing a semi-invariant T cell receptor (TCR), which recognizes microbially derived small intermediate molecules from the riboflavin biosynthetic pathway. However, in addition to their TCR-mediated functions they can also be activated in a TCR-independent manner via cytokines including IL-12, -15, -18, and type I interferon. Emerging data suggest that they are expanded and activated by a range of viral infections, and significantly that they can contribute to a protective anti-viral response. Here we describe methods used to investigate these anti-viral functions in vivo in murine models. To overcome the technical challenge that MAIT cells are rare in specific pathogen-free laboratory mice, we describe how pulmonary MAIT cells can be expanded using intranasal bacterial infection or a combination of synthetic MAIT cell antigen and TLR agonists. We also describe protocols for adoptive transfer of MAIT cells, methods for lung homogenization for plaque assays, and surface and intracellular cytokine staining to determine MAIT cell activation.

Key words Virus, MAIT cell, Flow cytometry, MR1-tetramer, Infection, Mouse

1 Introduction

MAIT cells are relatively recently described innate-like lymphocytes, with similarities to the invariant natural killer T (iNKT) and γδ T cell subsets [1–4]. They are the most abundant innate-like population in the lungs in humans [5] though relatively rare in specific pathogen-free mice [6] and show a striking evolutionary conservation between diverse species of mammals [7]. MAIT cells express a semi-invariant T cell receptor (TCR), which recognizes microbially derived small molecule intermediates from the riboflavin biosynthetic pathway [1, 4, 8, 9]. These molecular intermediates exist only in microbes but not in mammals, and therefore constitute a signature of microbial infection. This property implicates MAIT cells in anti-bacterial host defense, and potentially also in other roles such as tissue repair [3]. However, in addition to their

Helen Kaipe and Isabelle Magalhaes (eds.), *MAIT Cells: Methods and Protocols*, Methods in Molecular Biology, vol. 2098, https://doi.org/10.1007/978-1-0716-0207-2_17,

TCR-dependent functions, they can be activated in a TCR-independent manner via cytokines including IL-12, -15, -18, and type I interferon [10–12]. Emerging data suggest that they are expanded and activated by a range of human viral infections including dengue, hepatitis C, and influenza virus [11, 13]. It was not clear from observational human studies whether this would lead to enhanced immune protection, or, conversely, contribute to immunopathology. To address this question, we conducted experimental influenza A virus challenge in vivo in mice and demonstrated that MAIT cells could contribute to a protective anti-viral response [12].

Here we describe the methods used to investigate these anti-viral functions in vivo in murine models. To overcome the technical challenge that MAIT cells are rare in specific pathogen-free laboratory mice, we describe (1) how pulmonary MAIT cells can be expanded using intranasal (i.n.) bacterial infection or a combination of synthetic MAIT cell antigen and TLR agonists as well as protocols for (2) adoptive transfer of MAIT cells, (3) viral preparation and infection of mice, (4) lung homogenization, (5) surface and intracellular cytokine staining to determine MAIT cell activation, and (6) plaque assays.

2 Materials

2.1 Reagents and Buffers

1. Antibodies are specified in Tables 1–4.
2. Collagenase medium: Roswell Park Memorial Institute medium (RPMI) containing 3 mg/mL collagenase III, 5 μg/mL DNase, and 2% fetal calf serum (FCS). Aliquots can be frozen at −20 °C.
3. Fluorescence activated flow cytometry (FACS) buffer: phosphate buffered saline (PBS), 2 mM EDTA, 0.5% bovine serum albumin (BSA). From a 500 mL bottle of PBS, add 40 mL to a 50 mL falcon containing 2.5 g BSA powder, vortex hard, then filter sterilize back into PBS bottle using a syringe through a 0.22-μm filter. Do not add azide as will be toxic to the cells.
4. Percoll (Density 1.13 g/mL) 40% and 70% solutions, pre-warmed to room temperature for each use.
5. RPMI with pen/strep: RPMI containing 100 μg/mL streptomycin and 100 U/mL penicillin.
6. Tris-based Ammonium Chloride (TAC)–HCl, pH 7.5 hypotonic red blood cell lysis buffer: 0.14 M NH_4Cl, 0.017 mM Tris (pH 7.5), then adjust pH to 7.2 with HCl (2 M). The solution is filter (0.22 μm) sterilized and kept at room temperature.

Table 1
Flow cytometry panel compatible with a three-laser BD Aria III flow cytometer, allowing identification and sorting of MR1-5-OP-RU-tetramer+ MAIT cells

Marker	Fluorophore	Laser	Standard dilution if staining in 1500 μL, amount in μL
CD45.2	FITC	Blue	3.75 μL 1:400
7AAD	7AAD	Blue or Yellow/Green	3.75 μL *titrate
CD19	PerCpCy5.5	Blue or Yellow/Green	7.5 μL 1:200
TCRβ	APC	Red	7.5 μL 1:200
MR1-5-OP-RU tetramer	BV421	Violet	7.5 μL 1:200

Make up volume to final 720 μL with FACS buffer

Table 2
Flow cytometry panel compatible with a three-laser BD Aria III flow cytometer, allowing optimal identification of MR1-5-OP-RU-tetramer+ MAIT cells using surface stains only

Marker	Fluorophore	Laser	Standard dilution if staining in 40 μL, amount in μL
CD45.2 (*see* **Note 1**)	FITC	Blue	1:200, 0.2
TCRβ	APC	Red	1:200, 0.2
CD19	PerCpCy5.5	Blue or Yellow/Green	1:200, 0.2
CD8	PE	Blue or Yellow/Green	1:800, 0.08
CD4	APC Cy7	Red	1:200, 0.2
MR1-5-OP-RU-tetramer	BV421	Violet	1:200, 0.2

Antibodies should be titrated by each laboratory

Table 3
Surface markers for flow cytometry panel compatible with a three-laser BD Aria III flow cytometer, allowing measurement of MR1-5-OP-RU-tetramer+ MAIT cell activation by intracellular cytokine staining

Marker	Stain	Laser	Standard dilution if staining in 40 μL, amount in μL
TCRβ	APC	Red	1:200, 0.25
CD19	PerCpCy5.5	Blue or Yellow/Green	1:200, 0.25
MR1-5-OP-RU-tetramer	BV421	Violet	1:200, 0.25

Table 4
Intracellular markers for flow cytometry panel for intracellular staining

Marker	Intracellular stain (*see* Note 2)	Laser	Standard dilution if staining in 50 μL, amount in μL
IFNγ	PE Cy7	Blue or Yellow/Green	1:400, 0.125
TNF	PE	Blue or Yellow/Green	1:300, 0.17
IL-17	PE or PECy7 or APC (depending on surface stains used)	Blue or Yellow/Green, Red	1:200, 0.25

7. Fixation buffer: 1% formaldehyde, 2% glucose in PBS. Fully dissolved solution is kept cold (+4 °C) and dark (aluminum foil wrapped) as formaldehyde is sensitive to light.
8. Media for growing MDCK cells: Dulbecco Modified Eagle Medium (DMEM) containing 2 mM L-glutamine, 1 mM MEM sodium pyruvate, 100 U/mL penicillin/streptomycin, and 10% heat-inactivated FCS.
9. Serum-free (SF) DMEM: Dulbecco Modified Eagle Medium (DMEM) containing 2 mM L-glutamine, 1 mM MEM sodium pyruvate, and 100 U/mL penicillin/streptomycin.
10. 2× Leibovitz's L-15 media for overlay, make 2× stock as it will be diluted 1:1 with agarose. For 1 L: Use 1 L sterile water. Remove 100 mL of the water but keep for later use. Add two 14 g packets of L-15 powdered media (kept at 40 °C). Add magnetic flea and stir for 4 h or more to ensure the powder is completely dissolved. Adjust pH to 6.8 using 1 M HCl. Then add the following to the medium.
    (a) 8 mL of 7% w/v $NaHCO_3$ prepared in Hanks Buffered Saline Solution (HBSS) (stored at 4 °C).
    (b) 800 μL 1 M HEPES buffer (pH 6.8).
    (c) 20 mL of 10,000 U/mL Pen/Strep.
    (d) Make up the volume to 1 L (using the 100 mL previously removed) and filter sterilize. Store at 4 °C. To reduce precipitation, aliquot into 50 mL tubes for storage.
11. 1 mg/mL trypsin: warm up trypsin powder for 30 min at room temperature (kept at +4 °C). Weigh out 10 mg of powder and dissolve in 10 mL of PBS. Filter using 0.45 μm filter. Aliquot aseptically into 220 μL/aliquot. Store at −20 °C. 200 μL will be added to 100 mL of overlay (50 mL L-15 and 50 mL agarose) for a final concentration of 2 μg/mL trypsin/well.
12. Salmonella: *Salmonella enterica*, serovar Typhimurium (attenuated strain BRD509) [14], stored at −20 °C in Luria-

Bertani (LB) broth with 50% glycerol, to prevent freezing at this temperature.

13. MR1-tetramers (5-OP-RU and 6-FP) are available from the NIH core tetramer facility, on application. Store in component parts at −80 °C until ready for use, at which point small aliquots can be tetramerized and stored at 4 °C for days to weeks. They should be reconstituted according to instructions supplied with the product. Typically a 5-μg aliquot of MR1-5-OP-RU monomer or MR1-6-FP monomer should be expanded to a total volume of 18 μL in Tris-buffered saline. About 6.8 μL of commercially available streptavidin-PE at 0.5 mg/mL should be made up to a total volume of 17 μL in Tris-buffered saline. Add 1/10 of the streptavidin-PE solution (1.7 μL) to the monomer solution every 10 min and pipette to mix, incubating at room temperature in the dark between steps. Repeat until all the streptavidin-PE solution has been added. This will give a final volume of 35 μL containing 0.143 μg/μL of tetramer. The tetramer should be titrated for use; typically 1:200–1:1000 dilutions are sufficient.
14. Madin-Darby Canine Kidney (MDCK) cells.
15. Live/Dead Fixable Aqua Dead Cell Stain Kit or Zombie Yellow Viability Stain Kit.
16. Brefeldin A.
17. Phorbol 12-myristate 13-acetate (PMA).
18. Ionomycin.
19. Trypsin–versene.
20. 1% Crystal Violet in 20% ethanol and dH_2O.
21. Flow cytometry compensation beads.
22. Flow cytometry 6 μm blank size calibration beads.
23. Fixation/permeabilization buffer and perm-wash buffer.
24. LB agar plates, containing 50 μg/mL streptomycin.
25. LB culture medium.
26. 2.4G2 (anti CD16/32) hybridoma cell culture supernatant.
27. Anti-CD4 (GK1.5) and anti-CD8 (53.762) monoclonal antibodies for depletion of adoptively transferred T cell subsets.
28. 1% Virkon or 10% Lysol or Hypochlorite (5000 ppm).
29. 80%w/v EtOH.
30. Hanks buffered saline solution (HBSS).
31. Isoflurane.

2.2 Plastic and Other Supplies

1. 1 and 10 mL syringes.
2. 26 G needles.

3. Dissection scissors.
4. 1 mL Eppendorf tubes.
5. 40 and 70 μm cell strainers.
6. 10 cm Petri dishes.
7. 10, 15, and 50 mL Falcon tubes.
8. 5 mL polypropylene or polycarbonate FACS tubes.
9. Flat-bottom 6-well (TC6) plates.
10. 96-well flat-bottom plates.
11. 96-well U- or V-bottom plates.

2.3 Equipment

1. Flow cytometer with capability for cell sorting, BD LSR Aria or equivalent.
2. Spectrophotometer capable of reading at 600 nm.
3. Hemocytometer and light microscope.
4. Animal anesthetic circuit capable of administering volatile inhalational anesthetics.
5. Shaking incubator.
6. Gaseous carbon dioxide and gas exposure chamber.
7. Benchtop mechanical roller for tubes.
8. Tissue homogenizer for disrupting tissue into single cell suspensions.

3 Methods

Personal protective equipment (PPE) should be worn at all times (gloves, lab coat, & eye protection) (*see* **Notes 3** and **4**).

3.1 MAIT Cell Expansion in Donor Mice

MAIT cells are rare in specific pathogen-free mice [6], typically comprising about 1×10^4 recoverable pulmonary MAIT cells in an infection-naïve adult C57BL/6 mouse. Therefore, for adoptive transfer experiments, the MAIT cell population should first be expanded using intranasal infection [15] or immunization (5-OP-RU with TLR agonists) [3, 15] (*see* **Note 5**). When planning the adoptive transfer experiment, estimate that one *S.* Typhimurium BRD509-infected mouse will yield 1–2 $\times 10^6$ sorted MAIT cells, which are enough for 10–20 recipient mice (10^5 MAIT cells/RAG2$^{-/-}\gamma$C$^{-/-}$ mouse in this case). Infect donor mice 7 days earlier than the adoptive transfer.

1. Two days before infection streak out a plate of *S.* Typhimurium BRD509 (an attenuated vaccine strain [14]) on LB agar plates,

containing 50 μg/mL streptomycin and incubate plates overnight at 37 °C.

2. The day before infection, pick a single colony under flame and inoculate to 10 mL LB culture medium with 50 μg/mL streptomycin and leave static at 37 °C (double contained if working with wild type/virulent SL1344 or equivalent strains) overnight.
3. On the day of infection, re-inoculate into fresh 10 mL pre-warmed LB culture medium with 0.5, 100, or 20 μL of overnight culture, under flame. This is to ensure an optimal optical density (O.D.) reading (bacteria in log phase growth) for preparing the inoculum later (*see* **Note 6**). The doubling time for *Salmonella* can vary between 0.5 and 1 h. Make the infection inoculum from culture with $O.D._{600nm}$ reading between 0.2 and 0.6. Calculate the required CFU of bacteria estimating 1 O.D. = 5–10 × 10^8 CFU (this constant needs to be established for individual labs). Dilute with PBS to a final concentration of 2 × 10^7 CFU/mL, allowing 50 μL inoculum/mouse, i.e., 10^6 CFU/50 μL/mouse.
4. Infect mice i.n. with 10^6 CFU *S.* Typhimurium BRD509 in 50 μL PBS under isoflurane anesthesia (*see* **Notes 7** and **8**).
5. Allow mice to recover and monitor mice for 7 days to allow the infection to take its course and MAIT cell frequencies to expand dramatically from 10^4 to 5 × 10^6 MAIT cells, or from 1% to 20–50% of all alpha-beta T cells [15] (*see* **Note 8**).

3.2 MAIT Cell Adoptive Transfer

1. 7 days or more after intranasal infection with *S.* Typhimurium, MAIT cells can be harvested (*see* **Note 9**). As MAIT cells are to be used for adoptive transfer, all procedures should be performed in a BSCII biosafety cabinet. All tools and reagents should be sterile.
2. Prewarm collagenase media and shaking incubator to 37 °C.
3. Mice should be euthanized (e.g., using a rising concentration of CO_2 with a second method to confirm death).
4. Open the diaphragm by cutting the rib cage to expose both the heart and lungs. Gently perfuse the right ventricle with 8–10 mL of ice-cold RPMI to dispense circulating blood. Perfuse using a 10-mL syringe and a 26-G needle. Efficient perfusion will result in lung inflation and a color change to pink/white.
5. Remove lungs using scissors to cut through the hilum and place into a 24-well plate containing ice-cold RPMI to transfer organs to the laboratory.
6. Chop lungs into fine pieces (*see* **Note 10**).

7. Place lung tissue into a 1-mL Eppendorf tube containing 1–2 mL/lung of pre-warmed collagenase medium. Incubate tubes on their sides in a shaking incubator at 37 °C, at 100–180 rpm, for 90 min.
8. During this time prepare Percoll gradients and antibody cocktails (*see* Table 1).
9. After 90 min pour digested tissue through a 70-μm cell strainer and force through into a Petri dish with the plunger from a 1-mL syringe. Rinse residual sample with extra FACS buffer for maximum MAIT cell yield. Cells from multiple lungs (if required) (*see* **Note 11**) are pooled into a single 50-mL Falcon tube with a total of 50 mL of sterile FACS buffer.
10. Centrifuge at 400 × *g* for 5 min to pellet the cells. Pour off supernatant (SN).
11. Resuspend cells in 20 mL 40% Percoll. **Underneath** this layer use a transfer pipette to layer 20 mL 70% Percoll (*see* **Note 12**). Centrifuge this gradient at 800 × *g* for 20 min at room temperature with the centrifuge brake OFF. Lymphocytes and other immune cells will form a visible interphase layer between the 40% and 70% Percoll post centrifugation.
12. During this centrifugation step, prepare single color controls. It is convenient to use part of a spleen forced through a 70-μm filter and resuspended in 5 mL TAC for 5 min at 37 °C, then washed once with 5 mL FACS buffer.
13. Collect the interphase between 40% and 70% Percoll into a fresh 50 mL Falcon and top up with FACS buffer to a total of 50 mL. Centrifuge at 400 × *g* for 5 min.
14. Pour off supernatant and resuspend in 5 mL FACS buffer, transferring to a 10 mL Falcon tube. Centrifuge at 400 × *g* for 5 min.
15. Resuspend all lung cells in 750 μL FACS buffer.
16. Block non-specific tetramer binding by adding 7.5 μL 2.4G2 (anti CD16/32) cell culture supernatant, containing MR1-6-FP tetramer [8, 16] (no fluorochrome, 1:100). Incubate at room temperature on a roller or bench rocker for 15 min.
17. For lungs from 5 mice, add 750 μL of staining cocktail (Table 1).
18. Cover in aluminum foil to protect fluorochromes from light and shake on roller for 30 min RT.
19. Wash with 10 mL of FACS wash. Centrifuge at 400 × *g* for 5 min. Pour off supernatant.
20. Wash again with 10 mL of FACS wash. Centrifuge at 400 × *g* for 5 min. Pour off supernatant.

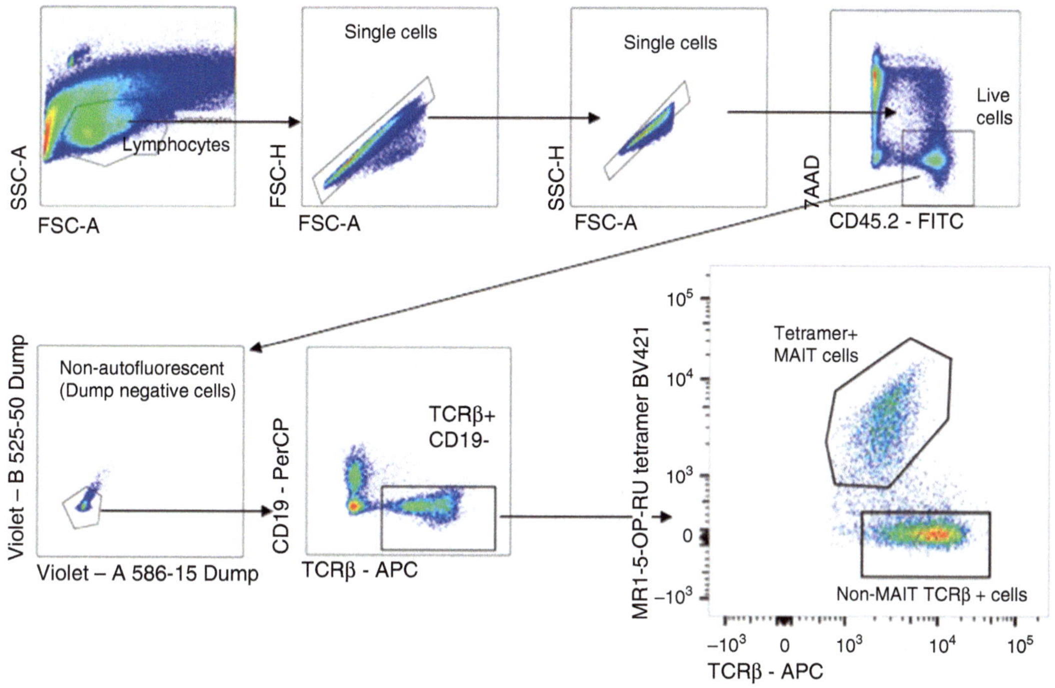

Fig. 1 Flow cytometry gating strategy for MR1-5-OP-RU-tetramer+ MAIT cells

21. Resuspend cells in 2 mL FACS wash and filter through 40 μm filter into non-pyrogenic FACS tubes.
22. Sort live MAIT cells (defined as $CD3^{+}CD45^{+}$MR1-5-OP-RU tetramer^{+} cells) (Fig. 1) into 3 mL of FCS in 15 mL Falcon tube. For detailed gating strategy, refer to [17]. Wash cells and adjust cell concentration to 5×10^5 cell/mL, allowing 10^5 in 200 μL for injection to each mouse.
23. Inject 10^5 cells into the tail vein of recipient mice using cells suspended in 200 μL PBS in a 1-mL syringe with a 26-G cannula after warming the mice for 5–15 min with appropriate monitoring.
24. To deplete residual non-MAIT T cells (*see* **Note 13**), inject recipient mice on days 2 and days 5 or 6 with 0.1 mg each of purified anti-CD4 (GK1.5) and anti-CD8 (53.762) monoclonal antibodies i.v.
25. Rest mice for a total of 2 weeks post adoptive transfer to allow MAIT cell populations to settle in the host.

3.3 Influenza A Virus Infection

1. Thaw virus vial rapidly in a 37 °C water bath until all ice crystals have melted.
2. Decontaminate the outer surface of the vial with 70% ethanol.

3. Perform serial dilutions in sterile PBS to achieve the desired inoculum. For example:
   (a) If titer of PR8 stock = 1.0×10^9 plaque-forming units (PFU)/mL, require 25 PFU/inoculum (*see* **Note 14**).
   (b) Volume of inoculum required for intranasal infection = 50 μL/inoculum.
   (c) 25 PFU/50 μL = 500 PFU/mL, dilution required from virus stock DF = 2,000,000.
      - Dilution 1: 1/100: 10 μL of virus stock +990 μL PBS.
      - Dilution 2: 1/100: 10 μL of dilution 1 + 990 μL PBS.
      - Dilution 3: 1/100: 10 μL of dilution 2 + 990 μL PBS.
      - Dilution 4: ½: (depends on volume required, e.g., 20 inoculations = 1000 μL) 500 μL of dilution 3 and 500 μL of PBS.
4. Mix virus with vortex before administering the inoculum.
5. Infect mice i.n. with 50 μL under isoflurane anesthesia (*see* **Note 7**).
6. Allow mice to recover and monitor mice for until recovery (typically 10 days) (*see* **Note 15**).

3.4 Lung Homogenization

1. Collect the lungs into 2 mL of RPMI supplemented with penicillin/streptomycin.
2. For homogenization, place the lung and the 2 mL of media into 10 mL falcon tubes with lids (*see* **Note 16**).
3. Prepare 10 or 15 mL Falcon tubes with 2 × 5 mL 80%w/v EtOH for cleaning the homogenization probe initially and 1 tube containing HBSS. For each group of samples, prepare further 1 × 5 mL EtOH and 1 × 5 mL HBSS, and for the final probe clean set up 2 × 5 mL of EtOH.
4. Homogenize the sample using a homogenizer, mounted on a retort stand with the probe set to medium for 30 s per sample. Keep samples on ice (*see* **Note 17**).
5. Centrifuge the samples at 1000 × *g* for 7 min.
6. Using a 1 mL pipette, carefully draw up approximately 1 mL (a little bit more is good) of supernatant, avoiding the pellet and fatty residue on top. Divide this volume into two 1.5 mL Eppendorf tubes. Store at −80 °C for subsequent plaque assays.

3.5 MAIT Cell Intracellular Cytokine Staining

To analyze MAIT cell frequencies and function during viral infection.

1. Prewarm collagenase media and shaking incubator to 37 °C.

2. Mice should be euthanized (e.g., using a rising concentration of CO_2 with a second method to confirm death).
3. Open the diaphragm by cutting the rib cage to expose both the heart and lungs. Gently perfuse the right ventricle with 8–10 mL of ice-cold RPMI to remove circulating blood. Perfuse using a 10-mL syringe and a 26-G needle. Proper perfusion will result in lung inflation and a color change to pink/white.
4. Remove lungs (*see* **Note 18**) using scissors to cut through the hilum and place into a 24-well plate containing ice-cold RPMI to transfer organs to the laboratory.
5. Chop lungs into very small pieces (*see* **Note 9**).
6. Place lung tissue into a 1-mL Eppendorf tube containing 0.5 mL/lung of pre-warmed collagenase/DNase medium. This should also contain, 0.5 μL of Brefeldin A (1:1000) (final concentration 3.0 μg/mL).
7. Incubate tubes on their sides in a shaking incubator at 37 °C, at 100–180 rpm, for 90 min.
8. After 90 min pour digested tissue through a 70-μm cell strainer and force through into Petri dish with the plunger from a 1-mL syringe. Rinse residual cells into a total of 10 mL FACS wash in 10 mL falcon tubes at RT. Centrifuge at 400 × *g* for 5 min at RT.
9. Resuspend in 2 mL per lung (*see* **Note 19**) of pre-warmed TAC lysis buffer at 37 °C. Vortex well, then place in a pre-warmed water bath at 37 °C. After 5 min neutralize by adding an equal volume of FACS buffer. Centrifuge at 400 × *g* for 5 min, RT.
10. Numbers of lung cells can be estimated using a hemocytometer or spectrophotometer (*see* **Notes 20** and **21**).
11. Transfer 100 μL containing 0.5–1 million cells to a 96-well U- or V-bottom plate format or into FACS tubes for staining, passing them through a 40-μm mesh (*see* **Note 22**).
    (a) 1 × 100 μL into a plate for surface stain (**steps 16–22**).
    (b) 2 × 100 μL (unstimulated and stimulated) to a second plate (*see* **Note 23**), and include a no Brefeldin control (**steps 12**, **13**, and **21–30**).

    In vitro stimulation phase:
12. Keep the cells for the surface stain on ice, while setting up PMA/Ionomycin stimulation to induce production of cytokines of interest.
    (a) PMA final concentration: 20 ng/mL.
    (b) Ionomycin 1 μg/mL.
    (c) 1000× stock Brefeldin A (final concentration 3.0 μg/mL).

13. Incubate for 3 h at 37 °C with 5% CO_2.

    Surface staining:

14. During stimulation phase perform surface staining for extracellular markers (*see* **Note 24**).
15. If performing Zombie Yellow vital staining wash cells with 1–2 mL PBS. Centrifuge at 400 × *g*, 5 min (or if using plate format wash twice with 200 μL FACS buffer centrifuging for 2 min at 400 × *g*). Resuspend in 20 μL PBS + 0.4 μL Zombie Yellow for 15 min.
16. Add 20 μL of 2.4G2 (anti-CD16/32) containing 0.2 μL of MR1-6-FP tetramer (no fluorochrome conjugate) to block non-specific binding. Incubate for 15 min dark, room temp.
17. Add surface cocktail (Table 2) using a cocktail made up in 10 μL FACS buffer. Pipette carefully to mix. Stain for 20–30 min at room temperature.
18. For single color controls use splenocytes or compensation beads.
19. Wash cells twice with 2 mL FACS buffer, centrifuging at 400 × *g* for 5 min (or if using plate format wash three times with 200 μL FACS buffer centrifuging for 2 min at 400 × *g*).
20. Resuspend cells in 100 μL FACS buffer (*see* **Note 25**). To enable estimation of absolute cell numbers, add a known number of calibration beads.
    (a) Vortex calibration beads hard. Dilute (1:10) counting beads in PBS before using. To each sample 25 μL of these diluted beads was added, and an additional 10 μL of beads were saved to be counted with a hemocytometer, giving a count of X in a large square, i.e., X × 10^4 beads/mL (which is X × 10 beads/μL, or X × 10 × 25 beads/sample). Typically add a total of 25,000 beads per sample.
    (b) When samples have been acquired on flow cytometer, these calibration beads can be detected using their FSC/SSC profile and the absolute number of cells of interest can be estimated using the following approach. Total number of MAIT cells per sample = Number of MAIT cells counted on flow cytometer × Number of beads added/Number of beads counted/proportion of total lung cell suspension actually used for staining.

    Intracellular staining:

21. After 3-h stimulation, continue processing the cells for intracellular staining. Resuspend into FACS tube with +1 mL PBS. Centrifuge at 400 × *g*, 5 min. (Alternatively, if in 96-well format resuspend in 100 μL PBS, centrifuge at 400 × *g* for 2 min and repeat.)

22. Resuspend in 20 μL PBS with 0.4 μL Zombie Yellow for 15 min.
23. Add 20 μL of 2.4G2 (anti CD16/32) SN containing 0.2 μL of unlabeled MR1-6-FP tetramer to block non-specific tetramer staining. Incubate for 15 min dark, room temperature.
24. Add surface cocktail (Table 3) using a cocktail made up in 10 μL FACS buffer. Pipette carefully to mix. Stain for 20–30 min at room temperature.
25. For single color controls use splenocytes or compensation beads. These may be available from being made up earlier in the protocol.
26. Wash cells once with 1 mL FACS buffer, centrifuging at 400 × *g* for 5 min (or if using plate format wash twice with 200 μL FACS buffer centrifuging for 2 min at 400 × *g*).
27. Resuspend in 200 μL of commercially available Fixation/Permeabilization solution and incubate for 30 min on ice.
28. Wash with 2 mL Perm Wash (diluted 1:9 with FACS buffer). Centrifuge at 400 × *g*, 5 min (or if using plate format wash twice with 200 μL Perm Wash centrifuging for 2 min at 400 × *g*).
29. Resuspend in 50 μL of Perm Wash containing intracellular cocktail (Table 4) and pipette carefully to mix. Incubate for 45 min or leave to stain overnight.
30. Wash cells with 2 mL Perm Wash (or if using plate format wash twice with 200 μL Perm Wash centrifuging for 2 min at 400 × *g*). Resuspend cells in 100 μL FACS wash. If cells are in plate format use a multichannel pipette to transfer them to 1.2 mL "bullet" cluster tubes for acquisition, or use a plate reader attachment with the cytometer.
31. Analyze cells on flow cytometer.

3.6 Viral Plaque Assay

Viral plaque assays are used to determine influenza viral titers. A diluted solution of egg-adapted Influenza A viruses/lung-infected tissue homogenates are applied to a six-well tissue culture dish containing a monolayer of Madin-Darby canine kidney (MDCK) cells. The infected MDCK cells grow under a semisolid overlay medium (agar) containing trypsin. A plaque is produced when a virus particle infects a cell, replicates, and then kills the cell. This process can be repeated several times as surrounding cells can be infected by newly replicated virus and killed. When visualized by eye, plaques appear as white spots. The assay is measured in PFU/mL.

Passaging MDCK cells:

1. Warm up MDCK cell media, trypsin–versene, and PBS at 37 °C.

2. Check the confluency of MDCK cells, aspirate the medium, add 10 mL of PBS, aspirate the medium, and repeat wash.
3. Discard PBS, add 2–3 mL of trypsin–versene (stored −20 °C) to MDCK monolayers, and incubate at 37 °C for 5 min. After 5 min tap the flasks, and incubate for longer if required (maximum 15 min).
4. In the meantime, add 15 mL of MDCK cell media to fresh T75 flasks.
5. Add MDCK cell media to a total volume of 10 mL to the trypsinized cells, and transfer cells to a 10-mL tube.
6. Count cells using a hemocytometer.
7. Set up multiple T75 flasks with different cell densities to determine the growth pattern of MDCK cells. Generally ~3–5 × 10^5 for 3-day split.
8. Incubate at 37 °C, 5% CO_2.

   Amplification of MDCK cells for plaque assay:
9. Warm up MDCK cell media, trypsin–versene, and PBS at 37 °C.
10. Check the confluency of MDCK cells, aspirate the medium, add 10 mL PBS, aspirate the medium, and repeat wash.
11. Discard PBS, add 2–3 mL of trypsin–versene (stored −20 °C) to MDCK monolayers, and incubate at 37 °C for 5 min. After 5 min tap the flasks, and incubate for longer if required (maximum 15 min).
12. In the meantime, add 40 mL of MDCK cell media to fresh T175 flasks. Set up one T175 flask of MDCK cells per ~4 plates for plaque assay. Each 6-well plate assays 3 viral dilutions (as dilutions are done in duplicate).
13. Add MDCK cell media to a total volume of 10 mL to trypsinized cells, and transfer cells to a 10-mL tube.
14. Count cells using a hemocytometer.
15. Add ~7–8 × 10^5 cells per T175 flask for 3-day culture.

    Seeding flat-bottomed 6-well tissue culture (TC6) plates for plaque assay:
16. Warm up MDCK cell media, trypsin–versene, and PBS at 37 °C.
17. Check the confluency of MDCK cells, aspirate the medium, add 10 mL of PBS, aspirate the medium, and repeat the wash.
18. Discard PBS, add 5–8 mL of trypsin–versene (−20 °C) to MDCK monolayers in T175 flask, incubate at 37 °C for 5 min. After 5 min, tap bottles, incubate for longer if required (maximum 15 min).

19. Add 17–20 mL of MDCK cell media to each flask (total 25 mL including 5–8 mL trypsin) to inhibit trypsin–versene.
20. Pool cells into one flask.
21. Count the cells, adjust the concentration to 3.3×10^5 cells/mL.
22. Add 3 mL of 3.3×10^5 cells/mL to each well of TC6 plates (~1×10^6/well), swirl plates gently to distribute cells evenly. Include a negative control plate.
23. Incubate cells at 37 °C, 5% CO_2 overnight. Aim for monolayers to be confluent in 6-well plates for assay.

    Plaque assay
24. Warm up SF-DMEM at 37 °C.
25. Prepare dilutions of samples to be titrated. This can be done in a 96-well flat-bottom plate.
26. Cells will be infected with 150 μL of each dilution in duplicate, so a minimum of 300 μL of each dilution is required. A 96-well flat-bottom plate can hold ~350 μL/well. 35 μL of samples would be added to 315 μL of media for a final volume of 350 (35 in 350 = 1:10 dilution). If a sample is to be plaqued neat, add 350 μL of sample to the first well.
27. Add 315 μL of SF-DMEM in each well of 96-well plate. A multichannel can be used and add 157 and 158 μL (total 315) to each well.
28. Add 35 μL of sample to the first well with media (tenfold dilution) and continue serial tenfold dilutions by transferring 35 μL across wells, changing tips between dilutions.
29. For titration of viral stocks use dilutions from 10^{-4} to 10^{-6}. 10^{-1} can be used as a positive control. Half-log dilutions can also be performed. For titration of mouse lung homogenates, generally:
    (a) Days 1–5: 10^{-1} to 10^{-3}.
    (b) Days 6–10: neat to 10^{-2} (*see* **Note 26**).
30. Wash MDCK cells with 1–2 mL of PBS/well.
31. Infect cells with 150 μL/well of the appropriate dilution, swirl gently to cover all cells and incubate at 37 °C, 5% CO_2 for 60 min, shake gently every 15 min.
32. In the meantime, prepare overlay media, best to start doing so in the beginning of the 1 h incubation of MDCK cells.
33. Weigh out 1.8 g of agarose into 200 mL glass bottle, add 100 mL of sterile water (1.8% agarose), and melt in microwave. Store in a 55 °C water bath.
34. Aliquot 50 mL of 2 × L-15 medium into 50 mL tubes (need 2 × 50 mL aliquots of 2 × L-15 medium and 1 bottle (100 mL) of 1.8% agarose) and store in 37 °C water bath.

35. Thaw trypsin–versene at 40 °C. Thaw a 200 μL aliquot/50 mL of 2 × L-15.
36. After 1 h incubation of MDCK cells with sample: Add 200 μL trypsin to each 50 mL tube of 2 × L-15.
37. Make overlay media by adding 100 mL of 2 × L-15 + trypsin solution to 100 mL of 1.8% agarose and mix well (*see* **Note 27**).
38. Add 3 mL of overlay medium/well and leave at RT until it sets.
39. Incubate upside-down at 37 °C, 5% CO_2 for 3 days. Plaques maybe visible by the end of day 2 and the plates can be incubated till day 4 if plaques are too small on day 3.
40. Count the plaques. This can be done by holding the plates against the light. Alternatively, remove agarose overlay and stain with crystal violet. To stain, cover the cells with a minimal amount of crystal violet solution for ~15 min. Rock plates if necessary to ensure even coverage. Gently wash off the crystal violet stain with water. Once fixed, stained, and dried, store plaques indefinitely for future analysis.
41. Calculate viral titer in PFU/mL: Average count of duplicate well × Dilution factor × (1000/150) = PFU/mL (multiply lung homogenate counts by 2 to give total viral load, as lungs were taken and homogenized in 2 mL RPMI).

4 Notes

1. Allow a little extra for pipetting wastage when making up antibody cocktails. Keep on ice and protect from light (e.g., with aluminum foil). Make up cocktails in FACS buffer, but for the intracellular stains these should be made up in Perm Wash buffer containing 0.1% Saponin.
2. Congenic markers could be reversed or other markers are used as appropriate to the mouse strains being used and to the specific experimental set-up.
3. Biological Hazards—*S.* Typhimurium BRD509 is a risk group 2 pathogen. Influenza A virus-PR8-strain (H1N1) is a lab adapted strain of IAV virus. Work should be risk assessed and we recommend controls that include but are not restricted to the following: Lab coat, safety glasses, and gloves should be worn when performing this protocol. Gloves should be removed or sterilized before exiting the biohazard hood. Solutions of Lysol (200 ppm) or hypochlorite (5000 ppm) should be accessible in case of a spill.
4. Decontaminate all pipette tips in 1% Virkon when working in the biohazard cabinet. After use, the biohazard hood should be

decontaminated by wiping down with 70% ethanol and by UV sterilization for 15 min before any further use. All waste and its container must be disposed as hazardous waste.

5. Pulmonary MAIT cells can be expanded using any source of 5-OP-RU and an appropriate TLR agonist [15, 17]. A systematic assessment of effective TLR agonists has shown strong MAIT cell expansion 7 days after intranasal inoculation with 76 pmol 5-OP-RU on days 1, 2, and 4 in combination with a single dose of agonist on day 1 to TLR3 (high molecular weight poly I:C), TLR4 (lipopolysaccharide from *E. coli*), TLR2/6 (FSL-1 (Pam2CGDPKHPKSF)), or TLR9 (CpG ODN1826), but not with agonists of TLR1/2 (Pam3CSK4), TLR2, TLR5, TLR7 [3]. Each inoculum should be instilled in 50 μL PBS. However, the requirement for accurate repeated inoculations can introduce significant variability in MAIT cell expansion. A simple, less costly on reagents and time, and equally effective, if not more so, is a single intranasal inoculation with *S.* Typhimurium BRD509 in 50 μL PBS.
6. Growth of bacteria is estimated by measuring the culture in a spectrophotometer at 600 nm. To do so fill a cuvette with fresh LB media, place in spectrophotometer, and use this to blank. Then take 500 μL of bacteria-containing broth and measure optical density. To calculate the inoculum dose, use the estimate that an $O.D._{600nm}$ of $1 = 5 \times 10^8$ CFU/mL.
7. Accurate intranasal inoculation depends critically on the depth of anesthesia. Administer isoflurane and observe breathing pattern until respiratory rate has decreased to approximately 100 breaths/min and is deep and relaxed. If insufficient depth is achieved mice will sneeze. If depth of anesthesia is too great (further slowing of respiratory rate and very deep breaths), then mice tend to spontaneously breath-hold and again, volume inhaled will be unreliable. Place 50 μL of inoculum onto the left nasal opening (if user is right-handed) using a P200 pipette, gradually ejecting the 50 μL over a few breath cycles until all has been inspired.
8. Intranasal *S.* Typhimurium is well tolerated in immunocompetent strains such as C57BL/6 and BALB/c with less than 5% of animals showing minor signs of illness (ruffled hair) within 1–2 days after infection. These animals fully recover after days 3–5. The lethal dose of *S.* Typhimurium BRD509 is $>2 \times 10^7$ CFU/mouse (wild-type C57BL/6 adult). Caution should be used in immunocompromised strains in which pilot experiments should be performed to confirm optimal safe inocula.
9. This MAIT cell expansion is long-lived [15], so donor mice can be prepared several weeks in advance.

10. The lungs can conveniently be chopped up using the back of an upturned Petri dish. Using fine forceps lift lungs from the RPMI in which they have been transferred, gently blot off excess liquid with tissue paper and place on the Petri dish. Use a large curved scalpel blade to repeatedly chop through the lungs at multiple angles for at least 60 s each until a very fine and homogeneous texture is achieved.
11. Typically this method will yield 1.5×10^6 pulmonary MAIT cells per mouse, so multiple mice may be required as donors, depending on the requirements of the experiment.
12. This will be sufficient for lungs from 8 mice.
13. If transferring cells into a $Rag2^{-/-}\gamma C^{-/-}$ mouse then low frequencies of "contaminating" conventional $CD4^+$ or $CD8^+$ T cells tend to expand more rapidly than the MAIT cells and produce artifacts (not obvious for other T-cell-deficient mice, e.g., $TCR\alpha^{-/-}$ or $RAG2^{-/-}$). As many MAIT cells are double-negative, it is possible to prevent this effect by repeated injections with T-cell-depleting anti-CD4 and anti-CD8 antibodies [17].
14. The PR8 strain of influenza virus is highly virulent in mice and only low inoculate are tolerated. The exact inoculum required for each experimental system will need to be carefully determined depending on the exact strain and batch of PR8 and the strain of mice, and local welfare and monitoring requirements. In our hands C57BL/6 mice receiving 100 PFU of A/PR/8/34 AF18 WCN experienced severe pneumonia in mice, characterized by parenchymal necrosis and infiltrates of macrophages, lymphocytes, and neutrophils, with 10–25% mortality due to welfare concerns or weight loss >20%.
15. Virally infected mice experience a transient viral illness with transient. Viral titers peak at day 3. Weight loss peaks at day 5–7 post infection, and there would be a significant weight gain expected by day 8 and resolving by day 10 post infection. Typically mice should be monitored and/or weighed daily for signs of ill health such as ruffled fur, hunched-up appearance, gait abnormalities, lethargy and loss of body condition for 10 days after challenge or till all the symptoms disappear and body weight returns to pre-challenge level. Monitoring can then return to twice weekly.
16. For many homogenization probes a wide tube is needed, such as the sterile, capped, round-bottom polypropylene tubes which are available.
17. The homogenizer generates a lot of heat at the probe tip. Samples should be kept on ice before and after homogenization, and the probe should be intermittently rested to cool down in ice-cold EtOH between groups of 5 or 10 samples.

Between samples or groups of samples clean the probe by running briefly in EtOH and then rinsing briefly in HBSS. Often connective tissue will clog the probe and this can be removed with large forceps. After use the probe tip should be sterilized.

18. Only approximately 2/7 of one lung is needed for intracellular cytokine staining, so the other lung, or other sections of lung, can be saved for viral titer estimation, histology, or other assays if required.
19. To clarify terminology there are two lungs in each animal, so "one lung" refers to all the 2 or 3 lobes in a single hemithorax. Due to the presence of the heart on the left side, the left lung is smaller with only 2 lobes.
20. Using a spectrophotometer saves time for large numbers of samples. To do this resuspend cell pellet in 1–2 mL PBS (or adjust according to pellet size/counts). Select $O.D._{600nm}$. Blank cuvette with 1 mL FACS wash/PBS. Measure $O.D._{600nm}$ with 200 μL samples + 800 μL PBS (5×). Calculate the number of cells: this is a simple linear relationship between O.D. and the number of cells, which can be derived by measuring a few cell counts in parallel on both the hemocytometer and the spectrophotometer.
21. An alternative is to resuspend the entire pellet in 700 μL of FACS buffer and take 200 μL into 96-well plate: this should contain approximately $1–1.5 \times 10^6$ cells, appropriate for staining.
22. To avoid using multiple filters, it is possible to buy large sheets of 40 μm mesh. A single rectangle can be cut which covers a whole plate. Using this, multiple cells can be pipette simultaneously with a multichannel pipette.
23. In round-bottom plates cells may clump so consider using flat-bottom plate for the stimulation step, especially if doing further steps in FACS tubes rather than staining in plate format.
24. While surface markers can be measured on the intracellularly stained cells, the most accurate measurement of MAIT cell frequencies will be obtained from immediate surface staining prior to stimulation, due to activation-induced downregulation of the TCR.
25. If cells are not to be acquired immediately, then they can instead be resuspended in 100 μL of fixation buffer and stored at 4 °C until required.
26. This may differ depending on virus and mouse strains.
27. The overlay media will start setting so proceed to the following steps quickly. Overlay media can be made in batches to assist with that.

Acknowledgments

This work was funded by grants to T.S.C.H. from the Wellcome Trust (104553/z/14/z, 211050/Z/18/z) and Project Grants 1062889 and 1120467 and Program Grant 1113293 from the National Health and Medical Research Council of Australia. B.W. was supported by the Royal Society (IE160540). A.J.C. is supported by a Future Fellowship from the Australian Research Council, FT1600100083. The research leading to these results has received funding from the People Programme (Marie Curie Actions) of the European Union's Seventh Framework Programme (FP7/2007-2013) under REA grant agreement number 608765. The content represents only the authors' views and not those of the European Commission. HW was supported by a Melbourne International Engagement Award (University of Melbourne).

References

1. Eckle SB, Corbett AJ, Keller AN et al (2015) Recognition of vitamin B precursors and byproducts by mucosal associated invariant T cells. J Biol Chem 290:30204–30211
2. Godfrey DI, Uldrich AP, Mccluskey J et al (2015) The burgeoning family of unconventional T cells. Nat Immunol 16:1114–1123
3. Hinks TSC, Marchi, E, Jabeen, M et al (2019) Activation and in vivo evolution of the MAIT cell transcriptome in mice and humans reveals tissue repair functionality. Cell Reports 28 (12):3249–3262.e5
4. Kjer-Nielsen L, Patel O, Corbett AJ et al (2012) MR1 presents microbial vitamin B metabolites to MAIT cells. Nature 491:717–723
5. Hinks TS, Zhou X, Staples KJ et al (2015) Innate and adaptive T cells in asthmatic patients: relationship to severity and disease mechanisms. J Allergy Clin Immunol 136:323–333
6. Rahimpour A, Koay HF, Enders A et al (2015) Identification of phenotypically and functionally heterogeneous mouse mucosal-associated invariant T cells using MR1 tetramers. J Exp Med 212:1095–1108
7. Tsukamoto K, Deakin JE, Graves JA et al (2013) Exceptionally high conservation of the MHC class I-related gene, MR1, among mammals. Immunogenetics 65:115–124
8. Corbett AJ, Eckle SB, Birkinshaw RW et al (2014) T-cell activation by transitory neo-antigens derived from distinct microbial pathways. Nature 509:361–365
9. Patel O, Kjer-Nielsen L, Le Nours J et al (2013) Recognition of vitamin B metabolites by mucosal-associated invariant T cells. Nat Commun 4:2142
10. Ussher JE, Bilton M, Attwod E et al (2014) CD161++ CD8+ T cells, including the MAIT cell subset, are specifically activated by IL-12 +IL-18 in a TCR-independent manner. Eur J Immunol 44:195–203
11. Van Wilgenburg B, Scherwitzl I, Hutchinson EC et al (2016) MAIT cells are activated during human viral infections. Nat Commun 7:11653
12. Wilgenburg BV, Loh L, Chen Z et al (2018) MAIT cells contribute to protection against lethal influenza infection in vivo. Nat Commun 9:4706
13. Loh L, Wang Z, Sant S et al (2016) Human mucosal-associated invariant T cells contribute to antiviral influenza immunity via IL-18-dependent activation. Proc Natl Acad Sci U S A 113:10133–10138
14. Hoiseth SK, Stocker BA (1981) Aromatic-dependent Salmonella typhimurium are non-virulent and effective as live vaccines. Nature 291:238–239

15. Chen Z, Wang H, D'souza C et al (2017) Mucosal-associated invariant T-cell activation and accumulation after in vivo infection depends on microbial riboflavin synthesis and co-stimulatory signals. Mucosal Immunol 10:58–68
16. Reantragoon R, Corbett AJ, Sakala IG et al (2013) Antigen-loaded MR1 tetramers define T cell receptor heterogeneity in mucosal-associated invariant T cells. J Exp Med 210:2305–2320
17. Wang H, D'souza C, Lim XY et al (2018) MAIT cells protect against pulmonary Legionella longbeachae infection. Nat Commun 9:3350

Chapter 18

MAIT Cells in Type 1 Diabetes Mouse Models

Isabelle Nel, Lucie Beaudoin, and Agnès Lehuen

Abstract

MAIT cells are unconventional T cells expressing a semi-invariant αβ TCR, and they recognize bacterial metabolites via the highly conserved MR1 protein. MAIT cells interact with gut microbiota and literature reports alterations of gut homeostasis in type 1 diabetes (T1D), suggesting the involvement of MAIT cells in T1D. Since NOD mice is a well-established mouse model of T1D, MAIT cells were studied in these mice to evaluate their potential involvement in disease development. This chapter describes the material and methods required to characterize MAIT cells and to determine their function in T1D mouse models.

Key words MAIT, MR1, T1D, NOD, Intestinal microbiota, Migration, Cytokine, Activation

1 Introduction

MAIT cells are unconventional T cells expressing a semi-invariant αβ T cell antigen receptor (TCR) [1, 2]. In mouse, the αβTCR is composed of an invariant α chain (Vα19 joined to Jα33) in association with a restricted set of TCRβ-chains (Vβ6 and Vβ8) [3, 4]. MAIT cells recognize a limited number of ligands via MR1 (MHC-related 1 protein), a highly conserved molecule in mammals [5]. MAIT cell ligands are mainly derived from riboflavin metabolite pathway synthetized by bacteria [6, 7]. For a long time, the identification of MAIT cells in mouse model was very difficult due to the lack of a specific antibody [4]. The discovery of the 5-OP-RU [8], a major activating MAIT cell ligand, has been a critical step leading to identification of MAIT cells via mMR1-Ag-loaded tetramer [9, 10]. 5-OP-RU is also a tool used for in vitro MAIT cell TCR-specific activation [11, 12]. Alterations of MAIT cells in metabolic [13–15] and liver diseases [16] lead our laboratory to study MAIT cells in T1D.

Type 1 diabetes (T1D) results from the destruction of insulin-secreting pancreatic β cells leading to chronic hyperglycemia. T1D is an autoimmune disease with multifactorial origins [17–19]. The NOD mouse model is widely used to study T1D physiopathology

Helen Kaipe and Isabelle Magalhaes (eds.), *MAIT Cells: Methods and Protocols*, Methods in Molecular Biology, vol. 2098, https://doi.org/10.1007/978-1-0716-0207-2_18,

since it shared many features with the human disease [18, 20]. Studies in humans and in mouse models report alterations of gut homeostasis and intestinal microbiota in T1D [21–24]. Since MAIT cells are abundant in mucosa including gut and are in interaction with microbiota [3, 25, 26], we hypothesized that MAIT cells could play a role in the physiopathology of T1D.

MAIT cells were firstly characterized in NOD mice at different stages of disease after isolation from various tissues (spleen, pancreatic lymph nodes (PLNs), mesenteric lymph nodes (MLNs), ileum, and liver) and identification via tetramer staining. Since MAIT cells in mouse models can be divided into three subsets (double negative, CD4+, or CD8+ cells), activation and maturation of MAIT cells were analyzed by surface flow cytometry staining in each of these subsets. The ability of MAIT cells to produce various cytokines (TNFα, IFNγ, IL-17, IL-22) and granzyme B (GzB) [27] was evaluated in tissues from NOD mouse via intracytoplasmic flow cytometry staining. Then to highlight a possible MAIT cell migration toward inflamed tissues, as suggested by results from T1D patients and NOD mice, transfer experiments of MAIT cells obtained after purification from CD45.2 NOD mouse were performed. Moreover, to assess the hypothesis that MAIT cell activation observed in the blood of T1D patients and in target tissues from NOD mice could result from translocation of bacterial compounds following alteration of intestinal permeability, TCR-specific activation of MAIT cells from NOD mice by intestinal contents and by 5-OP-RU as control was realized. The very low frequency of circulating MAIT cells in NOD mice does unfortunately not allow their characterization. Finally, study of diabetes incidence in NOD mice (overexpressing Vα19 or depleted in MR1) as compared to littermate NOD mice and in streptozotocin-induced disease *MR1*$^{-/-}$ B6-MAIT^{CAST} mice emphasizes a global protective effect of MAIT cells in T1D [11, 28]. The main tools and methods introduced above, allowing the study of MAIT cell involvement in T1D physiopathology, are the topic of this chapter.

2 Materials

2.1 Cell Preparation for MAIT Cell Analysis and Isolation

1. Phosphate-buffered saline (PBS).
2. Buffer for cell isolation (CI buffer): RPMI 1640 containing 5% (v/v) fetal calf serum (FCS). Use at +4 °C.
3. Buffer for intestine wash (IW buffer): HBSS without Ca^{2+} and Mg^{2+}, containing 10 mM HEPES. Use at room temperature (RT).
4. Buffer for intestine predigestion (IPD buffer): IW buffer with 5 mM EDTA, 5% (v/v) FCS and 1 mM DTT freshly added before each digestion. Use at RT.

5. Buffer for intestine digestion (ID buffer): HBSS with Ca^{2+} and Mg^{2+}, containing 10 mM HEPES and 5% (v/v) FCS. Use at RT.
6. Red blood cell (RBC) lysis solution (Tris-buffered ammonium chloride solution): 90% v/v 0.16 M NH_4Cl + 10% v/v 0.17 M Tris–HCl, pH 7.65. Adjust to pH 7.2 with HCl. Prewarm at RT.
7. For pancreas digestion: Collagenase P from *Clostridium histolyticum* 0.8 mg/mL in DMEM (Dulbecco's modified eagle's medium) (*see* **Note 1**).
8. For intestine digestion: Enzymes D, R, and A are prepared according to manufacturer's instructions from the Lamina Propria Dissociation Kit mouse (Miltenyi Biotec).
9. Nylon cell strainers of 40 and 100 μm.
10. Ficoll: After being autoclaved 5 min at 110 °C, 40% stock solution of Ficoll 400 in Eurocollins is then diluted to make 23%, 20%, and 11% solutions in Eurocollins. Prewarm at RT before use.
11. Cell dissociation solution is nonenzymatic. Prewarm at 37 °C.
12. Percoll™ used at 40% and 80% after dilution in CI buffer. Prewarm at RT.
13. Petri dish (3 cm in diameter).
14. Falcon tubes of 14, 15, and 50 mL.
15. Plunger (1 mL).
16. Forceps.
17. Clamp.
18. Dissection scissors.
19. Needles 18G and 23G.
20. Large bench top shaking incubator for continuous rotation (180 rpm).

2.2 Flow Cytometry for MAIT Cells

1. FACS buffer: Phosphate-buffered saline (PBS) without calcium and magnesium pH 7.2, 2% (v/v) FCS, 0.1% (w/v) NaN_3. Use at +4 °C.
2. RPMI medium: RPMI 1640 containing 10% (v/v) FCS and 1% (v/v) Penicillin-Streptomycin (10,000 UI/mL).
3. 5-OP-RU-loaded mMR1 tetramer (to specifically identify MAIT cells) or Ac-6-FP-loaded mMR1 tetramer (negative control) coupled to PE (phycoerythrin) fluorochrome for optimal detection sensitivity provided by National Institutes of Health tetramer core facility and by James McCluskey and Jamie Rossjohn. Store at +4 °C.

4. Phorbol 12-myristate 13-acetate (PMA): 1 mg/mL in ethanol. Ionomycin: 0.5 mg/mL in ethanol. Brefeldin A (BA): 1 mg/mL in ethanol. All reagents are aliquoted and stored at −20 °C.
5. Cytofix/Cytoperm Kit (BD Pharmingen).
6. Anti-mouse mAbs: Anti-Fcγ receptor (2.4G2); anti-CD19 (6D5); anti-CD4 (GK1.5); anti-CD69 (HI-2F3); anti-CD25 (PC61); anti-IL-22 (5164); anti-CD45 (30F11); anti-TCR-β (H57); anti-CD8α (53-6.7); anti-CD44 (IM7); anti-IFNγ (XMG1.2); anti-TNF (MP6-XT22) and anti-IL-17A (TC11-18H10) and anti-GzB (NGZB).

2.3 Mouse MAIT Cell Purification and Migration In Vivo

1. Injection buffer: PBS without calcium and magnesium pH 7.2, 2% (v/v) FCS. Use at +4 °C.
2. Dynabeads Untouched Mouse T Cells Kit (Invitrogen by Thermo Fisher Scientific).
3. Magnets: Dynal MPC-15 for 1–15 mL or MPC-50 for 15–50 mL samples.
4. Anti-mouse mAb: anti-CD45.2 (104).

2.4 In Vitro MAIT Cell Activation (5-OP-RU/Fecal Supernatant Dilution)

1. Culture medium: DMEM Glutamax, 10% (v/v) FCS, 10 mM HEPES, 1 mM Sodium pyruvate, 1× MEM NEAA and 1% (v/v) Penicillin-Streptomycin (10,000 UI/mL).
2. WT3-MR1^{+} cell line (mycoplasma free) provided by O. Lantz [12] (*see* **Note 2**).
3. 5-A-RU (10 mmol/L) provided by O. Lantz [12]. Store at −80 °C (*see* **Note 3**).
4. Methylglyoxal solution 40%.
5. Acetyl-6-formylpterin (Ac-6-FP) 1 mM.
6. 0.22 μM filter.
7. Petri dish (3 cm in diameter).
8. 1.5-mL Eppendorf tube.
9. Tissue culture 96-well plate, flat bottom.
10. Clear V-bottom 96-well plate.

2.5 Mouse Strains ($MR1^{-/-}$ and Vα19 Transgenic Mice)

Among all the mouse models allowing the study of T1D physiopathology, the NOD mouse model is the most used. However, diabetes can also be induced by multiple low-dose streptozotocin injection in C57BL/6 mice or in genetically susceptible mice by infection with Coxsackievirus B (CBV4).

NOD mice usually develop diabetes in 12–15 weeks. The characterization of MAIT cells at different stages of disease development (early, prediabetic, and at the onset of diabetes) represents a first step to evaluate MAIT cell involvement in T1D physiopathology [11].

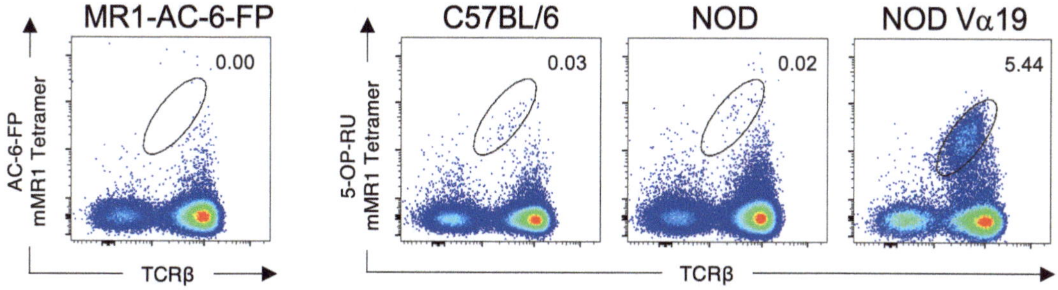

Fig. 1 Frequency of MAIT cells in different mouse strains. MAIT cells are identified by 5-OP-RU mMR1 tetramer staining. Ac-6-FP mMR1 tetramer is used as negative control. MAIT cell frequency among T cells from spleen is evaluated in adult female mice from various strains (C57BL/6, NOD, and NOD Vα19)

1. MR1 deficient mice: these mice do not express the MR1 molecule and therefore they do not contain any MAIT cells. NOD MR1-deficient mice were generated by 15 successive backcrosses of *Mr1*$^{-/-}$ C57BL/6J mice onto NOD background. Transmission of Idd susceptibility genes close to MR1 was verified.
2. Vα19 transgenic mice: these mice overexpress the Vα19-Jα33 invariant chain as a transgene and therefore they contain a higher frequency of MAIT cells (*see* Fig. 1). NOD Vα19 transgenic mice were obtained directly after microinjection of the *Vα19* transgene into NOD embryos.

3 Methods

3.1 Cell Preparation for MAIT Cell Analysis and Isolation (See Note 4)

Mouse abdomen is opened immediately after mouse death.

3.1.1 Mouse Spleen and Lymph Nodes

1. Drop tissue to be processed into a 40-μm cell strainer put on a Petri dish (3 cm in diameter) containing 5 mL of CI buffer.
2. Crush tissue with a plunger and collect cells in a 15-mL Falcon tube.
3. Wash cells by adding 9 mL of CI buffer and centrifuge at 500 × *g* for 5 min at +4 °C.
4. Resuspend cell pellet from spleen in 1 mL of RBC lysis solution and incubate 2 min at RT.
5. Wash cells in 10 mL CI buffer and centrifuge at 500 × *g* for 5 min at +4 °C. Resuspend the pellet in 1 mL of CI buffer and count live cells.

3.1.2 Mouse Liver

1. Infuse 10 mL of ice-cold PBS with 23G needle through the portal vein of mice and cannulate the inferior vena cava with 18G needle to eliminate peripheral blood lymphocyte contamination (*see* **Note 5**).
2. Transfer the liver into a 40-μm cell strainer put on a 50-mL Falcon tube containing 5 mL of CI buffer.
3. Crush the liver with a plunger and wash with 5 mL of CI buffer. Repeat several times until entire dissociation of the liver. Complete the Falcon tube up to 50 mL and centrifuge at 50 × *g* for 5 min at +4 °C (*see* **Note 6**).
4. Transfer supernatant into a 50-mL Falcon tube and centrifuge at 400 × *g* for 10 min at +20 °C.
5. Resuspend cell pellet in 10 mL of 40% Percoll diluted in CI buffer.
6. Prepare two tubes with 5 mL of 80% Percoll.
7. Meticulously deposit 5 mL of cell suspension on 5 mL of 80% Percoll diluted in RPMI to perform a Percoll density gradient into two 14-mL Falcon tubes (*see* **Note 7**). Centrifuge at 850 × *g* for 25 min at +20 °C with no brake.
8. Discard the layer on the top of Percoll of the two 14-mL Falcon tubes to avoid contaminations during the next step.
9. Collect cells contained in the interface between the layers of the two 14-mL Falcon tubes and resuspend them in 12 mL of CI Buffer. Homogenize cell suspension to dilute Percoll. Centrifuge at 805 × *g* for 5 min at +20 °C.
10. Resuspend the cell pellet in 2 mL of RBC lysis solution and incubate 2 min at RT.
11. Wash cells in 10 mL of CI buffer (centrifuge at 500 × *g* for 5 min at +4 °C), resuspend the pellet in 1 mL of CI buffer, and count live cells.

3.1.3 Mouse Pancreatic Islets

1. Preparation of pancreas (*see* **Note 5**). Displace duodenum and liver to uncover bile duct. Clamp off the bile duct at its entrance to duodenum. Free the upper part of bile duct by cutting it off from liver. Inject 3 mL of collagenase P solution through common bile duct into the pancreas. Remove and transfer the distended pancreas into a 50-mL Falcon tube containing 2 mL of collagenase P solution. Incubate for 10 min at 37 °C in a water bath and vigorously tap three times (*see* **Note 8**).
2. Isolation of pancreatic islets. Resuspend the tissue in about 30 mL of CI buffer. Perform the following steps at +2 to 8 °C (*see* **Note 9**). Centrifuge at 1800 × *g* for 1 min, discard supernatant. Resuspend the pellet in 10 mL of CI buffer and

transfer the cell pellet in 14 mL polystyrene round-bottom Falcon tube (*see* **Note 10**). Wash two times with 10 mL of CI buffer (centrifuge at 1800 × *g* for 1 min). After last wash, completely remove the supernatant and dry the pellet on absorbent paper. Resuspend the cell pellet in 1 mL of the 40% stock solution of Ficoll 400. Meticulously deposit on the top of this Ficoll suspension a density gradient of 23% (1.5 mL), 20% (1.5 mL), and 11% (1.5 mL) Ficoll solutions. Centrifuge at 1065 × *g* for 17 min at RT with no brake. The islets are collected from both interfaces between layers of 11% and 20% Ficoll solutions and between layers of 20% and 23% Ficoll solutions. Wash the islets in 10 mL of CI buffer (centrifuge at 805 × *g* for 5 min at RT) (*see* **Note 11**).

3. Isolation of single cells from pancreatic islets. Wash the islets with 10 mL of PBS (centrifuge at 805 × *g* for 5 min at +4 °C). Resuspend the pellet in 1 mL of cell dissociation solution nonenzymatic and incubate 10 min at 37 °C in a water bath. Dissociate the clumps using the pipette then add 10 mL of CI buffer. Wash the cell suspension (centrifuge at 500 × *g* for 5 min at +4 °C), resuspend the pellet in 1 mL of CI buffer, and count live cells.

3.1.4 Mouse Lamina Propria (According to the Manufacturer's Instructions Miltenyi Biotec)

1. Drop the intestine on a Petri dish (3 cm in diameter) containing 5 mL of IW buffer.
2. Hold the intestine with forceps and wash with 10 mL of IW buffer using a syringe to clean off feces.
3. Remove with forceps the residual fat tissue and the Peyer's patches.
4. Longitudinally cut the intestine (*see* **Note 12**) and transfer it into a 50-mL Falcon tube containing 20 mL of IPD buffer. Close the tube tightly.
5. Horizontally incubate the sample for 20 min at 37 °C under continuous rotation (180 rpm).
6. Vortex the tube during 10 s and empty the tube on a 100-μm cell strainer put on a 50-mL Falcon tube to eliminate the buffer (*see* **Note 13**).
7. Transfer the tissue remaining on the cell strainer into a new 50-mL Falcon tube containing 20 mL of IPD buffer (*see* **Note 14**).
8. Repeat **steps 5** and **6**. Transfer the tissue remaining on the cell strainer into a new 50-mL Falcon tube containing 20 mL of IW buffer.
9. Repeat **steps 5** and **6**.

10. Prepare the enzyme mix by adding 100 μL of Enzyme D, 50 μL of Enzyme R, and 12.5 μL of Enzyme A into a tube containing 2.35 mL of pre-heated ID buffer. Gently mix.
11. Quickly dry with absorbent paper the tissue obtained from **step 9** (*see* **Note 15**) and cut it into small fragments with dissection scissors in the upper part of a 15-mL Falcon tube (*see* **Note 16**).
12. Add 2.5 mL of the enzyme mix prepared in **step 11** and close the tube tightly.
13. Horizontally incubate the sample for 20 min at 37 °C under continuous rotation (180 rpm).
14. Add 10 mL of CI buffer in the tube and vigorously resuspend the tissue.
15. Transfer the sample into a 100-μm cell strainer put on a 50-mL Falcon tube and crush the tissue with a plunger. Rinse the cell strainer with 10 mL of CI buffer.
16. Centrifuge the cell suspension at 300 × *g* for 5 min at RT. Completely discard supernatant by aspiration.
17. Resuspend the cell pellet in 5 mL of 40% Percoll diluted in CI Buffer. Meticulously deposit the solution on 5 mL of 80% Percoll contained in a 14-mL Falcon tube. Centrifuge at 850 × *g* for 25 min at +20 °C with no brake (*see* **Note 17**).
18. Collect cells contained in the interface between the layers and resuspend them in 12 mL of FACS buffer. Centrifuge at 805 × *g* for 5 min at 4 °C. Completely discard the supernatant by aspiration.
19. Resuspend lamina propria lymphocytes in 1 mL of CI buffer and count live cells.

3.2 Flow Cytometry for MAIT Cells

3.2.1 Surface Staining for Mouse MAIT Cells

1. Stain 2 × 10^6 cells with 20 μL of 5-OP-RU-loaded mMR1 tetramer or with 20 μL of Ac-6-FP-loaded mMR1 tetramer diluted at the optimal concentration in FACS buffer (*see* **Note 18**). Incubate 1 h, protected from light, at RT.
2. Wash cells with 500 μL of FACS buffer (centrifuge at 500 × *g* for 5 min at +4 °C) and block Fcγ receptors by adding 20 μL of 2.4G2 mAb (100 μg/mL) for 15 min, protected from light, at +4 °C (*see* **Note 19**).
3. Wash cells with 500 μL of FACS buffer (centrifuge at 500 × *g* for 5 min at +4 °C) and stain the cells with 20 μL of the mix of mAbs composed by anti-CD45 and anti-CD19 (to exclude other cell populations, *see* the gating strategy Fig. 2), and by anti-TCR-β (to determine MAIT cell frequency among T cells, *see* Fig. 3) and other surface molecule-specific mAbs for 15 min, protected from light, at +4 °C (*see* **Note 20**).

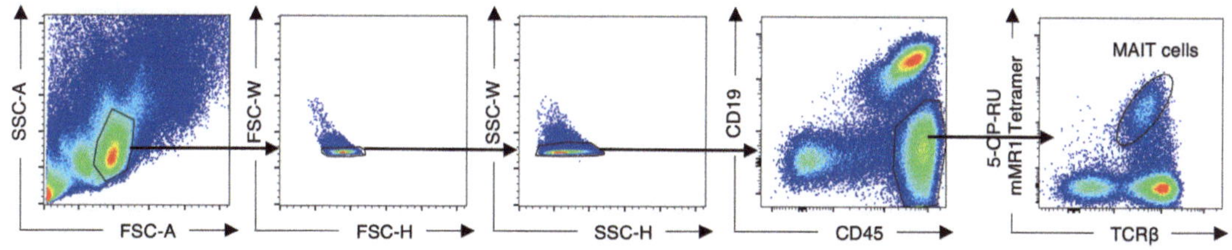

Fig. 2 Gating strategy to identify MAIT cells. Lymphocytes are separated from other cells obtained from the spleen of a NOD Vα19 Tg mouse according to their size and their granularity. After exclusion of doublets, MAIT cells are identified as CD45+, CD19−, TCRβ+ cells that recognize 5-OP-RU mMR1 tetramer

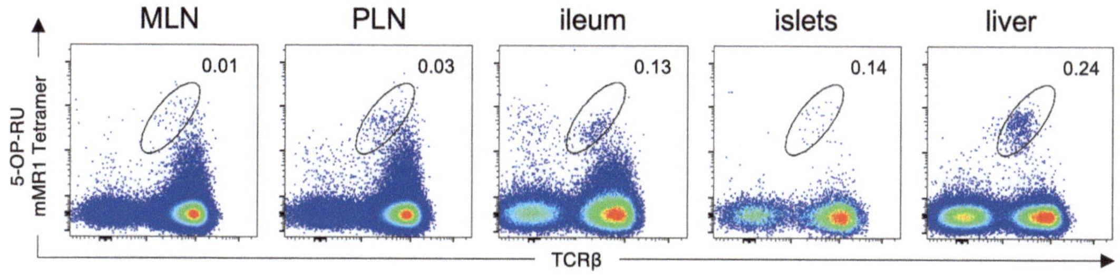

Fig. 3 Frequency of MAIT cells in different tissues from NOD mice. MAIT cell frequency among T cells is evaluated in mesenteric lymph nodes, pancreatic lymph nodes, ileum, pancreatic islets, and liver of an adult female NOD mouse

4. Wash cells with 500 μL of FACS buffer (centrifuge at 500 × *g* for 5 min at +4 °C) and resuspend the pellet in 300 μL of FACS buffer.

3.2.2 Intracytoplasmic Staining

1. To detect intracytoplasmic cytokines production (*see* Fig. 4), 5 × 10^6 mouse cells are activated during 4 h with PMA (10 ng/mL) and ionomycin (1 μg/mL) in the presence of BA (10 μg/mL) at 37 °C in RPMI medium in incubator.
2. Wash 5 × 10^6 cells with 500 μL of FACS buffer (centrifuge at 500 × *g* for 5 min at +4 °C). After surface staining, wash cells with 500 μL of FACS buffer (centrifuge at 500 × *g* for 5 min at +4 °C) and resuspend the cell pellet in 350 μL of Cytofix. After vortex agitation, incubate cells protected from light for 8 min at +4 °C.
3. Wash cells with 500 μL of FACS buffer (centrifuge at 500 × *g* for 5 min at +4 °C).
4. Wash cells with 500 μL of Cytoperm buffer (centrifuge at 500 × *g* for 5 min at +4 °C).
5. Stain cells with 20 μL of the mix of anti-cytokine-specific mAbs diluted in Cytoperm buffer. Incubate for 30 min, protected from light, at +4 °C.

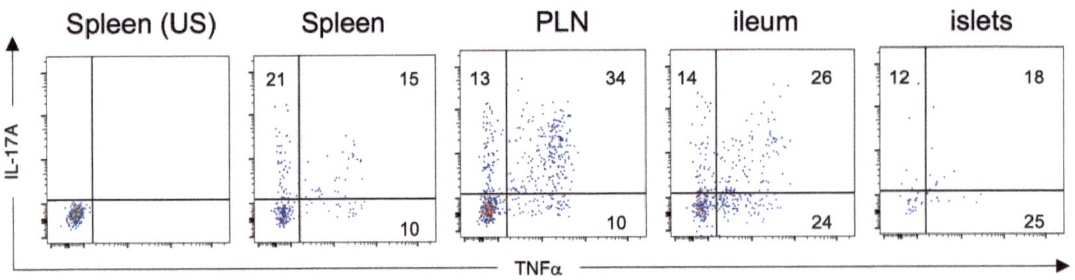

Fig. 4 Intracellular staining of MAIT cells in different tissues from NOD mice. IL-17A and TNFα secretion by MAIT cells in spleen, pancreatic lymph nodes, ileum, and pancreatic islets of an adult female NOD mouse after in vitro stimulation with PMA ionomycin or without stimulation (US)

6. Wash with 500 μL of Cytoperm buffer and resuspend the pellet in 300 μL of FACS buffer.

3.3 Mouse MAIT Cell Purification and Migration In Vivo

CD45.2+ MAIT cells are obtained from Vα19-transgenic *Trac*$^{-/-}$ NOD donor mice.

3.3.1 Enrichment of Mouse MAIT Cells by Negative Selection

1. Prepare a single-cell suspension from spleen using the previously described methods (*see* Subheading 3.1.1). Centrifuge the cell suspension at 500 × *g* for 5 min and resuspend the cell pellet in 5 mL of CI buffer. Count cells and adjust to obtain 1×10^7 cells in 100 μL.
2. For isolation of mouse MAIT cells, use Dynabeads Untouched Mouse T Cells Kit according to manufacturer's instructions (*see* **Note 21**).

3.3.2 Transfer of MAIT Cells into Recipient Micc (CD45.1 NOD Mice)

4×10^6 purified MAIT cells contained in 100 μL of injection buffer are intravenously injected into CD45.1 NOD mice. The migration is evaluated by detecting CD45.2+ MAIT cells in target tissues by flow cytometry staining 5 days after injection

3.4 In Vitro MAIT Cell Activation (5-OP-RU/ Fecal Supernatant Dilution) Designed by O. Lantz [12]

1. At least 2 h before activation, coat 10^5 WT3-MR1$^+$ cells in 100 μL of culture medium onto a 96-flat-well plate. Leave at 37 °C in cell culture incubator.
2. Feces preparation should be performed on ice. In a 3-cm Petri dish, collect all the feces contained in the colon or caecum from one mouse. Transfer the feces in 1.5-mL Eppendorf tube and resuspend in 1 mL of cold-PBS. Vigorously vortex and centrifuge at 11,000 × *g* for 5 min at +4 °C. Next steps have to be done under a culture hood. Harvest supernatant in a 1.5-mL Eppendorf and pass it through 0.22-μM filter (*see* **Note 22**).
3. If necessary, add 10 μmol/L Ac-6-FP on WT3 MR1$^+$ 1 h before MAIT cell activation (*see* **Note 23**).
4. Generation of the 5-OP-RU stock solution extemporaneously. 5-A-RU reacts with methylglyoxal to form 5-OP-RU. Dilute

methylglyoxal in sterile H_2O to obtain a solution at 10 mmol/L. In a 0.5-mL Eppendorf tube, mix 2 µL of 5-A-RU with 4 µL of 10 mmol/L methylglyoxal and incubate for 5 min, protected from light, at RT. Dilute 2 µL of the 5-OP-RU solution obtained (at 3.3 mmol/L) in 658 µL of culture medium to obtain the 5-OP-RU stock solution at 0.01 mmol/L. Prepare various solutions of 5-OP-RU with culture medium to obtain a range of final concentrations in culture from 0.12 nmol/L until 1000 nmol/L with an added volume of 5-OP-RU that does not exceed 20 µL.

5. Add the ligand (5-OP-RU) or 20 µL of fecal supernatant in each culture well.
6. MAIT cells are obtained from Vα19-transgenic *Trac*$^{-/-}$ NOD donor mice and are purified as in Subheading 3.3.1. Centrifuge the suspension at 500 × *g* for 5 min and resuspend pellet in 5 mL of culture medium. Count cells and adjust to obtain 5×10^6 cells in 1 mL. Deposit 60 µL of MAIT cell solution in each culture well. Incubate in cell culture incubator for 20 h.
7. The following day, resuspend MAIT cells with a pipette, then transfer MAIT cells in a new clear V-bottom 96-well plate. Centrifuge at 500 × *g* for 5 min. Wash MAIT cells in 180 µL of FACS Buffer (centrifuge at 500 × *g* for 5 min). MAIT cells are ready to be stained as in Subheading 3.2.1. MAIT cell activation is evaluated by CD25 and CD69 surface marker expression using flow cytometry staining (*see* Fig. 5).

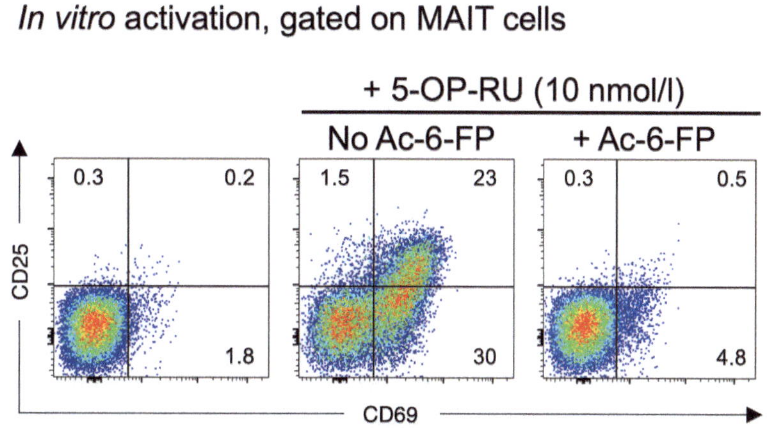

Fig. 5 MR1-dependent MAIT cell activation by 5-OP-RU. Surface expression of the activation markers, CD25 and CD69, on MAIT cells is measured by flow cytometry staining after stimulation overnight in the presence, or not, of 5-OP-RU (10 nmol/L). Addition of Ac-6-FP prevents 5-OP-RU binding to MR1 thereby inhibiting MAIT cell activation

4 Notes

1. The optimal concentration of each Collagenase P batch has to be determined.
2. WT3-MR1$^+$ cell line is WT3 cell line that stably overexpresses mouse MR1 molecules. These adherent cells need to be split using trypsin treatment.
3. 5-A-RU is an unstable molecule that can be affected by light or temperature change. This molecule has to be aliquoted to avoid excessive defrosting.
4. To optimize MAIT cell isolation from pancreatic islets and liver, both organs are obtained from two different mice. However, it is possible to isolate MAIT cells from pancreas and liver from the same mouse. In that case, the pancreas cannot be perfused but is only cut in little fragments, then digested with collagenase P solution under manual agitation for 10 min at 37°. MAIT cells from spleen, lymph nodes, and lamina propria can be indifferently isolated from mice in the same time as liver or pancreas.
5. For perfusion of liver and pancreas, the use of a binocular loupe (10×) is recommended.
6. The slow centrifugation enables to get rid of heavier cells obtained from liver as hepatocytes and other resident hepatic cells that would be also contained in the pellet.
7. To optimize collect of the interface between the layers of Percoll in Subheading 3.1.2, **step 9**, it is more efficient to use two 14-mL Falcon tubes as one 50-mL Falcon tube in Subheading 3.1.2, **step 7**.
8. Heat improves enzyme efficiency and mechanic step helps dissociation of the tissue.
9. Enzyme efficiency decreases consecutively to its dilution. However, a low temperature is highly recommended for the following steps to avoid degradation of pancreatic islets by dissociation enzyme.
10. Sometimes, during injection, collagenase P solution does not diffuse in all the pancreas. If some fragments are remaining, discard them because they could disturb the isolation of pancreatic islets.
11. To improve islets purity, pick islets under binocular loupe.
12. Even if the manufacturer recommends cutting intestine in fragments, we advise keeping the whole intestine and only cut it longitudinally to make easier the handling.
13. IPD buffer contains intraepithelial cells that can be collected after centrifugation, if needed.

14. Instead of doing Subheading 3.1.4, **step 6**, tissue can be directly collected with forceps and transferred in a new 50-mL Falcon tube with 20 mL of IPD buffer.
15. Just absorb exceeding humidity, be careful to not let tissue becoming dried.
16. Tissue can also be cut with scissors in a Petri dish or on an aluminum leaf but loss of tissue will be reduced by cutting the tissue in the 15-mL Falcon tube that will be directly used for the next step.
17. In comparison with the manufacturer's instructions Miltenyi Biotec, we perform this supplementary Percoll to enhance lymphocytes enrichment.
18. Ac-6-FP is a non-agonist ligand of MAIT cells, derived from folate metabolism pathway. Ac-6-FP is used as control for tetramer staining because its strong affinity for MR1 leading to a stable complex that is not recognized by MAIT cells [29].
19. 2.4G2 mAb against Fcγ receptors avoid nonspecific staining.
20. The amount of mAb is determined by titrating each of them in separate staining.
21. Dynabeads Untouched Mouse T Cells Kit purifies T lymphocytes by negative selection. Since cell suspension used is obtained from the spleen from Vα19-transgenic *Trac*$^{-/-}$ NOD donor mice, T cells resulting from the negative selection are highly enriched in MAIT cells.
22. Fecal supernatant samples can be prepared in advance and stored at −80 °C or at experiment time.
23. MR1 blocking by Ac-6-FP is a way to show that MAIT cell activation is TCR-dependent.

Acknowledgments

We thank M. Salou for help in analysis of in vitro MAIT cell activation (5-OP-RU/fecal supernatant dilution); the mouse and Cybio facilities of the Cochin Institute. This work was supported by grants from INSERM, CNRS, Laboratoire d'Excellence consortium Inflamex ANR-11-IDEX-0005-02, and the Fondation pour la Recherche Médicale (FRM grant number DEQ20140329520, EQU201903007779 to A.L.), Fondation Francophone pour la Recherche sur le Diabète to A.L., EFSD/JDRF/Lilly to A.L., Aide aux Jeunes Diabétiques fellowship to I.N., Agence National de la Recherche (ANR-17-CE14-0002-01 Diab1MAIT to A.L.).

Author Contributions: I.N. and L.B. wrote the review and A.L. supervised the writing of the review.

References

1. McWilliam HEG, Birkinshaw RW, Villadangos JA, McCluskey J, Rossjohn J (2015) MR1 presentation of vitamin B-based metabolite ligands. Curr Opin Immunol 34:28–34
2. Ussher JE, Klenerman P, Willberg CB (2014) Mucosal-associated invariant T-cells: new players in anti-bacterial immunity. Front Immunol 5:450
3. Treiner E, Duban L, Bahram S, Radosavljevic M, Wanner V, Tilloy F et al (2003) Selection of evolutionarily conserved mucosal-associated invariant T cells by MR1. Nature 422(6928):164–169
4. Lantz O, Legoux F (2018) MAIT cells: an historical and evolutionary perspective. Immunol Cell Biol 96(6):564–572
5. Lamichhane R, Ussher JE (2017) Expression and trafficking of MR1. Immunology 151 (3):270–279
6. Kjer-Nielsen L, Patel O, Corbett AJ, Le Nours J, Meehan B, Liu L et al (2012) MR1 presents microbial vitamin B metabolites to MAIT cells. Nature 491(7426):717–723
7. Kjer-Nielsen L, Corbett AJ, Chen Z, Liu L, Mak JY, Godfrey DI et al (2018) An overview on the identification of MAIT cell antigens. Immunol Cell Biol 96(6):573–587
8. Corbett AJ, Eckle SBG, Birkinshaw RW, Liu L, Patel O, Mahony J et al (2014) T-cell activation by transitory neo-antigens derived from distinct microbial pathways. Nature 509 (7500):361–365
9. Rahimpour A, Koay HF, Enders A, Clanchy R, Eckle SBG, Meehan B et al (2015) Identification of phenotypically and functionally heterogeneous mouse mucosal-associated invariant T cells using MR1 tetramers. J Exp Med 212 (7):1095–1108
10. Reantragoon R, Corbett AJ, Sakala IG, Gherardin NA, Furness JB, Chen Z et al (2013) Antigen-loaded MR1 tetramers define T cell receptor heterogeneity in mucosal-associated invariant T cells. J Exp Med 210 (11):2305–2320
11. Rouxel O, Da Silva J, Beaudoin L, Nel I, Tard C, Cagninacci L et al (2017) Cytotoxic and regulatory roles of mucosal-associated invariant T cells in type 1 diabetes. Nat Immunol 18(12):1321–1331
12. Soudais C, Samassa F, Sarkis M, Le Bourhis L, Bessoles S, Blanot D et al (2015) In vitro and in vivo analysis of the gram-negative bacteria-derived riboflavin precursor derivatives activating mouse MAIT cells. J Immunol (Baltimore, MD, 1950) 194(10):4641–4649
13. Rouxel O, Lehuen A (2018) Mucosal-associated invariant T cells in autoimmune and immune-mediated diseases. Immunol Cell Biol 96(6):618–629
14. Touch S, Assmann KE, Aron-Wisnewsky J, Marquet F, Rouault C, Fradet M et al (2018) Mucosal-associated invariant T (MAIT) cells are depleted and prone to apoptosis in cardiometabolic disorders. FASEB J Off Publ Fed Am Soc Exp Biol. https://doi.org/10.1096/fj.201800052RR
15. Magalhaes I, Pingris K, Poitou C, Bessoles S, Venteclef N, Kiaf B et al (2015) Mucosal-associated invariant T cell alterations in obese and type 2 diabetic patients. J Clin Invest 125 (4):1752–1762
16. Hegde P, Weiss E, Paradis V, Wan J, Mabire M, Sukriti S et al (2018) Mucosal-associated invariant T cells are a profibrogenic immune cell population in the liver. Nat Commun 9 (1):2146
17. Atkinson MA, Eisenbarth GS, Michels AW (2014) Type 1 diabetes. Lancet Lond Engl 383(9911):69–82
18. Lehuen A, Diana J, Zaccone P, Cooke A (2010) Immune cell crosstalk in type 1 diabetes. Nat Rev Immunol 10(7):501–513
19. Bluestone JA, Herold K, Eisenbarth G (2010) Genetics, pathogenesis and clinical interventions in type 1 diabetes. Nature 464 (7293):1293–1300
20. Anderson MS, Bluestone JA (2005) The NOD mouse: a model of immune dysregulation. Annu Rev Immunol 23:447–485
21. Wen L, Ley RE, Volchkov PY, Stranges PB, Avanesyan L, Stonebraker AC et al (2008) Innate immunity and intestinal microbiota in the development of Type 1 diabetes. Nature 455(7216):1109–1113
22. Kostic AD, Gevers D, Siljander H, Vatanen T, Hyötyläinen T, Hämäläinen A-M et al (2015) The dynamics of the human infant gut microbiome in development and in progression toward type 1 diabetes. Cell Host Microbe 17 (2):260–273
23. Sapone A, de Magistris L, Pietzak M, Clemente MG, Tripathi A, Cucca F et al (2006) Zonulin upregulation is associated with increased gut permeability in subjects with type 1 diabetes and their relatives. Diabetes 55(5):1443–1449
24. Alam C, Bittoun E, Bhagwat D, Valkonen S, Saari A, Jaakkola U et al (2011) Effects of a germ-free environment on gut immune regulation and diabetes progression in non-obese

diabetic (NOD) mice. Diabetologia 54 (6):1398–1406

25. Franciszkiewicz K, Salou M, Legoux F, Zhou Q, Cui Y, Bessoles S et al (2016) MHC class I-related molecule, MR1, and mucosal-associated invariant T cells. Immunol Rev 272 (1):120–138
26. Koay H-F, Gherardin NA, Enders A, Loh L, Mackay LK, Almeida CF et al (2016) A three-stage intrathymic development pathway for the mucosal-associated invariant T cell lineage. Nat Immunol 17(11):1300–1311
27. Le Bourhis L, Martin E, Péguillet I, Guihot A, Froux N, Coré M et al (2010) Antimicrobial activity of mucosal-associated invariant T cells. Nat Immunol 11(8):701–708
28. Shimamura M, Huang Y-Y, Goji H, Endo S, Migishima R, Yokoyama M (2011) Regulation of immunological disorders by invariant Vα19-Jα33 TCR-bearing cells. Immunobiology 216 (3):374–378
29. Eckle SBG, Birkinshaw RW, Kostenko L, Corbett AJ, McWilliam HEG, Reantragoon R et al (2014) A molecular basis underpinning the T cell receptor heterogeneity of mucosal-associated invariant T cells. J Exp Med 211 (8):1585–1600

Chapter 19

Magnetic Bead-Based Enrichment of Murine MAIT Cells

Samantha J. Winter and Andreas Krueger

Abstract

Mucosal-associated invariant T cells (MAIT) are abundant in humans, comprising up to 40% of liver T cells and 10% of peripheral blood T cells. However, understanding MAIT cell biology is hampered by the fact that they are fundamentally rare in standard laboratory mouse strains, such as C57BL/6. The discovery of MAIT cell ligands and recent advances in MR1-tetramer technology has provided a means for detecting murine MAIT cells, but low frequencies still hinder precise characterization. Here we describe how to accurately isolate rare MAIT cells from murine lymphocyte populations using MR1-tetramer technology combined with magnetic bead enrichment. Isolated MAIT cells can be used for downstream characterization or functional analysis.

Key words MAIT cells, Tetramers, T cells, Magnetic bead enrichment, Flow cytometry

1 Introduction

MAIT cells constitute a substantial proportion of T cells in tissues and circulation in humans. They are characterized by expression of a semi-invariant T cell receptor (TCR) comprised of an invariant TCRα chain (Vα7.2Jα33/20/12 in humans and Vα19Jα33 in mice) and a restricted set of TCRβ chains [1, 2]. They recognize vitamin B metabolites, such as 5-(2-oxopropylideneamino)-6-D-ribityl-aminouracil (5-OP-RU) in the context of the non-classical MHC class I molecule MHC class I-related 1 (MR1) [3, 4]. Having such a huge presence in the human peripheral T cell compartment warrants in-depth biological characterization of MAIT cells. However, research into MAIT cell development and function has been hampered by low cell frequencies in conventional laboratory mice (e.g., there are fewer than 10,000 MAIT cells in the murine C57BL/6 thymus) [5]. Despite the recent development of tetramer reagents to faithfully identify MAIT cells by flow cytometry, low abundance is associated with typical technical difficulties, such as background staining, and with problems to unambiguously identify MAIT cell developmental stages and subsets. Development

Helen Kaipe and Isabelle Magalhaes (eds.), *MAIT Cells: Methods and Protocols*, Methods in Molecular Biology, vol. 2098, https://doi.org/10.1007/978-1-0716-0207-2_19, © Springer Science+Business Media, LLC, part of Springer Nature 2020

of MAIT cells has recently been broken down into three stages based on differential expression of surface markers CD24 and CD44 [6]. MAIT cell subsets are characterized by differential expression of the transcription T-bet and RORγt to identify MAIT1 and MAIT17 cells, respectively [7, 8]. It remains a particular challenge to dissect mechanisms of MAIT cell development and function using genetic tools that are accompanied with further loss of MAIT cell numbers. Examples of these tools include mouse models lacking the key innate lymphocyte transcription factor PLZF (encoded by Zbtb16) or miRNAs [6, 9].

To circumvent problems associated with low abundance of MAIT cells, we have provided a detailed overview of MAIT cell isolation from murine organs using MR1-5-OP-RU tetramers combined with magnetic bead enrichment. To improve staining,

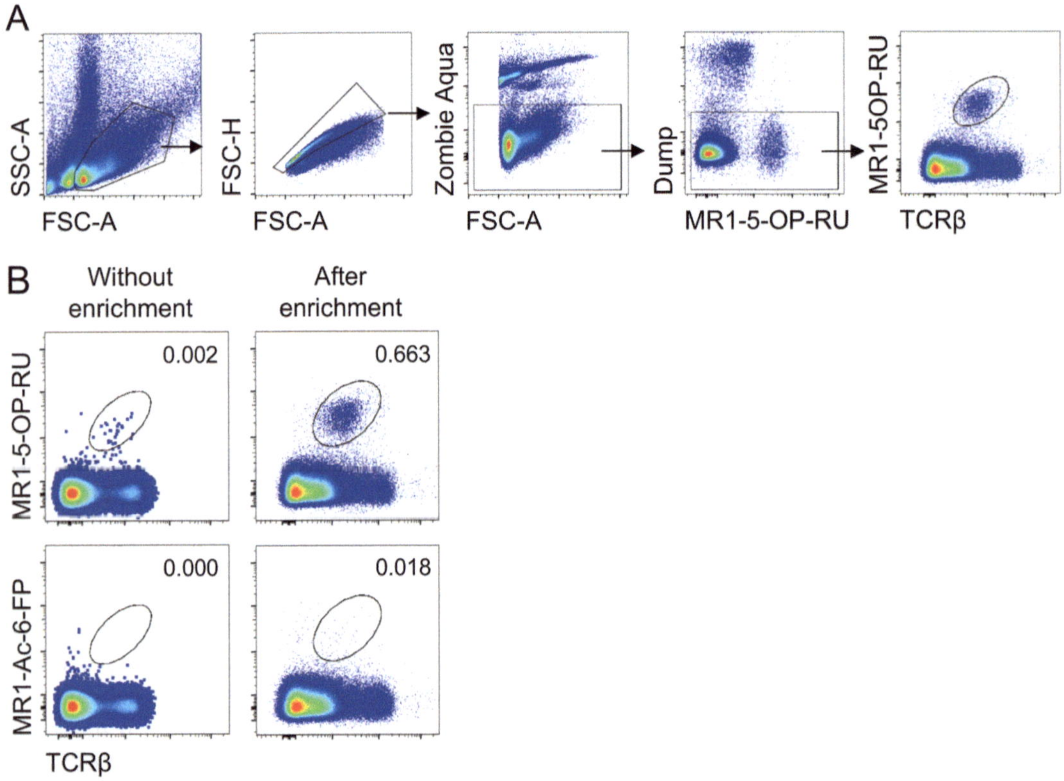

Fig. 1 (**a**) MAIT cell gating strategy. To identify MAIT cells, lymphocytes are gated based on FSC-A and SSC-A, followed by doublet exclusion. Singlets are then selected for living cells (Zombie Aqua^{-}) and lineage^{+} cells are removed using a dump channel containing the CD11b, CD11c, CD19, and B220 antibodies (PE-Cy7 conjugated). MAIT cells are then identified as the MR1-5-OP-RU-APC and TCRβ-BV421 double positive population. (**b**) MAIT cells before and after enrichment. Left column is representative of MR-1-5-OP-RU^{+} MAIT cells (top row) in a single thymus without enrichment. The right column shows the total amount of MR1-5-OP-RU^{+} MAIT cells (top row) recovered from three pooled thymi after enrichment. Plots in the bottom row are representative of tetramer specificity both before (left) and after (right) enrichment using the MR1-Ac-6-FP negative control tetramer. Numbers adjacent to gates indicate frequencies in percentages

cells that non-specifically bind MR1-tetramers can be removed via staining with biotin labeled antibodies which are detected by anti-streptavidin magnetic beads and removed through magnetic separation. Following this, MAIT cells labeled with MR1-5-OP-RU tetramers can be positively selected for using anti-fluorochrome magnetic beads, specific for the same fluorochrome coupled to MR1-5-OP-RU tetramers and subsequent magnetic separation. Isolated MAIT cells can be used directly for flow cytometric analysis or downstream applications (Fig. 1). Interestingly, we and others discovered that enrichment of MAIT cells via magnetic beads retains differences in MAIT cell frequencies observed between wild-type and genetically engineered mouse strains [6, 9]. One underlying reason may be that MAIT cell enrichment by tetramers is comparatively inefficient, not unlike enrichment of iNKT cells via a similar protocol [10], which in turn could be due to the low affinity of the MR1-TCR interaction when compared to typical antigen–antibody interactions.

2 Materials

Keep all solutions and reagents cool during MAIT cell isolation.

2.1 Buffers and Reagents

1. FACS buffer: PBS containing 3% fetal calf serum (FCS). Store at 4 °C.
2. Red blood cell (RBC) lysis buffer.
3. MACS buffer: PBS containing 2 mM EDTA, 0.5% BSA. This reagent can be store bought (Miltenyi Biotec, Bergisch-Gladbach, Germany) or self-prepared. Store at 4 °C.
4. B cell antibodies: Biotin anti-mouse CD19 (clone 6D5) and B220 (clone RA3-6B2).
5. Fc Block (rat anti-mouse CD16/32, clone 2.4G2; used purified or directly as hybridoma supernatant).
6. MR1-5-OP-RU and MR1-Ac-6-FP tetramers. Currently available through the NIH Tetramer Core Facility (APC-conjugated).
7. Zombie Aqua viability dye.

2.2 Magnetic Beads and Columns

1. MACS anti-Streptavidin MicroBeads.
2. MACS LD columns.
3. MACS anti-APC MicroBeads.
4. MACS LS columns.

2.3 Equipment and Supplies

1. 70 μm filter.
2. 2 mL syringe plunger.
3. 30 μm MACS filters.
4. 15 mL Falcon tubes.
5. Magnetic MACS separator (e.g., QuadroMACS™).

3 Methods

To preserve cell integrity throughout MAIT cell isolation, work fast and keep samples on ice. After initial tetramer staining, shield cells from light for the remainder of protocol to maintain optimal fluorescence.

1. Harvest organs of interest (*see* **Note 1**) and isolate lymphocytes via appropriate protocol. For lymphoid organs such as thymus, spleen, and lymph nodes, samples can be extracted and immediately mashed through a 70-μm filter using the plunger from a 2-mL syringe.
2. RBCs are then lysed (spleen or well-vascularized organs) by incubating samples for 10 min on ice in RBC lysis buffer. For samples that do not require RBC lysis (e.g., thymus and lymph nodes), skip to **step 4**.
3. Samples are then centrifuged at 300 × *g*, 4 °C, 5 min, and the supernatant is discarded. To remove traces of RBC lysis buffer, wash samples once more in FACS buffer, centrifuge at 300 × *g*, 4 °C, 5 min and discard supernatant.
4. **Steps 4–7** are optional (*see* **Note 2**). To remove B cells, resuspend samples and stain with biotin-conjugated anti-CD19 and anti-B220 antibodies for 15 min on ice. Wash samples with MACS buffer, centrifuge at 300 × *g*, 4 °C, 5 min and resuspend in residual volume (~200 μL when working with 15 mL Falcon tubes).
5. Add 50 μL of MACS anti-Streptavidin MicroBeads to each sample (*see* **Note 3**), vortex and incubate for 15 min at 4 °C. Wash samples, centrifuge at 300 × *g*, 4 °C, 5 min and resuspend in 2 mL of MACS buffer.
6. Prepare MACS LD Columns by placing them onto an appropriate magnetic MACS separator. Place a 30-μm MACS filter on top of each column and rinse both with 2 mL of MACS buffer (*see* **Note 4**).
7. Once prepared, each sample is loaded onto an individual filter/column and the flow through is collected. Rinse columns with 2 × 1 mL of MACS buffer and collect the flow through. Flow through contains unlabeled cells which have been depleted of B cells. Centrifuge at 300 × *g*, 4 °C, 5 min and resuspend in residual buffer.

8. To block Fc receptors, add Fc Block to samples (1:5 to 1:10 dilution if using hybridoma supernatant, 10 μg/mL if using purified antibody), vortex, and incubate for 10 min on ice. Without washing, proceed to tetramer staining.
9. Before staining, centrifuge tetramer stocks at 10,000 × *g*, 4 °C, 5 min to remove aggregates (*see* **Note 5**).
10. To stain MAIT cells, add MR1-5-OP-RU-APC tetramer to samples (*see* **Note 6**), vortex, and incubate in the dark at room temperature for 40 min. Negative control tetramer MR1-Ac-6-FP-APC can also be stained at this stage using the same procedure (*see* **Note 7**).
11. Wash once with MACS buffer, centrifuge at 300 × *g*, 4 °C, 5 min and resuspend in residual volume.
12. For each sample add 50–100 μL of MACS anti-APC MicroBeads (*see* **Notes 3** and **8**) and vortex. Incubate for 15 min at 4 °C in the dark.
13. Wash once with MACS buffer, centrifuge at 300 × *g*, 4 °C, 5 min and resuspend in 3 mL of MACS buffer.
14. Prepare MACS LS columns by placing them onto an appropriate magnetic MACS separator. Rinse columns with 3 mL of MACS buffer.
15. Once prepared, load each sample onto an individual LS column and wait for the column reservoir to empty. Once emptied rinse column with 3 mL of MACS buffer, wait for the column to empty and repeat two more times to remove unlabeled cells.
16. To collect labeled cells enriched for the tetramer$^+$ population, remove the column from the magnet and place into a 15-mL Falcon tube. Load 5 mL of MACS buffer onto column and using the plunger provided, plunge the buffer through the reservoir and into the tube. Centrifuge samples at 300 × *g*, 4 °C, 5 min and resuspend in residual volume.
17. Following enrichment stain with surface antibodies for 15 min at 4 °C in the dark (*see* **Note 9**). Wash with PBS, centrifuge at 300 × *g*, 4 °C, 5 min and resuspend in residual volume.
18. For live/dead discrimination, stain cells with Zombie Aqua in PBS for 15 min at room temperature and in the dark. Wash with FACS buffer, centrifuge (300 × *g*, 4 °C, 5 min) and resuspend in residual volume (*see* **Note 10**).
19. Cells can be used for flow cytometric analysis or downstream intracellular staining. An example of MAIT cells analysis by flow cytometry is provided in Fig. 1.

4 Notes

1. In mice, MAIT cells are generally quite rare which can make detection even with tetramer enrichment difficult. Depending on your purposes, it may be necessary to pool organs from multiple mice to isolate enough MAIT cells for statistical analysis. Our laboratory routinely pools cells from thymi, combined peripheral lymph nodes or spleens from three young adult C57BL/6 mice.
2. B cells tend to non-specifically bind to MR1-5-OP-RU tetramers. To circumvent this, it is possible to remove B cells through magnetic depletion as described in **steps 3–6**. If not performing a magnetic B cell depletion, it is crucial to include a dump channel which includes B cell markers, to exclude them from analysis.
3. We recommend titrating the number of MACS Microbeads needed to prevent wasting expensive reagents.
4. When performing magnetic depletion or enrichment over columns, it is important to ensure that columns do not become clogged to achieve adequate depletion/enrichment. To circumvent this, it is important to filter samples directly before loading onto columns as well as using MACS buffer that has been degassed to remove small bubbles that can negatively affect column flow through.
5. Aggregates that form in tetramer stocks may non-specifically bind to cells. To remove these aggregates, it is recommended to centrifuge tetramers at high speed in a cooled benchtop microfuge before use.
6. It is essential to titrate tetramers before use to find the optimal concentration required per sample. Generally, they can be used at quite low concentrations, and it is recommended to start titrations with an initial dilution between 1:500 and 1:1000. Tetramer preparations obtained by our laboratory routinely gave good results at dilutions between 1:1500 and 1:3000.
7. Due to the rarity of MAIT cells, it is important to include negative controls to determine tetramer staining specificity. The currently accepted negative control is the MR1-Ac-6-FP tetramer. This can either be stained simultaneously with MR1-5-OP-RU as a dual-tetramer stain (e.g., MR1-5-OP-RU-APC with MR1-Ac-6-FP-PE) or alongside MR1-5-OP-RU in a separate sample.
8. This protocol calls for the use of MR1-5-OP-RU-APC tetramers and therefore MACS Anti-APC MicroBeads. It is also possible to use MR1-5-OP-RU tetramers coupled to various fluorochromes if the corresponding MicroBeads are used for enrichment purposes (e.g., MR1-5-OP-RU-PE with Anti-PE MicroBeads).

9. Stage 1 thymic MAIT cells that have recently gone through selection can be distinguished by their high expression of CD24 [6]. CD24 antibody binding appears to be prevented in the presence of EDTA, therefore to adequately stain for CD24 EDTA-containing MACS buffer must be washed away from samples before surface staining.
10. Alternate live/dead discrimination dyes can be used at **step 18**. We recommend Zombie Aqua dye due to its compatibility with downstream intracellular staining methods. Zombie Aqua staining needs to occur in the absence of proteins such as BSA or FCS and should therefore be carried out on cells washed with PBS.

Acknowledgments

MR1-5-OP-RU tetramer technology was developed jointly by Dr. James McCluskey, Dr. Jamie Rossjohn, and Dr. David Fairlie, and the material was produced by the NIH Tetramer Core Facility as permitted to be distributed by the University of Melbourne. This work was funded by grants from the German Research foundation (DFG) KR2320/5-1 and SFB902-B15 (to A.K.).

References

1. Tilloy F, Treiner E, Park S-H et al (1999) An invariant T cell receptor α chain defines a novel TAP-independent major histocompatibility complex class Ib–restricted α/β T cell subpopulation in mammals. J Exp Med 189(12):1907–1921. https://doi.org/10.1084/jem.189.12.1907
2. Reantragoon R, Corbett AJ, Sakala IG et al (2013) Antigen-loaded MR1 tetramers define T cell receptor heterogeneity in mucosal-associated invariant T cells. J Exp Med 210(11):2305–2320. https://doi.org/10.1084/jem.20130958
3. Corbett AJ, Eckle SBG, Birkinshaw RW et al (2014) T-cell activation by transitory neo-antigens derived from distinct microbial pathways. Nature 509(7500):361–365. https://doi.org/10.1038/nature13160
4. Kjer-Nielsen L, Patel O, Corbett AJ et al (2012) MR1 presents microbial vitamin B metabolites to MAIT cells. Nature 491(7426):717–723. https://doi.org/10.1038/nature11605
5. Cui Y, Franciszkiewicz K, Mburu YK et al (2015) Mucosal-associated invariant T cell-rich congenic mouse strain allows functional evaluation. J Clin Invest 125(11):4171–4185. https://doi.org/10.1172/JCI82424
6. Koay H-F, Gherardin NA, Enders A et al (2016) A three-stage intrathymic development pathway for the mucosal-associated invariant T cell lineage. Nat Immunol 17(11):1300–1311. https://doi.org/10.1038/ni.3565
7. Rahimpour A, Koay HF, Enders A et al (2015) Identification of phenotypically and functionally heterogeneous mouse mucosal-associated invariant T cells using MR1 tetramers. J Exp Med 212(7):1095–1108. https://doi.org/10.1084/jem.20142110
8. Salou M, Legoux F, Gilet J et al (2019) A common transcriptomic program acquired in the thymus defines tissue residency of MAIT and NKT subsets. J Exp Med 216(1):133–151. https://doi.org/10.1084/jem.20181483
9. Winter SJ, Kunze-Schumacher H, Imelmann E et al (2019) MicroRNA miR-181a/b-1 controls MAIT cell development. Immunol Cell Biol 97(2):190–202. https://doi.org/10.1111/imcb.12211
10. Ziętara N, Łyszkiewicz M, Witzlau K et al (2013) Critical role for miR-181a/b-1 in agonist selection of invariant natural killer T cells. Proc Natl Acad Sci U S A 110(18):7407–7412. https://doi.org/10.1073/pnas.1221984110

INDEX

Helen Kaipe and Isabelle Magalhaes (eds.), *MAIT Cells: Methods and Protocols*, Methods in Molecular Biology, vol. 2098, https://doi.org/10.1007/978-1-0716-0207-2, © Springer Science+Business Media, LLC, part of Springer Nature 2020